Advances in Forensic Applications of Mass Spectrometry

Advances in Forensic Applications of Mass Spectrometry

Edited by

Jehuda Yinon

CRC PRESS

Boca Raton London New York Washington, D.C.

Library of Congress Cataloging-in-Publication Data

Advances in forensic applications of mass spectrometry / edited by Jehuda Yinon.
p. cm.
Includes bibliographical references and index.
ISBN 0-8493-1522-0
1. Mass spectrometry—Forensic applications. 2. Chemistry, Forensic. I. Yinon, Jehuda.

HV8073.5.A38 2003
363.25′62—dc22 2003055691

This book contains information obtained from authentic and highly regarded sources. Reprinted material is quoted with permission, and sources are indicated. A wide variety of references are listed. Reasonable efforts have been made to publish reliable data and information, but the author and the publisher cannot assume responsibility for the validity of all materials or for the consequences of their use.

Direct all inquiries to CRC Press LLC, 2000 N.W. Corporate Blvd., Boca Raton, Florida 33431.

Visit the CRC Press Web site at www.crcpress.com

International Standard Book Number 0-8493-1522-0
Library of Congress Card Number 2003055691
Printed in the United States of America 1 2 3 4 5 6 7 8 9 0
Printed on acid-free paper

Introduction

The continuing developments in analytical instrumentation during the last several years have had a major influence on the forensic laboratory. Among the instrumental methods used by the forensic analyst, the mass spectrometer has become the method of choice. While in the early days of forensic mass spectrometry GC/MS with electron ionization (EI), and later with chemical ionization (CI), were the only mass spectrometry techniques used, today a variety of MS techniques can be found in many forensic laboratories. Those include mainly LC/MS with electrospray ionization (ESI) or atmospheric pressure chemical ionization (APCI) and tandem mass spectrometry (MS/MS) in triple quadrupole or ion trap configurations. These techniques enable the detection and identification of trace components in complex mixtures at a high sensitivity as well as analysis of nonvolatile and thermally labile compounds. The incorporation of new mass spectrometry techniques in the forensic laboratory paved the road for new applications in forensic analysis.

The purpose of this book is to assemble in one volume forensic applications using novel mass spectrometry techniques which result in lower detection limits and more reliable identification. In addition, the book describes some new applications which were made possible due to this advanced instrumentation.

Two chapters have been devoted to the detection and analysis of drugs of abuse and their metabolites in body fluids (forensic toxicology). While GC/MS was the main method used for this purpose, LC/MS is now playing a major role in this important area and complements GC/MS.

Testing of athletes for drugs of abuse in major sporting events has become routine. GC/MS and LC/MS have been used for drug testing of urine and blood samples, which provides short-term information. A special chapter has been devoted to the detection of doping agents in hair by mass spectrometry, as hair testing has a larger surveillance window and can provide a long-term history.

The application of stable isotope ratios has become relevant for forensic purposes with the development of more sensitive and accurate isotope ratio mass spectrometers. The chapter dealing with this subject describes the methodology, the instrumentation, and several forensic applications, including origin identification of drugs and explosives and isotope tagging.

For many years, GC/MS has been the major method used for identification of accelerants in fire debris. MS/MS now provides an additional dimension of identification. GC/MS/MS has the capability to isolate and identify ignitable liquid residues in the presence of background products such as pyrolysates that may mask the target analytes of interest. A chapter devoted to this subject consists of two subchapters written by two different authors who are using GC/MS/MS with a different approach for the identification of accelerants in fire debris.

LC/MS, with electrospray ionization (ESI) and atmospheric pressure chemical ionization (APCI), are now being used for the analysis of explosives in postblast residues. This chapter describes both ionization techniques and their application in explosive residue analysis.

The mass spectrometer in its various configurations is now an integral part of every forensic laboratory. The number of possible forensic applications of this analytical technique is limited only by the imagination of the forensic chemist.

Finally, I would like to thank the contributing authors who have made the publication of this book possible. Many thanks are due to authors and publishers for permission to reproduce copyrighted material.

— Jehuda Yinon

The Editor

Dr. Jehuda Yinon is currently a visiting professor of forensic science at the National Center for Forensic Science (NCFS), University of Central Florida (UCF), Orlando, Florida.

Dr. Yinon received his B.Sc. and M.Sc. degrees in electrical engineering from the Technion, Israel Institute of Technology (Haifa, Israel), and his Ph.D. in chemistry from the Weizmann Institute of Science (Rehovot, Israel).

Dr. Yinon was a senior research fellow at the Weizmann Institute of Science, Rehovot, Israel, for 35 years, until 2000. He was a research associate (1971 to 1973) and a senior research associate (1976 to 1977) at Caltech's Jet Propulsion Laboratory (Pasadena, California), and spent sabbatical leaves as a visiting scientist at the National Institute of Environmental Health Sciences (Research Triangle Park, North Carolina) (1980 to 1981), the EPA Environmental Monitoring Systems Laboratory (Las Vegas, Nevada) (1988 to 1989), and the University of Florida, Gainesville (1993 to 1994).

Dr. Yinon's main research interests are applications of mass spectrometry techniques in forensic science with emphasis on analysis of explosives and the application of novel technologies to the detection of explosives.

Dr. Yinon has authored, coauthored, and edited 7 books and published over 100 papers in the scientific literature. He is a member of the American Academy of Forensic Science, the American Chemical Society, the American Society for Mass Spectrometry, and the Israel Society for Mass Spectrometry. He is a regional editor of Forensic Science Review and is on the editorial boards of the *Journal of Energetic Materials*, *Environmental Forensics Journal*, and *Modern Mass Spectrometry.*

Contributors

José R. Almirall
Department of Chemistry and International Forensic Research Institute
Florida International University
Miami, Florida

Maciej J. Bogusz
Department of Pathology and Laboratory Medicine
King Faisal Specialist Hospital and Research Centre
Riyadh, Saudi Arabia

Vincent Cirimele
Institut de Medicine Legale
Universite Louis Pasteur
Strasbourg, France

Pascal Kintz
Institut de Medicine Legale
Universite Louis Pasteur
Strasbourg, France

Ray H. Liu
Graduate Program in Forensic Science
Department of Justice Sciences
University of Alabama at Birmingham
Birmingham, Alabama

Hans H. Maurer
Department of Experimental and Clinical Toxicology
Institute of Experimental and Clinical Pharmacology and Toxicology
University of Saarland
Homburg (Saar), Germany

Wolfram Meier-Augenstein
Queen's University Belfast
Environmental Engineering Research Group
School of Civil Engineering
Belfast, U.K.

Jeannette Perr
Department of Chemistry and International Forensic Research Institute
Florida International University
Miami, Florida

Dale Sutherland
Activation Laboratories Ltd.
Ancaster, Ontario, Canada

Marion Villain
Institut de Medicine Legale
Universite Louis Pasteur
Strasbourg, France

Jehuda Yinon
National Center for Forensic Science
University of Central Florida
Orlando, Florida

List of Abbreviations

2-ADNT — 2-amino-4,6-dinitrotoluene
4-ADNT — 4-amino-2,6-dinitrotoluene
2-D — two-dimension
6-MAM — 6-monoacetylmorphine
A — amphetamine
AC — acetylation
AC — acetylcodeine
ACE — angiotensin converting enzyme
ACN — acetonitrile
ACS — activated charcoal strip
AGC — automatic gain control
AN — ammonium nitrate
APCI — atmospheric pressure chemical ionization
APE — atom% excess
API — atmospheric pressure ionization
APPI — atmospheric pressure photoionization
ASTM — American Society for Testing and Materials
AT_1 — angiotensin II AT_1 receptor
BDB — benzodioxolybutanamine
BMA — benzphetamine
BSIA — bulk stable isotope analysis
BU — buprenorphine
BUG — buprenorphine glucuronide
BZE — benzoylecgonine
BZP — *N*-benzylpiperazine
C4 — composition 4
C6G — codeine-6-glucuronide
CAD — collisionally activated dissociation
CAM — crassulacean acid metabolism
CE — capillary electrophoresis
CF — continuous flow
CI — chemical ionization
CID — collision-induced dissociation

CL-20 — hexanitrohexazaisowurtzitane
Cod — codeine
CSIA — compound specific isotope analysis
DAD — diode array detection
DAM — diacetylmorphine
DC — direct current
DHEA — dehydroepiandrosterone
DHM — dihydromorphine
DHT — dihydrotestosterone
DLI — direct liquid introduction
DMT — *N,N*-dimethyltryptamine
DNA — deoxyribonucleic acid
DNB — dinitrobenzene
DNT — dinitrotoluene
DUID — driving under the influence of drugs
E — ephedrine
EA — elemental analysis
ECG — ecgonine
EDDP — 2-ethylidene-1,5-dimethyl-3,3-diphenylpyrrolidine
EGDN — ethylene glycol dinitrate
EI — electron ionization
EM — electron multiplier
EMDP — 2-ethyl-5-methyl-3,3-diphenyl-1-pyrroline
EME — ecgonine methyl ester
ESI — electrospray ionization
EU — extraction uniformity
FC — Faraday cup
FID — flame ionization detector
FT-ICR — Fourier transform ion cyclotron resonance
GC — gas chromatography
GC/C — gas chromatography/conversion interface
GC/FID — gas chromatograph/flame ionization detector
GCxGC — two-dimension gas chromatography
GC/HRMS — gas chromatography/high resolution mass spectrometry
GC/MS — gas chromatography/mass spectrometry
GC/MS/MS — gas chromatography/tandem mass spectrometry
GC/PID — gas chromatograph/photoionization detector
GHB — gamma-hydroxybutyrate
H3G — hydromorphine-3-glucuronide
Hexyl — hexanitrodiphenylamine
HFB — heptafluorobutyration
HFB — heptafluorobutyrate

HMTD — hexamethylenetriperoxidediamine
HMX — 1,3,5,7-tetranitro-1,3,5,7-tetrazacyclooctane
HNS — hexanitrostilbene
HPLC — high performance liquid chromatography
HRcGC — high resolution capillary gas chromatography
HRMS — high resolution mass spectrometry
HYC — hydrocodone
HYM — hydromorphone
ID — internal diameter
IDT — isotope dilution technique
IFD — ion fingerprint detection
ILA — isotope-labeled analog
ILR — ignitable liquid residue
IOC — International Olympic Committee
IRMS — isotope ratio mass spectrometry
IS — internal standard
ISCIRA — internal standard carbon isotope ratio analysis
LC/MS — liquid chromatography/mass spectrometry
LC/MS/MS — liquid chromatography/tandem mass spectrometry
LLE — liquid–liquid extraction
LOD — limit of detection
LOQ — limit of quantitation
LSD — lysergic acid diethylamide
K9 — canine
M — morphine
M3G — morphine-3-glucuronide
M6G — morphine-6-glucuronide
MA — methamphetamine
MBDB — *N*-methyl-benzodioxolylbutanamine
MBHFA — *N*-methyl-*N*-trimethylsilylheptafluorobutyramide
MBHFBA — *N*-methyl-bis-heptafluorobutyramide
mCPP — 1-(3-chlorophenyl)piperazine
MDA — methylenedioxyamphetamine
MDBP — *N*-(3,4-methylenedioxybenzyl)piperazine
MDE — methylenedioxyethylamphetamine
MDEA — methylenedioxyethylamphetamine
MDMA — methylenedioxymethamphetamine
MDPPP — 3′,4′-methylenedioxy-alpha-pyrrolidinopropiophenone
ME — methylation, methylated
MeOPP — 1-(4-methoxyphenyl)piperazine
MF — matrix factor
MOPPP — 4′-methoxy-alpha-pyrrolidinopropiophenone

MPHP — 4′-methyl-alpha-pyrrolidinohexaphenone
MRM — multiple reaction monitoring
MS — mass spectrometry
MS/MS — tandem mass spectrometry
MSTFA — *N*-methyltrimethylsilyltrifluoroacetamide
NBU — norbuprenorphine
NG — nitroglycerin
NICI, NCI — negative-ion chemical ionization
NMT — *N*-methyltryptamine
NorCod — norcodeine
Nor-K — nor-ketobemidone
Nor-M — normorphine
NQ — nitroguanidine
NSAID — non-steroidal anti-inflammatory drug
NTO — 5-nitro-2,4-dihydro-3H-1,2,4-triazol-3-one
ODS — octadecylsilica
PB — particle beam
PBI — particle beam ionization
PDB — PeeDee belemnites (standard)
PETN — pentaerythritol tetranitrate
PFP — pentafluoropropionylation
Pholcod — pholcodine
PID — photoionization detector
PMA — paramethoxyamphetamine
PPP — alpha-pyrrolidinopropiophenone
Q — single stage quadrupole mass analyzer
QIT — quadrupole ion trap
qIT — quadrupole/ion trap mass analyzer
QQQ — triple stage quadrupole mass analyzer
qTOF — quadrupole/time-of-flight mass analyzer
RDX — 1,3,5-trinitro-1,3,5-triazacyclohexane
RF — radio frequency
RIA — radioimmunological assay
RT — retention time
SB — secobarbital
SD — standard deviation
SIM — selected ion monitoring
SMOW — standard mean ocean water
SPE — solid-phase extraction
SPME — solid-phase microextraction
SRM — single reaction monitoring
SSI — sonic spray ionization

SSRI — selective serotonin reuptake inhibitor
STA — systematic toxicological analysis
TC — thermal conversion
TCA — tricyclic antidepressants
TFA — trifluoroacetate
TFA — trifluoroacetic acid
TFA — trifluoroacetylation
TFMPP — 1-(3-trifluoromethylphenyl)piperazine
THC — tetrahydrocannabinol
TLC — thin layer chromatography
TMC — trimethoxycocaine
TMS — trimethylsilylation
TMSI — trimethylsilyl imidazole
TNB — 1,3,5-trinitrobenzene
TNC — trinitro-*m*-cresol
TNT — 2,4,6-trinitrotoluene
TOF — time-of-flight
TS — thermospray
TSQ — triple stage quadrupole
UV — ultraviolet

Table of Contents

Screening for Drugs in Body Fluids by GC/MS

1

HANS H. MAURER

Contents

0-8493-1522-0/04/$0.00+$1.50

1.1 Introduction

In forensic toxicology, proof of abuse of illegal drugs or of murder by poisoning are important tasks. Furthermore, drugs, which may reduce the penal responsibility of a defendant, or which may reduce the fitness to drive a car, must be monitored in body fluids or tissues.

In clinical toxicology, the diagnosis or the definite exclusion of an acute or chronic poisoning is of great importance. Furthermore, patients addicted to alcohol, medicaments, or illegal drugs have to be monitored. For determination of clinical death as a prerequisite for explantation of organs, the presence of drugs which may depress the central nervous system must be analytically excluded. The compliance of patients can be monitored by determination of the prescribed drugs. Finally, monitoring of drugs with a narrow therapeutic range can be performed by the clinical toxicologist. Similiar problems arise in forensic toxicology.

In doping control, the use or abuse of drugs that may stimulate the build-up of muscles, enhance endurance during competition, lead to reduction of body weight, or that may reduce pain caused by overexertion must be monitored, typically in urine.

An efficient toxicological analysis is the basis of competent toxicological judgement, consultation, and expertise. The choice of methods in analytical toxicology depends on the problems to be solved. Usually, the compounds to be analyzed are unknown. Therefore, the first step is the identification of the compounds of interest which can then be quantified, e.g., in plasma. The screening strategy of systematic toxicological analyses (STA) must be very extensive because several thousands of drugs or pesticides have to be considered. It often includes screening and confirmatory tests. If only a single drug or category has to be monitored, immunoassays can be used for screening in order to differentiate between negative and presumptively positive samples. Positive results must be confirmed by a second independent method that is at least as sensitive as the screening test and that provides the highest level of confidence in the result. Without doubt, GC/MS, especially in the full-scan electron ionization (EI) mode, is still the reference method for confirmation of positive screening tests.[1–21] Nevertheless, LC/MS has also been applied for screening and confirmation of particular drugs or drug classes, especially in blood.[22–31]

The two-step strategy, immunoassay screening and MS confirmation, is employed only if those drugs or poisons have to be determined that are scheduled, e.g., by law or by international organizations, and for which immunoassays are commercially available. If this is not the case, the screening strategy must be more extensive, because several thousands of drugs or pesticides are on the market worldwide.[32] For these reasons, STA procedures are necessary that allow the simultaneous detection of as many toxicants in biosamples as possible. As already mentioned, GC/MS procedures are most often used today. HPLC coupled to diode array detectors (DAD)[33–39] have also been described for general screening purposes, but the specificity is lower than that of full-scan EI MS. Valli et al.[40] combined GC/MS blood screening with urine REMEDi testing and Saint-Marcoux et al.[26] combined GC/MS, HLPC-DAD, and LC/MS. LC/MS procedures in this field were several times reviewed in the last years, e.g., by Maurer,[41] Van Bocxlaer et al.,[42] Marquet,[28] and Bogusz.[31,43]

Most of the STA procedures cover basic (and neutral) drugs, which include the majority of toxicants. For example, most of the psychotropic drugs have basic properties like neurotransmitters. Nevertheless, some classes of acidic drugs or drugs producing acidic metabolites, like cardiovascular drugs such as angiotensin converting enzyme (ACE) inhibitors and angiotensin II AT_1 receptor blockers, dihydropyridine-type calcium channel blockers (metabolites), diuretics, coumarin anticoagulants, antidiabetics of the sulfonylurea type, barbiturates, or nonsteroidal antiinflammatory drugs (NSAIDs), are relevant to clinical and forensic toxicology or doping. Therefore, GC/MS screening procedures are described here not only for detection of basic and neutral, but also acidic drugs in biosamples. After the unequivocal identification, reliable quantification of the drugs can also be performed by GC/MS, especially if stable isotopes are used as internal standards (for example, see References 44, 45), or in case of rather polar or unstable compounds, by LC/MS, even using universal internal standards like deuterated trimipramine.[22,24,25,46] However, whatever technique is used, quantification procedures must be validated according to international guidelines, which have recently been reviewed by Peters and Maurer.[47]

1.1.1 Matrices for Drug Screening

Blood (plasma, serum) is the sample of choice for quantification. However, if the blood concentration is high enough, screening can also be performed herein. This is especially advantageous if only blood samples are available and/or the procedures allow simultaneous screening and quantification.[6,9,32,48,49] In driving under the influence of drugs (DUID) cases, blood analysis is even mandatory. A GC/MS screening procedure has been described for about 100 acid, neutral, and basic drugs in horse plasma.[50] Methods for postmortem drug analysis have been reviewed recently.[51]

GC/MS analysis of drugs in alternative matrices like hair,[8,19,52] sweat and saliva,[7,19] meconium,[53] or nails[54] have also been described, but a comprehensive screening for a series of various drugs has not yet been described in alternative matrices, probably because the concentrations are too low for full-scan GC/MS detection. Negative-ion chemical ionization (NICI) allows markedly lowering the detection limits,[20,24,44] but this technique is not suitable for comprehensive screening because the analytes must contain an electronegative moiety and the NICI mass spectra are less informative and reproducible than EI spectra.

In conclusion, urine is still the sample of choice for comprehensive screening for, and identification of, unknown drugs or poisons, mainly because concentrations of drugs are relatively high in urine and the samples can be taken noninvasively.[1,2,4,16,18,32] However, the metabolites of these unknowns must be identified, additionally or even exclusively. In (horse) doping control, urine is also the common sample for screening.[55]

1.1.2 Sample Preparation

Suitable sample preparation is an important prerequisite for GC/MS analysis in biosamples. It may involve cleavage of conjugates, isolation, and derivatization, preceded or followed by cleanup steps. Cleavage of conjugates can be performed by fast acid hydrolysis or by gentle but time-consuming enzymatic hydrolysis.[2] However, the enzymatic hydrolysis of acyl glucuronides (ester glucuronides of carboxy derivatives like NSAIDs) may be hindered due to acyl migration,[56] an intramolecular transesterification at the hydroxy groups of the glucuronic acid, which leads to β-glucuronidase-resistant derivatives. If the analysis must be finished within a rather short time as in emergency toxicology, it is preferable to cleave the conjugates by rapid acid hydrolysis.[57–62] Alkaline hydrolysis is only suitable for cleavage of ester conjugates. However, the formation of artifacts during chemical hydrolysis must be considered.[63] A compromise over both cleavage techniques is the use of a column packed with immobilized glucuronidase/arylsulfatase. It combines the advantages of both methods — the speed of acid hydrolysis and the gentle cleavage of enzymatic hydrolysis.[64] Acyl glucuronides, e.g., of acidic drugs are readily cleaved under the conditions of extractive alkylation (alkaline pH, elevated temperature) and need no extra cleavage step.[65–69]

Isolation can be performed by liquid–liquid extraction (LLE) at a pH at which the analyte is nonionized or by solid-phase extraction (SPE) preceded or followed by cleanup steps. Sample pretreatment for SPE depends on the sample type: whole blood and tissue (homogenates) need deproteinization and filtration/centrifugation steps before application to the SPE columns, whereas for urine usually a simple dilution step and/or centrifugation is satisfactory. Whatever SPE column is used, the analyst should keep in mind that there are large differences from batch to batch, and that comparable sorbents from different manufacturers may also lead to different results.[70] Therefore, use of a suitable internal standard (e.g., deuterated analytes) is recommended.

Solid-phase microextraction (SPME) is becoming a modern alternative to SPE and LLE. SPME is a solvent-free and concentrating extraction technique especially for rather volatile analytes. It is based on the adsorption of the analyte on a stationary phase, coating a fine rod of fused silica. The analytes can be desorbed directly in the GC injector. Fast GC/MS procedures for screening, e.g., for benzodiazepines,[71] for barbiturates,[72] clozapine,[73] or for drugs of abuse[74–82] have been published in recent years.

Extractive alkylation has been proved to be a powerful procedure for simultaneous extraction and derivatization of acidic compounds.[65–69,83] The acidic compounds are extracted at pH 12 as ion pairs with a phase-transfer catalyst (tetrahexyl ammonium iodide, THA^+I^-) into the organic phase (toluene). In the organic phase, the phase-transfer catalyst could easily be solvated due to its lipophilic hexyl groups, whereas poor solvation of the anionic analytes leads to a high reactivity with the alkylation (most often methylation) reagent alkyl (methyl) iodide. Part of the phase-transfer catalyst can also reach the organic phase as an ion pair with the iodide anion formed during the alkylation reaction or with anions of the urine matrix. Therefore, the remaining part had to be removed to prevent a loss of the GC column's separation power and to exclude interactions with analytes in the GC injection port. Several SPE sorbents and different eluents have been tested for efficient separation of the vestige of the phase-transfer catalyst from the analytes. A diol sorbent yielded best reproducibility and recovery under the described conditions. Further advantages of such SPE columns were easy handling, commercial availability, and that they had not to be manually prepared as described by Lisi et al.[84]

Derivatization steps are necessary if relatively polar compounds containing, e.g., carboxylic, hydroxy, primary, or secondary amino groups are to be determined by GC/MS, and/or if electronegative moieties (e.g., halogen atoms) have to be introduced into the molecule for sensitive NICI detection. The following procedures are typically used for basic compounds: acetylation (AC), trifluoroacetylation (TFA), pentafluoropropionylation (PFP), heptafluorobutyration (HFB), trimethylsilylation (TMS), or for acidic compounds: methylation (ME), extractive methylation, PFP, TMS or *tert*-butyldimethylsilylation. Further details on derivatization methods can be found in References 57, 63, and the pros and cons of derivatization procedures were discussed in a review of Segura et al.[85]

1.2 Screening for Drugs in Blood, Serum, or Plasma by GC/MS

GC/MS procedures have been published for blood screening, mainly of drugs of abuse[6,86–89] because they have to be monitored or confirmed after immunoassay prescreening, e.g., in DUID cases.[21,90] An automated screening procedure for barbiturates, benzodiazepines, antidepressants, morphine, and cocaine in blood after SPE and TMS has been developed.[9,48] The sample preparation consists of SPE and TMS derivatization, both automated using an HP PrepStation. The samples are directly injected by the PrepStation and analyzed by full-scan GC/MS. Using macros, peak identification, and the reporting of results are also automated. This fully automated procedure takes about 2 h, which is acceptable for forensic drug testing or doping control but not for emergency toxicology. Automation of the data evaluation is a compromise between selectivity and universality. If the exclusion criteria are chosen too narrowly, peaks may be overlooked. If the window is too large, a series of proposals is given by the computer, which have to be revised by the toxicologist. In order to extend this limited blood screening, one working group recommended to combine it with a urine screening using the REMEDI black box system.[40] Maurer[57] has described a rather comprehensive plasma screening procedure based on a standard LLE after addition of the universal internal standard trimipamine-d_3. This universal extract can be used for GC/MS as well as for LC/MS[25] screening, identification, and quantification. The GC/MS screening is based on mass chromatography[24,46] using macros for selection of suspected drugs,[91] followed by identification of the unknown spectra by library search.[92] The selected ions for screening in plasma (and gastric content) have recently been updated by the author's coworkers using experiences from their daily routine work with this procedure, and they are summarized in Table 1.1. The drugs or poisons which can be detected in plasma after therapeutic or toxic dosage are listed in the author's handbook.[32] Of course, further compounds can be detected,[93] if they are present in the extract, volatile in GC, and their mass spectra are contained in the reference libraries.[92,94–96]

The plasma screening procedure is illustrated in Figure 1.1 to Figure 1.6. Figure 1.1 shows mass chromatograms corresponding to fragment ions typical for analgesics of a plasma extract. Generation of the mass chromatograms can be started by clicking the corresponding pull-down menu which executes the user-defined macros.[91] Three major peaks appear, the identity of which is confirmed by comparison of the underlying full mass spectrum with reference spectra. Figure 1.2 shows the unknown mass spectrum underlying peak 1 (upper part), the reference spectrum (biomolecule linoleic acid, middle part), and the structure and the hit list found by library search in Reference 92 (lower part). Figure 1.3 and Figure 1.4 show the same for peaks 2 (oxycodone) and 3 (biomolecule cholesterol). Figure 1.5 shows mass chromatograms corresponding to fragment ions typical for sedative–hypnotics generated from the same data file. The peaks 1 and 3 indicate the same compounds as indicated in Figure 1.1. Figure 1.6 shows the mass spectrum underlying peak 4 (upper part), the reference spectrum (meprobamate,

Table 1.1 The Selected Target Ions for Screening in Plasma (and Gastric Content)

Target Ion (*m/z*)	Psychotropics
58	Amitriptyline, Citalopram, Clomipramine, Dosulepin, Doxepin, Fluoxetine, Imipramine, Trimipramine, Venlafaxine, Alimemazine, Chlorpromazine, Chlorprothixene, Levomepromazine, Promazine, Prothipendyl, Triflupromazine, Zotepine, etc.
61	Trimipramine-d_3 (internal standard)
70	Clopenthixol, Fluphenazine, Maprotiline, Opipramol, Paroxetine, Perazine, Prochlorperazine, Trazodone, Trifluoperazine, Zuclopenthixol
72	Methadone, Promethazine
86	Licocaine, Metoclopramide
98	Amisulpride, Biperiden, Mepivacaine, Sulforidazine, Sulpiride, Thioridazine
112	Melperone
195	Mirtazapine
	Antidepressants/Neuroleptics
100	Amfebutamone, Moclobemide, Viloxazine
123	Bromperidol, Droperidol, Haloperidol, Pipamperone, Trifluperidol
210	Quetiapine
242	Olanzapine
243	Clozapine
276	Fluvoxamine
297	Trimipramine-d_3 (internal standard)
303	Nefazodone, Verapamil
329	Nifedipine
	Benzodiazepines
239	Lorazepam, Triazolam
269	Nordazepam, Prazepam
281	Nitrazepam
283	Chlordiazepoxide, Diazepam
300	Clobazam, Temazepam
305	Lormetazepam
308	Alprazolam
310	Midazolam
312	Flunitrazepam, Midazolam
315	Bromazepam, Triazolam
	Sedative–Hypnotics
83	Meprobamate
156	Pentobarbital
105	Etomidate
167	Diphenhydramine, Doxylamine
163	Propofol
172	Thiopental
180	Ketamine
235	Methaqualone, Zolpidem

Table 1.1 (*Continued*) The Selected Target Ions for Screening in Plasma (and Gastric Content)

Target Ion (*m/z*)	Sedative–Hypnotics
248	Zaleplon
261	Zopiclone
	Anticonvulsants
102	Valproic acid
113	Ethosuximide
180	Phenytoin
185	Lamotrigine
190	Primidone
193	Carbamazepine
204	Phenobarbital
280	Clonazepam, Phenprocoumon
288	Tetrazepam
394	Brotizolam
	Analgesics
120	Salicylic acid
151	Paracetamol
161	Ibuprofen
188	Tramadol artifact
217	Metamizol
230	Naproxen, 8-Chlorotheophylline
231	Propyphenazone, Flupirtine
299	Codeine
214	Diclofenac
139	Indomethacin

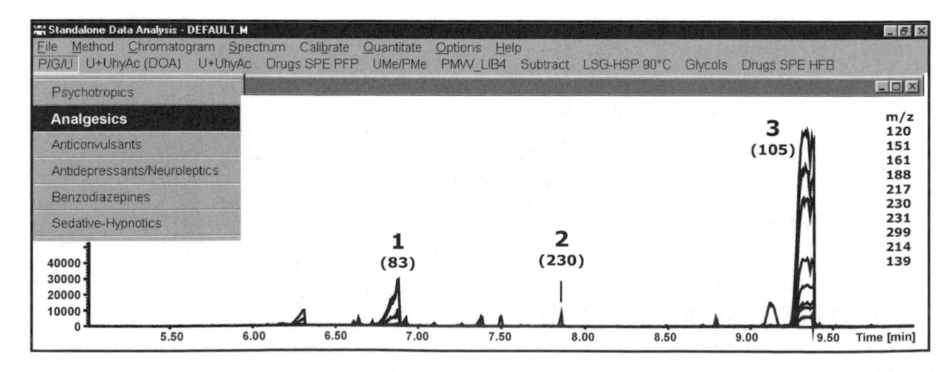

Figure 1.1 Mass chromatograms corresponding to fragment ions typical for analgesics of a plasma extract.

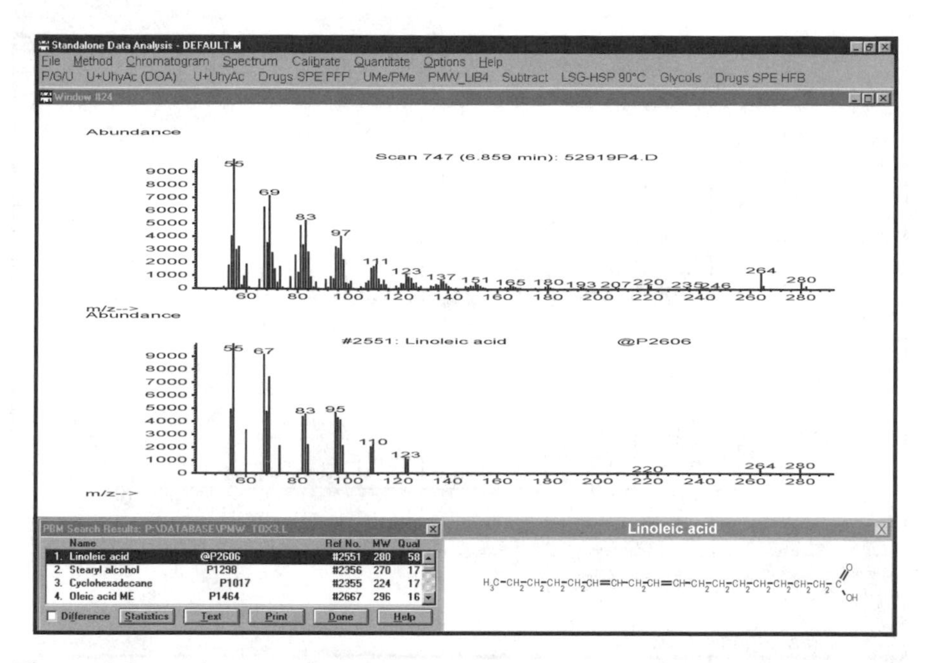

Figure 1.2 Unknown mass spectrum underlying the peak 1 in Figure 1.1 (upper part), the reference spectrum (middle part), and the structure and the hit list found by library search in Reference 92 (lower part).

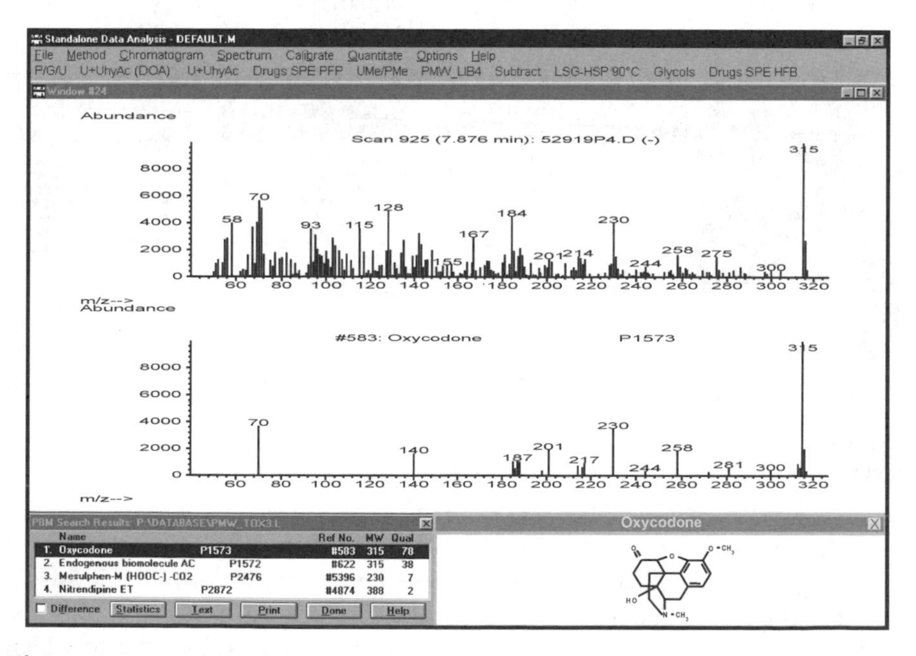

Figure 1.3 Unknown mass spectrum underlying the peak 2 in Figure 1.1 (upper part), the reference spectrum (middle part), and the structure and the hit list found by library search in Reference 92 (lower part).

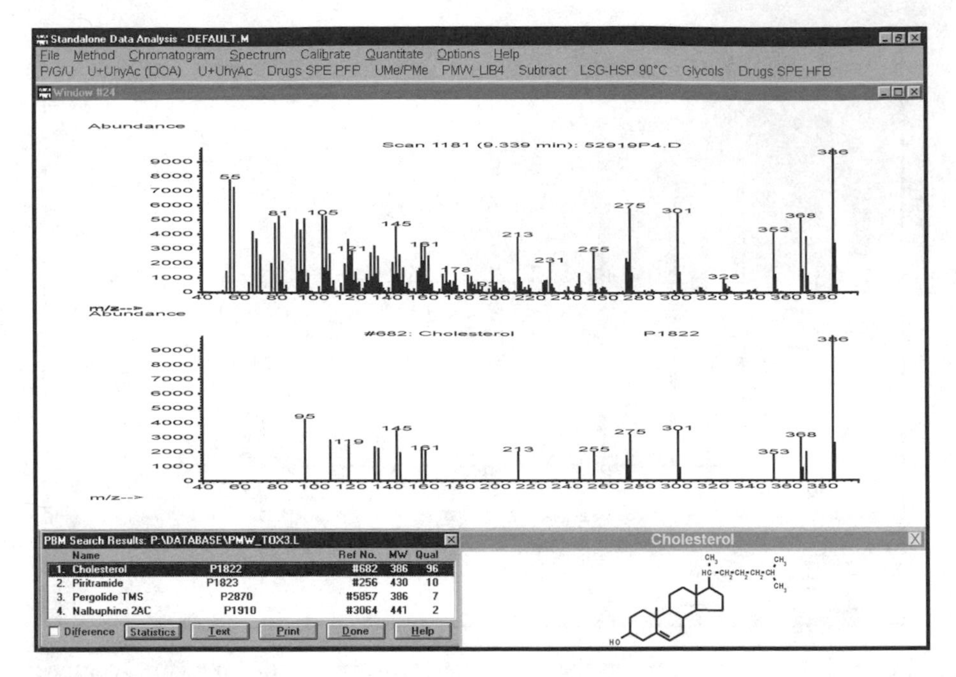

Figure 1.4 Unknown mass spectrum underlying the peak 3 in Figure 1.1 (upper part), the reference spectrum (middle part), and the structure and the hit list found by library search in Reference 92 (lower part).

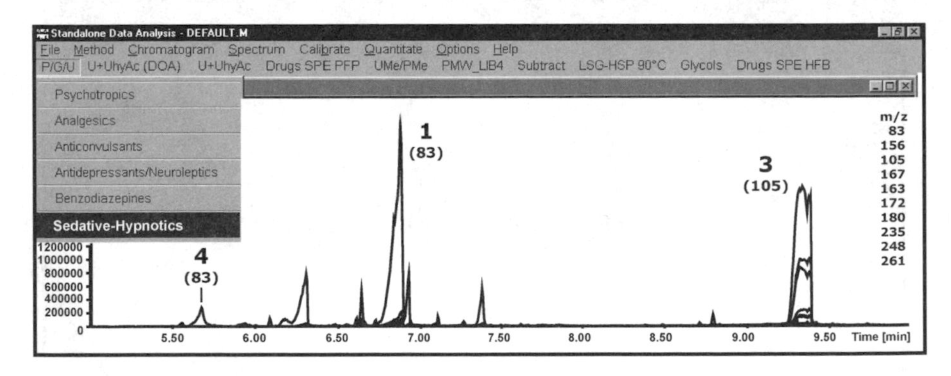

Figure 1.5 Mass chromatograms corresponding to fragment ions typical for sedative–hypnotics generated from the same data file as used for Figure 1.1.

middle part), and the structure and the hit list found by library search in Reference 92 (lower part). The appearance of peaks of biomolecules illustrate that the mass chromatograms are only the screening process, and only the comparison of the full spectra leads to a confirmed result. In order to widen the screening window, comprehensive urine screening by full-scan GC/MS

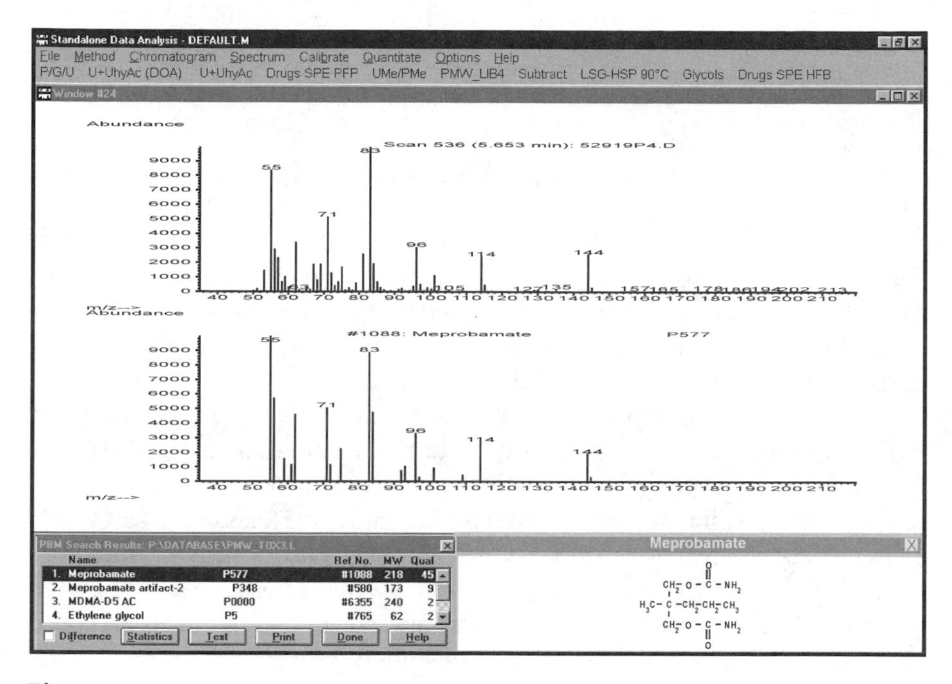

Figure 1.6 Unknown mass spectrum underlying the peak 4 in Figure 1.5 (upper part), the reference spectrum (middle part), and the structure and the hit list found by library search in Reference 92 (lower part).

detecting several thousand compounds is strictly recommended and will be described in 1.4.2.2, using the urine of the same case.

More or less comprehensive screening procedures for drugs in blood, serum, or plasma have been described, mainly using HPLC-DAD.[17,26,33,35,37,38,97–104] HPLC coupled with a single stage or tandem mass spectrometer (LC/MS, LC/MS/MS) is becoming more and more a routine apparatus, especially in blood and plasma analysis.[22,25,41,46,105] However, before establishing LC/MS screening procedures in routine work, several limitations should be kept in mind, as stated by all experts in this field.[24,26,28,106–108] The spectral information of electrospray ionization (ESI) and/or atmospheric pressure chemical ionization (APCI) spectra is limited, compared to EI mass spectra, and they can vary considerably between the apparatus. Another important problem for ESI is the reduction of the ionization of a compound (ion suppression) due to coeluting compounds (e.g., matrix) because in these cases a relevant toxicant might be overlooked,[109] resulting, in the worst case, in the patient's death. In the author's opinion and experience, analytes that are volatile in GC should be screened for, using full-scan GC/MS. Nevertheless, LC/MS is an excellent completion for screening, library-assisted identification, and quantification of unstable, low-dosed, and/or rather polar compounds, especially in plasma.[22–31]

1.3 Screening for Drugs in Alternative Matrices by GC/MS

There are several advantages of alternative matrices over urine and blood for drugs of abuse testing. They can be collected rather easily and noninvasively with low risk of adulteration. In addition, using hair, the window of drug detection is dramatically extended to weeks, months, or even years.[19] The great advances in analytical techniques today allow us to monitor drugs in such matrices. Several reviews on drug testing in alternative matrices have been published in the last years.[7,8,10,12–15,19,53,110] Most procedures published in last 2 years are still based on GC/MS; for example, see References 75, 76, 78, 111–120. However, as already mentioned above, it must be kept in mind that, until now, no really comprehensive screening procedure has been described for alternative matrices as would be necessary in general unknown cases, and that the pharmacokinetic interpretation is often difficult.

1.4 Screening for Drugs in Urine by GC/MS

1.4.1 Screening Procedures for Detection of Particular Drug Classes in Urine by GC/MS

As already discussed above, urine still remains the standard for comprehensive screening, especially in general unknown cases in clinical and forensic toxicology and doping control. Procedures are described not for single drugs but for all, or at least several, drugs of a pharmacological or chemical class. For toxicological screening procedures in urine, the detectability of the parent compound is of minor value if the concentrations of the metabolites are much higher in urine than those of the parent drug and if the metabolites can be detected by the procedure. Most of the toxicologically relevant drugs are rather lipophilic substances that undergo extensive metabolism. As pure substances of the metabolites are usually not available, it is necessary to control the quality of the screening procedures using urine samples of volunteers or in-patients treated with a known dose of the drug. The procedure should be sufficiently sensitive to detect therapeutic concentrations at least over a 12 to 24 h period after ingestion. Papers on the detection time of drugs of abuse in urine have recently been reviewed.[121] An interesting experimental approach for the validation of qualitative chromatographic methods and its application in an antidoping control laboratory was recently described.[122]

1.4.1.1 *Screening Procedures for Detection of Particular Drug Classes of Basic or Neutral Drugs and/or Their Metabolites in Urine*

1.4.1.1.1 Central Stimulants. Central stimulants are drugs of abuse as well as doping agents. Immunoassays are available for amphetamine and some of its derivatives as well as for cocaine metabolites. In the past, series of papers have described confirmation of amphetamines or cocaine after suitable isolation and derivatization and GC/MS.[5,123] SPME is a more recent alternative for rapid, sensitive, and solvent-free determination of amphetamines in urine.[80,124] However, such confirmation of positive immunoassay results for amphetamines should not be limited to detection of amphetamine and methamphetamine by GC/MS in the selected-ion monitoring (SIM) mode unless the testing program forbids analysis of analytes not specifically put into the testing program. Amphetamine and methamphetamine derived medicaments must be considered as part of a proper interpretation of amphetamine-positive results. Detection of the hydroxy metabolites may help extend the time of differentiation of illicit amphetamine abuse from intake of medicaments.[125–132] However, in the late phase of excretion after intake of amphetamine-derived medicaments, amphetamine, or methamphetamine are often the only metabolites which can be detected in urine. In such urine samples, differentiation of illicit amphetamine or methamphetamine intake from the intake of such medicaments may not be possible. In recent years, review articles have been published which discussed these problems.[5,133,134]

1.4.1.1.2 Designer Drugs (Rave Drugs). The amphetamine-derived designer drugs MDA, MDMA, and MDE as well as BDB and MBDB have gained great popularity as "rave drugs."[135] They produce feelings of euphoria and energy and a desire to socialize. Nichols coined the term entactogens for this new group of drugs.[136] Although the designer drugs had the reputation of being safe, several experimental studies in rats and humans and epidemiological studies indicated risks to humans. Recent papers and reviews describe the current knowledge on hepatotoxicity[137–141] and neurotoxicity,[142] psychopathology and abuse potential of such designer drugs.[135,142–156] Screening procedures are needed because these drugs may lead to more or less severe poisonings[138] and driving impairment.[21,157] Commercial immunoassays are availabe for urine (and blood) testing, but not all amphetamine IAs are suitable for detection of its methylenedioxy derivatives.[158–162] There is no IA available for testing the new piperazine-derived designer drugs which have been found on the illicit market as a new group of designer drugs.[45,60,163–167] Shulgin mentioned BZP as a "pure stimulant" and TFMPP as an "active hallucinogen" in his book *Pikhal*.[168] BZP is said to produce similar effects as dexamphetamine[169] and to act as a central serotoninomimetic.[170]

Again, GC/MS procedures were described for confirmation of amphetamine-derived designer drugs after extraction and derivatization[2,171–175] or SPME,[80,176] as well as for screening for and differentiation of all these drugs.[61,94,172] The latter procedure is part of a very comprehensive screening procedure (cf. 1.4.2.2) and the only one which also covers the new piperazine-derived designer drugs *N*-benzylpiperazine (BZP, "A2"),[166] *N*-(3,4-methylenedioxybenzyl)piperazine (MDBP, MDBZP), 1-(3-trifluoromethylphenyl)piperazine (TFMPP), 1-(3-chlorophenyl)piperazine (mCPP), and 1-(4-methoxyphenyl)piperazine (MeOPP).[60]

Alpha-pyrrolidinophenone derivatives like *R,S*-alpha-pyrrolidinopropiophenone (PPP), *R,S*-4′-methyl-alpha-pyrrolidinopropiophenone (MPPP), *R,S*-4′-methyl-alpha-pyrrolidinohexanophenone (MPHP), *R,S*-4′-methoxy-alpha-pyrrolidinopropiophenone (MOPPP), and *R,S*-3′,4′-methylenedioxy-alpha-pyrrolidinopropiophenone (MDPPP) are new designer drugs which have appeared on the illicit drug market.[177–181] Meanwhile, most of these substances are scheduled in the German Controlled Substances Act. So far, little information about the dosage as well as the pharmacological and toxicological effects of the pyrrolidinophenones is available. The chemical structures of all the alpha-pyrrolidinophenones are closely related to alpha-aminopropiophenone anorectics like amfepramone, drugs of abuse like cathinone/methcathinone, and antidepressants like bupropion and might, therefore, evoke similar effects including dopamine-release and sympathomimetic properties.[182–185] The metabolism and toxicological detection of such designer drugs were recently studied.[178–181,186–188] Unfortunately, these drugs cannot be detected by common screening procedures due to the zwitterionic structure of their metabolites. SPE showed good results because mixed-mode SPE has proven to be suitable for the extraction even of their zwitterionic metabolites.[181] Common trimethylsilylation leads to good GC properties.

1.4.1.1.3 Hallucinogens. Cannabis (marijuana) is the most frequently abused illegal drug around the world. Lysergide (LSD) and phencylidine are more rarely abused, but the margin of "therapeutic" safety is smaller. Again, immunoassays are available for these hallucinogens and confirmation by GC/MS or LC/MS (especially for LSD[189–191]) has been common for many years[1]; only a few recent papers cover tetrahydrocannabinol (THC) urine testing. For example, the need for coping with a big series of urine workplace testing leads to the development of high-throughput methods.[192] Another new aspect concerns the differentiation of therapeutic intake of synthetic THC (dronabinol/marinol as antiemetic under cytostatic treatment) and cannabis abuse. 11-Nor-δ^9-tetrahydrocannabivarin-9-carboxylic acid, the urinary metabolite of the natural component of most cannabis products,

δ^9-tetrahydrocannabivarin, was found to be a marker for marijuana or a related product use.[193]

1.4.1.1.4 Opioid (Narcotic) and Other Potent Analgesics. Opioid analgesics, often called narcotics, are widely used to reduce severe pain, especially in a postoperative state and in the final state of cancer diseases. These are also abused because of their euphoriant and anxiolytic effects. While opioid medicaments are usually misused by medical staff, heroin is widely abused by drug addicts. If heroin is not available the addicts often take opioid medicaments. For legal reasons, the use of heroin must analytically be differentiated from an intake of other opioids. Therefore, 6-monoacetylmorphine, the only heroin specific metabolite, must be detected in biosamples. Several papers have been published on the detection of 6-monoacetylmorphine, employing SPE or LLE and derivatization by TFA, PFP, propionylation, or by TMS.[1]

Screening procedures have been described for the detection of most of the opioids and other potent analgesics after acid or enzymatic hydrolysis with LLE or SPE at pH 8 to 9, followed by TMS or AC.[1,194–197] The latter procedure is part of a very comprehensive screening procedure (cf. 1.4.2.2) and the only one which also covers most of the synthetic opioids. The antitussive pholcodine was found to cross-react with opiate immunoassays, but it lacks opioid potency. Pholcodine is partly hydrolyzed to morphine by hydrochloric acid, so careful enzymatic hydrolysis must be used if differentiation in urine is needed.[198–200] In recent years, buprenorphine and oxycodone testing became more and more of interest. Since a few years ago, buprenorphine has been used, besides methadone and levomethadol, for heroin substitution therapy as an effective means of decreasing illicit heroin use, crime, HIV risk, and death, and in improving employment and social adjustment.[201] Urine testing can be performed by GC/MS,[83,202,203] while blood testing requires LC/MS, especially for precise quantification.[204–206]

Oxycodone has very high abuse potential because it is highly effective when taken orally, is often easily available, and has a high degree of consistent potency. Opiate tests might be negative for oxycodone. Therefore, urine specimens must be analyzed specifically for oxycodone, preferably by GC/MS after enzymatic hydrolysis, LLE or SPE, and derivatization by AC, TMS, PFP, or HFB.[1,194,207–210]

Recently, an improved GC/MS method for the simultaneous identification and quantification of opiates in urine was reported.[194] In this method, methoxyamine was used after enzymatic hydrolysis to form methoxime derivatives of the keto-opiates, which were extracted using solid-phase columns and derivatized with propionic anhydride/pyridine. This method demonstrated acceptable precision, lack of cross-interference from other opioids, short analysis time of about 6.5 min, and a small sample volume of 2.0 ml urine. However, LC/MS is the preferred technique for quantification.[211] It has the further advantage that the pharmacologically active morphine-6-glucuronide can also be monitored.

1.4.1.1.5 Nonopioid Analgesics. Nonopioid analgesics are widely used as over-the-counter drugs. Patients with chronic pain often misuse these drugs without control by physicians. Although perceived to be safe drugs, they may lead to severe toxic effects in case of acute overdosage or in case of chronic abuse. They are also misused in doping of humans and horses. Therefore, they may be encountered in clinical and forensic toxicological analysis, as well as in doping control. Analysis of such nonopioid analgesics was reviewed by Kraemer and Maurer.[212] GC/MS was described to be suitable for screening for, and confirmation of, nonopioid analgesics.[213,214] However, the more acidic compounds like the NSAIDs, can be better detected in acidic screening procedures (cf. 1.4.1.2.8).

1.4.1.1.6 Anticonvulsants. Anticonvulsants are usually not abused, but they relatively often lead to accidental, iatrogenic, or suicidal poisonings. They may impair the ability to drive a car or to work with machines. For these reasons a screening procedure for anticonvulsants is necessary. Immunoassays are available for drug monitoring of particular drugs in plasma. As the class of anticonvulsants is chemically heterogeneous, there is no immunoassay suitable for screening of the whole class of drugs.

Many papers have been published for quantification of anticonvulsants by HPLC, LC/MS, or GC/MS,[46,215–217] but there is still only one for screening and identification in urine by GC/MS.[218] Newer anticonvulsants like gabapentin, lamotrigine, oxcarbazepine, or valpromide can also be detected by this procedure.[32]

1.4.1.1.7 Benzodiazepines. Benzodiazepines are used as tranquilizers, hypnotics, anticonvulsants, or muscle relaxants and belong to the most frequently prescribed drugs. They may impair the ability to drive a car or to operate machines, and they may lead to addiction or severe poisonings, especially in combination with alcohol. Therefore, screening for benzodiazepines is necessary in clinical, forensic, and occupational toxicology.[39] Immunoassays are available for screening, but they need off- or on-line cleavage of conjugates to avoid a high percentage of false negatives. Also, for GC/MS confirmation or screening, cleavage of conjugates is mandatory. Both, enzymatic hydrolysis[39] or acid hydrolysis, are used. The latter cleaves the benzodiazepines to benzophenone derivatives or analogues,[219–221] which can sensitively be detected by GC/MS after derivatization, e.g., by AC. However, some benzodiazepines lead to common benzophenones, so that the interpretation of the result may be difficult.[32,219] An alternative to this workup is enzymatic hydrolysis, SPE and silylation.[222] The sensitivity of GC/MS procedures, e.g., for detection of low-dosed benzodiazepines, can be markedly improved by using the the NICI mode.[20,24] This technique is very suitable for benzodiazepine analysis in blood and alternative matrices[20] if LC/MS is not available.[24]

1.4.1.1.8 Sedative–Hypnotics. Sedative–hypnotic drugs are one of the largest groups of drugs, and they can be divided into barbiturates, benzodiazepines (already separately discussed), zopiclone and zolpidem, diphenhydramine and others, including meprobamate, methaqualone, chloral hydrate, and clomethiazole. They are widely used for the treatment of insomnia, anxiety, and convulsive disorders, as well as for anesthetic and preanesthetic treatment. Because of their central nervous and respiratory depressant effects, they may cause, alone or in combination with other drugs and/or ethanol, severe poisoning for which treatment is necessary. Furthermore, they may impair driving ability and the fitness to work with machines, even after therapeutic doses. In particular, barbiturates and benzodiazepines may lead to drug dependence, and they are misused by heroin addicts to ease the withdrawal symptoms from heroin or to augment the effects of "weak heroin." For all these reasons, sedative–hypnotics may be encountered in clinical or forensic toxicological analysis.

Use of barbiturates has been markedly decreased in recent years. However, some of them, like phenobarbital and its precursor primidone, are still used as anticonvulsants for which drug monitoring is necessary. Thiopental is widely used as a short-term intravenous anesthetic. Thiopental and its metabolite pentobarbital are often monitored prior to diagnosis of brain death. A confirmation or screening and identification can be performed by GC/MS in the EI, full-scan, or SIM mode. Barbiturates have only weakly acidic properties and can be detected in screening and confirmation procedures for basic and neutral drugs (e.g., after acid hydrolysis, LLE, and AC[223]), as well as in corresponding procedures for acidic drugs (e.g., after extractive methylation[68]).

Zopiclone and zolpidem have been found to interact with the omega-1 receptor subtype belonging to the $GABA_A$ receptor. They have rapid onset of action and short elimination half-life. Unlike benzodiazepines, they have weak myorelaxant and anticonvulsant effects. They are more and more prescribed as hypnotics instead of benzodiazepines. Diphenhydramine is clinically used as an antihistaminic, antiemetic, and sedative–hypnotic drug. Immunoassays for screening for zopiclon, zolpidem or diphenhydramine, meprobamate, methaqualone, and clomethiazole are not commercially available. Screening for and identification of these drugs can be performed by GC/MS (acid hydrolysis, LLE, and AC).[58,223,224] Chloral hydrate, the hydrate of trichloroacetaldehyde, is still used as a sedative–hypnotic, especially in pediatrics. Trichloroethanol is the main pharmacologically active principle of chloral hydrate therapy, and should therefore be included in the analysis. The usual screening procedures do not cover chloral hydrate or trichloroethanol. Therefore, the Fujiwara reaction for halogenated hydrocarbons is recommended as a qualitative test.[225] However, only chloral hydrate itself leads to the red reaction product, whereas the main metabolite, trichloroethanol, produces only a yellow color which cannot be differentiated from typical urine color. Chloral hydrate and/or trichloroethanol can be determined in urine or plasma by simple GC/MS.[226–228] Further procedures for sedative–hypnotics were reviewed by Kraemer and Maurer.[229]

1.4.1.1.9 Antidepressants. Tricyclic antidepressants (TCA), monamine oxidase inhibitors, and/or the newer selective serotonin reuptake inhibitors (SSRIs) are in use for the pharmacotherapy of depression. Among these, the SSRIs are less toxic since they do not show significant noradrenergic or anticholinergic properties. However, even if they are considered to be safe and well tolerated, they may multiply — by inhibition of cytochrome P450 metabolism — the pharmacological and toxic effects of other drugs like TCAs or anticonvulsants.[230,231] Therefore, they may be encountered in clinical and forensic cases. Methods for quantification have been reviewed.[232–234] In addition, immunoassays are available, but due to different cross-reactivity and potency, the TCA taken must be identified before calculation of the plasma level. Therefore, a systematic toxicological screening must cover these drugs.[58,59,235]

1.4.1.1.10 Neuroleptics of the Phenothiazine and Butyrophenone Type. Neuroleptics (phenothiazines, butyrophenones, and atypics) have suppressing effects on the consciousness and the respiration. Most exhibit anticholinergic effects in the autonomic nervous system. As overdosage may lead to severe poisonings, fast diagnosis is required. Immunoassays are not available, so screening must be performed directly by GC/MS.[1,58,236] Only risperidone cannot be detected by GC/MS; LC/MS is required.[24] Screening, library-assisted identification, and validated quantification of fifteen neuroleptics and three of their metabolites in plasma have recently been described using LC/MS with atmospheric pressure chemical ionization.[22]

1.4.1.1.11 Antihistamines (Histamine H_1-Receptor Blockers). Blockers of the histamine H_1-receptor (antihistamines) are used as allergy remedies, nonprescription hypnotics, and in combination with other drugs such as cold medicines. Antihistamines, often combined with other drugs or alcohol, reduce the ability to drive a car or to work with machines, and they often are the cause of poisonings. Therefore, screening for antihistamines is necessary in clinical, forensic, and occupational toxicology. Identification of 50 H_1-blockers and their metabolites in urine has been described. This procedure allows rapid and specific detection and differentiation of therapeutic concentrations of alkanolamine-, alkylamine-, ethylenediamine-, piperazine-, and phenothiazine antihistamines (H_1-blockers). They are integrated in a systematic GC/MS screening procedure.[1,224,237–239] Simultaneous screening and quantitation of 18 antihistamine drugs in blood has been described using liquid chromatography ionspray tandem MS.[30]

1.4.1.1.12 Antiparkinsonian Drugs. Antiparkinsonian drugs may lead to severe poisonings because of their central depressive and anticholinergic properties. Toxicological detection of these drugs has been described as part of a systematic screening and confirmation procedure.[1,132,240] Selegiline is of special interest since it is metabolized to *R*(–)-methamphetamine and *R*(–)-amphetamine, which interfere with immunoassays for amphetamines.[5,132,133,241] Enantioselective procedures help with differentiation of selegiline and amphetamine/methamphetamine ingestion in urine[132] and blood.[44]

1.4.1.1.13 Beta-Blockers (β-Adrenoceptor Blockers). β-Adrenoceptor blockers, conveniently named beta-blockers, are widely used, and therefore they are frequently encountered in clinical and forensic analysis as well as in doping control.[242] For both indications, GC/MS urine screening procedures have been published using enzymatic or acidic hydrolysis, followed by various derivatization procedures.[1,243–249] Leloux et al.[248] studied the effectiveness of three derivatization procedures: TFA of the amino group and TMS of the hydroxy group, twofold TFA, and *n*-butylboronylation to form a cyclic boronate. The combination of N-TFA with O-TMS proved to be the best procedure. Detection of beta-blockers in human urine by GC/MS-MS has been described and the implications have been discussed for doping control.[250]

1.4.1.1.14 Antiarrhythmics (Class I and IV). Antiarrhythmics may lead to severe cardiac and central nervous disorders if overdosed. As the symptoms of such overdosing are similiar to symptoms of poisonings with other drugs or to symptoms of internal or neurological diseases, a toxicological analysis may be of great importance for diagnosis. Before quantification in plasma, the drug must first be identified, preferably within a systematic screening procedure.[1,244,251]

1.4.1.1.15 Laxatives. Abuse of laxatives may lead to serious disorders like hypokalemia or chronic diarrhea. Toxicological screening should be performed before extensive diagnostic work is started.[252] Laxative use is common among adolescents with anorexia nervosa, and the risk of associated medical complications increases over time.[253] GC/MS procedures have been described for this purpose.[1,254,255] For the detection of the anthraquinone glycosides, which cannot easily be analyzed by GC/MS, high-performance thin-layer chromatographic (HPTLC) methods[256,257] as well as HPLC-DAD methods have been developed.[258]

1.4.1.2 Screening Procedures for Detection of Particular Drug Classes of Acidic Drugs and/or Their Metabolites in Urine

Some classes of acidic drugs, or drugs which are metabolized to acidic compounds like the cardiovascular drugs ACE inhibitors and AT-II blockers, dihydropyridine calcium channel blockers, diuretics, coumarin anticoagulants, antidiabetics of the sulfonylurea type, barbiturates, or NSAIDs, are relevant to clinical and forensic toxicology or doping. Therefore, these acidic drugs should also be monitored, ideally in one procedure.

For STA of acidic drugs and/or their metabolites, gas chromatographic procedures (GC with MS or other detectors), liquid chromatographic procedures (LC with DAD or other detectors), thin-layer chromatographic (TLC with different detection modes) or capillary electrophoretic (CE) procedures have been used.[16] As in STA, a broad range of unknown compounds — even in unknown combinations — must be screened, differentiated, and identified; the separation must be as powerful and universal as possible, and the detection modes must be of the highest specificity and universality. In most papers, the GC/MS coupling was applied.[9,36,55,65–67,72,214,259–269] As acidic compounds are too polar for sensitive GC separation, derivatization by alkylation or silylation is required. Extractive alkylation has proved to be a powerful procedure for simultaneous extraction and derivatization of many acidic compounds.[16,65–69,84]

1.4.1.2.1 ACE Inhibitors and Angiotensin II AT_1 Receptor Blockers. ACE inhibitors are widely used in the treatment of hypertension and congestive heart failure. AT_1 blockers, a new drug class, are used for the same indication. In case of poisoning, ACE inhibitors or AT_1 blockers may lead to severe cardiovascular disorders like hypotension and shock. For diagnosis or for differential diagnostic exclusion of such poisoning, a screening procedure is necessary for the detection of these drugs in urine. ACE inhibitors have a free carboxylic acid group. A further carboxylic group is formed by hydrolysis of the ethyl esters during metabolism and/or sample preparation. The pharmacologically active dicarboxylic acids, the so-called "prilates," are used for parenteral application. AT_1 blockers have also acidic properties, resulting from a carboxylic acid function and/or from the tetrazole ring. Only one screening procedure for ACE inhibitors and AT-II blockers has been published that allows the detection of therapeutic concentrations of most of the ACE inhibitors and for the AT_1 blocker valsartan in human urine samples after extractive methylation.[66] If necessary, the drugs can be quantified in plasma using GC/MS[270–274] or GC/MS-NICI.[24,270,275]

1.4.1.2.2 Anticoagulants of the Hydroxycoumarin Type. Anticoagulants of the 4-hydroxycoumarin type are used as therapeutics or as rodenticides of the so-called first generation. Coumarins of the second generation, the so-called "superwarfarins," are very potent rodenticides and therefore very low-dosed. For the differential diagnosis of unclear coagulopathies, which may occur after ingestion of therapeutic or rodenticide coumarins, screening is needed. Several screening procedures published in recent years are described. The relatively polar coumarin derivatives can be sufficiently separated by reversed phase chromatography. DADs were applied for screening of coumarin anticoagulants of the first generation and for indanedione anticoagulants.[36,276] As coumarins have fluorescent properties, fluorescence detection was used with at least ten times better sensitivity.[277–280] For determination of "superwarfarins," HPLC with fluorescence detection was necessary.[280] GC/MS was used for the detection of the 4-hydroxycoumarin anticoagulants of the first generation and their metabolites in urine after extractive methylation.[65] Derivatization was essential for sensitive GC/MS detection of these vinylogous carboxylic acids and their metabolites (aniline/anilide derivatives or phenols). Only alcoholic hydroxy groups could not be methylated due to their lower nucleophilicity, but this fact did not markedly influence the sensitivity. If necessary, the identified anticoagulants can be quantified in plasma by HPLC [276,277] or TLC.[281]

1.4.1.2.3 Calcium Channel Blockers of the Dihydropyridine Type. Calcium channel blockers, formerly named calcium antagonists, cover three main types: the phenylalkylamines (e.g., verapamil), the benzothiazepines (e.g., diltiazem), and the dihydropyridines (e.g., nifedipine). They are used in the treatment of cardiac dysrhythmias, angina, and/or hypertension. In case of overdose, they may lead to severe cardiovascular disorders like hypotension and shock, possibly resulting in life-threatening situations. For diagnosis or, even more important, for differential diagnostic exclusion of poisoning, a screening procedure is necessary for the detection of these drugs in urine prior to quantification in plasma. The phenylalkylamines and the benzothiazepines and their metabolites can be detected within the STA procedure for basic and neutral compounds.[1,251] The dihydropyridines are excreted only in minor amounts as parent compounds.[282–296] Most of the urinary metabolites are acidic compounds, so that they can be detected by screening procedures for acidic drugs and/or metabolites.[67] If necessary, quantification in plasma can be performed using GC/MS,[297–300] GC,[291,301–304] HPLC,[305–308] LC/MS-MS,[309,310] or modifications of these procedures. Enantioselective determination has been reviewed by Tokuma and Noguchi.[311]

1.4.1.2.4 Diuretics. Diuretics are misused mainly to reduce body weight. The resulting hypokalemia may lead to severe cardiac disorders. Toxicological screening for diuretics should be performed before extensive diagnostic work is started. Diuretics are also misused for doping purposes and therefore, they have been banned by the IOC. For both indications, screening is necessary.[1,312] Besides recent LC/MS procedures,[313–315] GC/MS procedures after (extractive) methylation are preferred,[84,316–318] which simultaneously cover most of the diuretics with series of other drugs relevant in clinical and forensic toxicology or doping.[32,65–69,92,319]

1.4.1.2.5 Antidiabetics of the Sulfonylurea Type. Antidiabetics of the sulfonylurea type have been used since the 1950s in the treatment of hyperglycemia in diabetes mellitus. Besides this therapeutic use, sulfonylureas are also misused. For differential diagnosis of unclear hypoglycemia, screening is necessary to allow differentiation between a surreptitious misuse of sulfonylureas or pathophysiological causes like insulinoma. Before exploratory surgery or even subtotal pancreatectomy, misuse of sulfonylurea drugs should analytically be excluded. Several LC or CE procedures have been published for screening, confirmation, and/or quantification.[320–322] GC/MS after extractive methylation allowed only detection of the sulfonamide part,[32] so that differentiaton was not possible. LC/MS is much better for screening and quantification of these drugs even in plasma.[25]

1.4.1.2.6 Barbiturates. As already mentioned in Section 1.4.1.1.8, barbiturates have only weakly acidic properties; they can be detected in screening and confirmation procedures for basic and neutral drugs[1,223] as well as acidic drugs.[265] In order to improve the GC/MS sensitivity, derivatization by ME,[90,259,265] ethylation,[260] or silylation[9] is preferred. SPME was also applied.[72]

1.4.1.2.7 Designer Drug of the Pyrrolidinophenone Type. As already mentioned in Section 1.4.1.1.2, the new designer drugs alpha-pyrrolidinophenone derivatives like PPP, MPPP, MPHP, MOPPP, and MDPPP are mostly excreted as acidic metabolites.[178–181] Common screening procedures did not cover such metabolites. Only mixed-mode SPE has proven to be suitable for the extraction even of these zwitterionic metabolites and showed good extraction yields.[181] Common trimethylsilylation leads to best GC/MS sensitivity.

1.4.1.2.8 Nonsteroidal Antiinflammatory Drugs (NSAIDs). Nonopiod analgesics are among the most commonly consumed over-the-counter preparations all over the world. Besides acetylsalicylic acid, paracetamol, and pyrazole derivatives, so-called NSAIDs, are used against acute and chronic pain, inflammation, and fever. Altough NSAIDs are perceived to be safe drugs, they may lead to severe toxic effects in case of acute overdosage or in case of chronic abuse. They are also misused in doping of humans and horses. Therefore, they may be encountered in clinical and forensic toxicological analysis, as well as in doping control. Analysis of such nonopioid analgesics has been reviewed by Kraemer and Maurer.[212]

NSAIDs are classified in arylacetic acid derivatives like indomethacin or diclofenac; arylpropionic acid derivatives like ibuprofen, naproxen, or ketoprofen; or oxicames like piroxicam. All these drugs have acidic properties due to (vinylogous) carboxyl groups. Many of the NSAIDs are chiral drugs but most often marketed as racemates. It is known that the enantiomers have different pharmacodynamic and pharmacokinetic properties. The anti-inflammatory activity of NSAIDs has been shown to be mainly evoked by the *S*-enantiomers.[323] However, this stereoselectivity of action is not manifest *in vivo*, due to the thus-far-unique unidirectional metabolic inversion of the chiral center from the inactive *R*(−)-isomers to the active *S*(+)-antipodes.[324] Nevertheless, series of enantioselective determination procedures have been published and reviewed by Davies[325] and Bhushan and Joshi.[326] Several GC/MS, LC, CE, or TLC procedures have been published in the last years for screening, confirmation, and/or quantification. Derivatization of NSAIDs before GC is recommended to improve chromatographic properties and to avoid thermal decarboxylation in the injection port of the GC. Most often, ME after extraction is used,[55,214,265,269] but extractive methylation has also been applied.[69] Silylation as an alternative for ME was studied for 26 NSAIDs.[267,268] The procedures of Laakkonen et al.[265] and Gaillard[36] allowed simultaneous detection of other acidic drugs like barbiturates. Simultaneous determination of 14 NSAIDs in plasma has been described using LC/MS.[327]

1.4.2 General Screening Procedures for Simultaneous Detection of Several Drug Classes in Urine by GC/MS

In so-called general unknown cases, comprehensive screening procedures are needed which cover as many drugs or poisons as possible. Only a few really comprehensive procedures have been published, mainly for urinalysis. As already mentioned, analytical quality criteria of the parent compound are of minor value if the concentrations of the metabolites are much higher in urine than those of the parent drug and if the metabolites are detected by the procedure. The procedure should be sufficiently sensitive to detect therapeutic concentrations at least over a 12 to 24 h period after ingestion.

Papers on the detection time of drugs of abuse in urine have recently been reviewed. [121] Three typical STA procedures have been published for urinalysis: one for the detection of doping-relevant stimulants, beta-blockers, beta-agonists, and narcotics after enzymatic hydrolysis, SPE, and combined TMS and TFA derivatization; [2] one for the detection of most of the basic and neutral drugs in urine after acid hydrolysis and LLE and AC, [1,57,172] recently modified and improved for newer drugs of interest; [58,61,62,126,139,166] and one for acidic drugs, poisons, and/or their metabolites in urine after extractive methylation. [65–69]

1.4.2.1 General Screening Procedure for Drug Classes Relevant in Doping after Enzymatic Hydrolysis, SPE, and Combined TMS and TFA Derivatization

A screening procedure was published for the detection of doping relevant stimulants, beta-blockers, beta-agonists, and narcotics after enzymatic hydrolysis, SPE, and combined TMS and TFA derivatization.[2] The time-consuming enzymatic cleavage of conjugates is acceptable for doping analysis since results do not have to be available as fast as in emergency toxicology. The authors did not focus their procedure on the detection of the metabolites, even if they could be detected in most cases for a longer time and more sensitively than their parent compounds. The chemical properties of the analytes allowed use of SPE with acceptable recoveries. However, the large differences from batch to batch should be kept in mind.[70] The combined TMS and TFA derivatization provided very good GC properties, but underivatized samples cannot be analyzed on the same GC/MS apparatus without changing the column.

This screening procedure, which is limited to some doping-relevant drugs, is also based on full scan GC/MS and mass chromatography for documentation of the absence of the corresponding drug. Such exclusion procedure is suitable for doping analysis since the prevalence for positives is low in contrast to clinical toxicology where the prevalence is high and where many more drugs must be detected or excluded.

1.4.2.2 General Screening for Most of the Basic and Neutral Drug Classes in Urine after Acid Hydrolysis, LLE, and AC

A screening method for detection of most of the basic and neutral drugs in urine after acid hydrolysis, LLE, and AC, has been developed, improved upon, and extended during the last few years. Cleavage of conjugates was necessary before extraction since part of the drugs and/or their metabolites were excreted into urine as conjugates. For studies on toxicological detection, rapid acid hydrolysis was performed to save time, which is relevant, e.g., in emergency toxicology. However, some compounds were destroyed or altered during acid hydrolysis.[57,59,328] Therefore, the standard procedure[63] had to be modified. Before extraction, half of the native urine volume was added to the hydrolyzed part. The extraction solvent used has proved to be very efficient in extracting compounds with very different chemical properties from biomatrices, so that it has been used for a STA procedure for basic and neutral analytes.[32,91,92,172,319] AC has proved to be very suitable for robust derivatization in order to improve the GC properties and thereby the detection limits of thousands of drugs and their metabolites.[32] The use of microwave irradiation reduced the incubation time from 30 to 5 min[59,329] so that derivatization should no longer be renounced due to time consumption.

This comprehensive full scan GC/MS screening procedure allows, within one run, the simultaneous screening and confirmation of the following categories of drugs: amphetamines,[130–132] designer drugs,[60,61,139,166,172,174] barbiturates and other sedative–hypnotics,[223] benzodiazepines,[219] opiates, opioids and other potent analgesics,[199,207] anticonvulsants,[218] antidepressants,[58,235] phenothiazine and butyrophenone neuroleptics,[236,330] nonopioid analgesics,[213,214] antihistamines,[224,237–239] antiparkinsonian drugs,[240] beta-blockers,[244] antiarrhythmics,[251,331] diphenol laxatives,[254] and, finally, herbal drugs like atropine, scopolamine,[32] lauroscholtzin, or protopine.[62] In addition, series of further compounds can be detected[93] if they are present in the extract and their mass spectra are contained in the used reference libraries.[92,94–96]

Eight to ten ions per category were individually selected from the mass spectra of the corresponding drugs and their metabolites identified in authentic urine samples. Table 1.2 summarizes these target ions, which have been updated and optimized. Generations of mass chromatograms can be started by clicking the corresponding pull-down menu which executes the user-defined macros.[91] The procedure is illustrated in Figure 1.7 to Figure 1.16. In Figure 1.7, mass chromatograms are depicted corresponding to fragment ions typical for opioids, indicating the peaks 5 to 9. Figure 1.8 to Figure 1.12 show the unknown mass spectra underlying peaks 5 to 9 (upper part, each), the reference spectra (middle part, each), and the structures and the hit lists found by library search in Reference 92 (lower part, each). As already discussed for the plasma screening, besides the opioids oxycodone and dihydrocodeine, a compound not belonging to the monitored drug class is indicated and could be identified as the muscle relaxant carisoprodol. Figure 1.13 shows mass chromatograms generated from the same data file, corresponding to fragment ions typical for nonopioid analgesics. Besides the already known compounds underlying the peaks 6, 7, and 9, peak 10 appears, which could be identified as acetaminophen (paracetamol [INN], Figure 1.14). This example illustrates again that the selective mass chromatograms provide only a more or less selective screening and only the comparison of the peak

Table 1.2 The Selected Target Ions for Screening in Urine after Acid Hydrolysis and Acetylation

Target Ion (*m/z*)	Psychotropics
58	Methamphetamine, MDMA, BDB, Psilocine, Tramadol, lots of antidepressives and neuroleptics
72	Ethylamphetamine, MBDB, MDE, Methadone, Promethazine, Beta-blockers
84	Methylphenidate, Nicotine
86	Amphetamine, Methylthioamphetamine, Cathinone
91	Amphetamine, Benzylpiperazine
98	Amisulpride, Mepivacaine, Nicotine, Sulpiride
100	Amfepramone, 58er AC, Prothipendyl
114	Ethylamphetamine, MBDB, MDE, Methadone, Promethazine, Beta-blockers
194	Caffeine
	Antidepressants
182	Nefazodone, Trazodone
190	Fluoxetine
192	Amineptine
195	Mirtazapine
215	Amitriptyline
234	Paroxetine
277	Maprotiline
290	Sertraline
292	Doxepin
361	Opipramol
	Phenothiazine Neuroleptics
141	Clopenthixol, Flupentixol, Fluphenazine, Perphenazine, Trifluperazine
185	Chlorprothixen, Flupenthixol, Perphenazine
203	Fluspirilen, Penfluridol, Pimozide
210	Quetiapine
221	Clopenthixol
260	Zotepine
284	Olanzapine
298	Clozapine
	Butyrophenone Neuroleptics
82	Benperidol
112	Melperone
123	Droperidol, etc.
165	Pipamperone
185	Moperone
189	Haloperidol
223	Trifluperidol
233	Bromperidol
235	Haloperidol

Table 1.2 (*Continued*) The Selected Target Ions for Screening in Urine after Acid Hydrolysis and Acetylation

	Benzodiazepines (Benzophenones)
111	Clonazepam, Lormetazepam
211	Clonazepam, Flunitrazepam, Nitrazepam
230	Chlordiazepoxide, Clorazepate, Lorazepam, Nordazepam, Oxazepam
245	Brotizolam, Diazepam, Temazepam
249	Bromazepam, Flurazepam, Tetrazepam
257	Clobazam
308	Alprazolam
340	Midazolam
357	Triazolam
	Barbiturates
83	Meprobamate, Vinylbital
117	(Methyl-)Phenobarbital
141	Amobarbital, Aprobarbital, Barbital, Butabarbital, Butobarbital, Crotylbarbital, Dipropylbarbital, Nealbarbital, Pentobarbital, Vinbarbital
157	Thiopental
167	Allobarbital, Butalbital, Secobarbital
207	Brallobarbital, Cyclobarbital
221	Heptabarbital, Hexobarbital, Methohexital
235	Methaqualone, Zolpidem
	Sedative–Hypnotics
83	Meprobamate
105	Etomidate
156	Pentobarbital
163	Propofol
167	Diphenhydramine
172	Thiopental
216	Ketamine
235	Methaqualone, Zolpidem
248	Zaleplon
261	Zopiclone
	Anticonvulsants
102	Valproic acid
113	Ethosuximide
146	Primidone
185	Lamotrigine
193	Carbamazepine
204	Phenobarbital
208	Phenytoin
241	Clonazepam
	Stimulants/Hallucinogens
82	Cocaine

Table 1.2 (*Continued*) The Selected Target Ions for Screening in Urine after Acid Hydrolysis and Acetylation

	Stimulants/Hallucinogens
94	Scopolamine
124	Atropine
140	Bupivacaine, Prolintane
162	MDMA, MDA, MDE
164	EA, MDMA, MDA
176	MBDB, MDE
178	BDB, MBDB, MDE
192	MMDA
250	Cafedrine, Fenetylline
	Opioids
111	Tilidine
138	Piritramide
245	Fentanyl
187	Pethidine
259	Pentazosine
327	Morphine, Pholcodine
341	Codeine, Naltrexone
343	Dihydrocodeine
359	Oxycodone
420	Buprenorphine
	Analgesics
120	Salicylic acid
151	Paracetamol
161	Ibuprofen
188	Morazone, Phenazone, Tramadol
217	Metamizol
230	Diclofenac
231	Propyphenazone
139	Indomethacin
308	Phenylbutazone
258	Flupirtine
	Cardiovascular Drugs
140	Atenolol
159	Metipranolol, Propranolol, Toliprolol, etc.
200	Betaxolol, Metipranolol, Metoprolol, Oxprenolol, Toliprolol, etc.
277	Gallopamil
282	Nifedipine
289	Verapamil
297	Nimodipine, Nitrendipine
303	Verapamil
347	Amlodipine

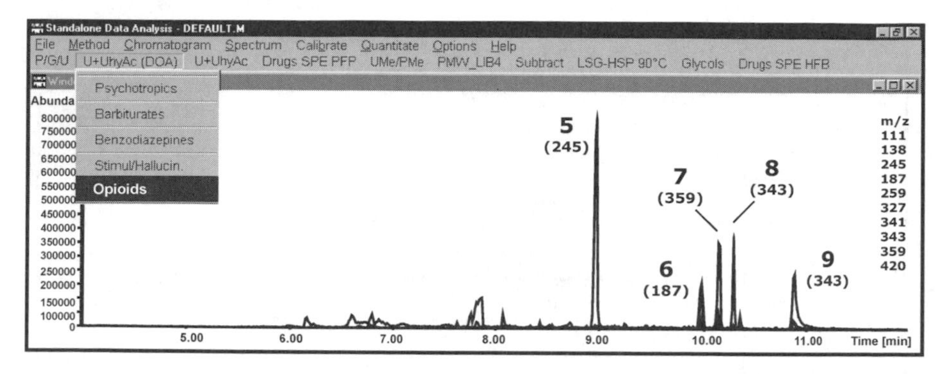

Figure 1.7 Mass chromatograms corresponding to fragment ions typical for opioids of a urine sample after acid hydrolysis, extraction, and acetylation.

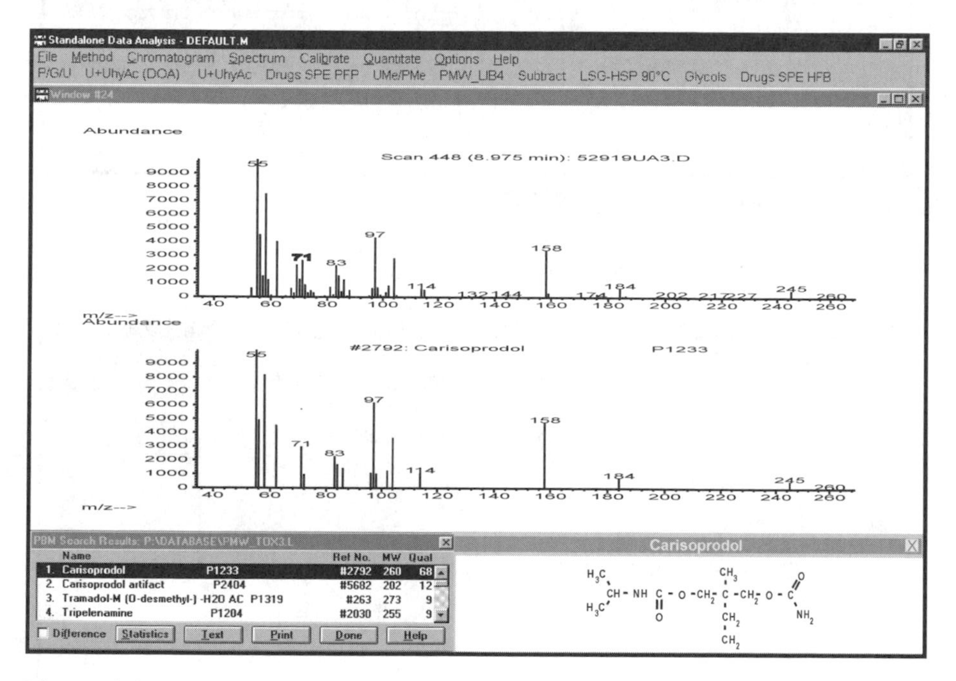

Figure 1.8 Unknown mass spectrum underlying the peak 5 in Figure 1.7 (upper part), the reference spectrum (middle part), and the structure and the hit list found by library search in Reference 92 (lower part).

underlying full mass spectrum allows the specific identification. Finally, Figure 1.15 shows mass chromatograms corresponding to fragment ions typical for sedative–hypnotics indicating peak 11, identified as meprobamate (Figure 1.16). In this case, urinalysis showed that the ingested drugs could be detected in urine in much higher amounts than in plasma (Figure 1.1 and Figure 1.5), showing that the detection window in urine is much wider that

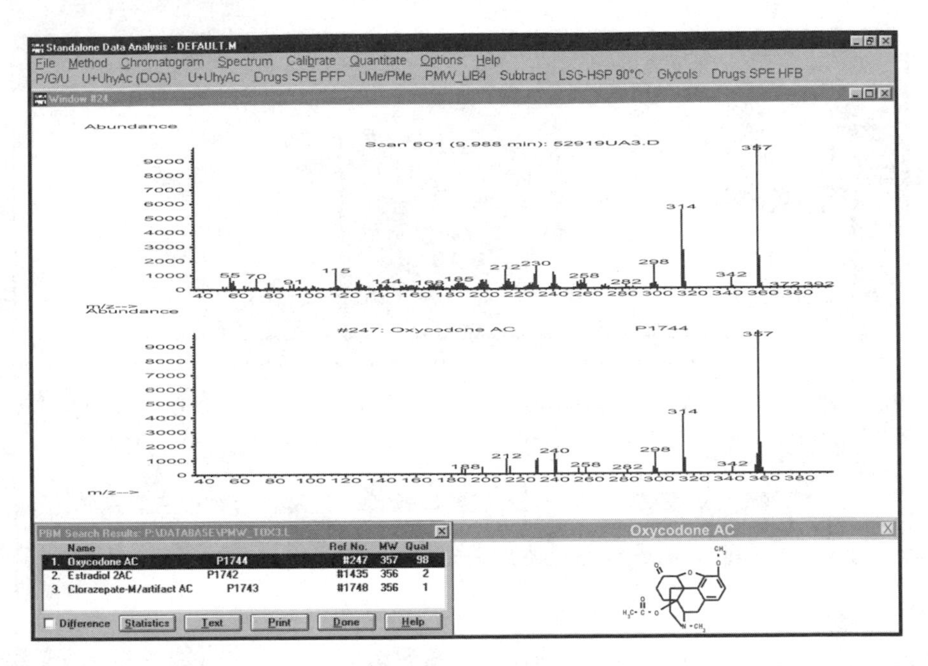

Figure 1.9 Unknown mass spectrum underlying the peak 6 in Figure 1.7 (upper part), the reference spectrum (middle part), and the structure and the hit list found by library search in Reference 92 (lower part).

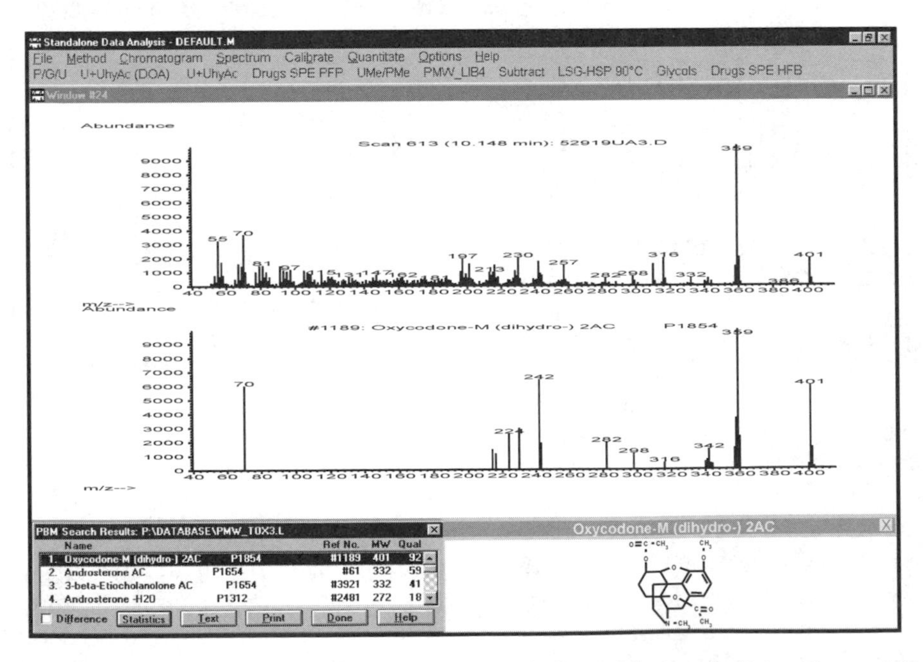

Figure 1.10 Unknown mass spectrum underlying the peak 7 in Figure 1.7 (upper part), the reference spectrum (middle part), and the structure and the hit list found by library search in Reference 92 (lower part).

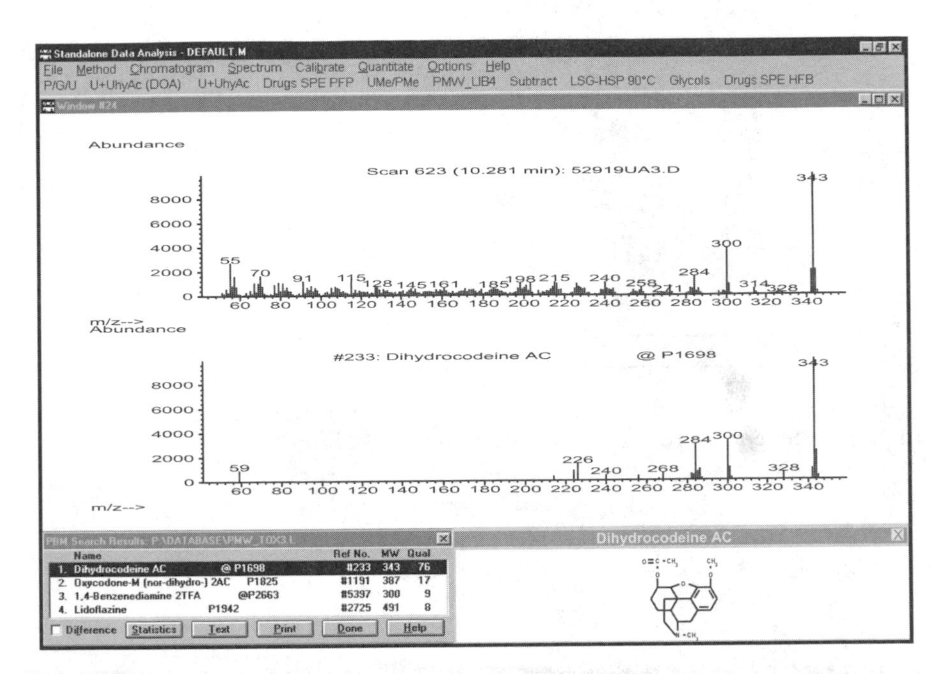

Figure 1.11 Unknown mass spectrum underlying the peak 8 in Figure 1.7 (upper part), the reference spectrum (middle part), and the structure and the hit list found by library search in Reference 92 (lower part).

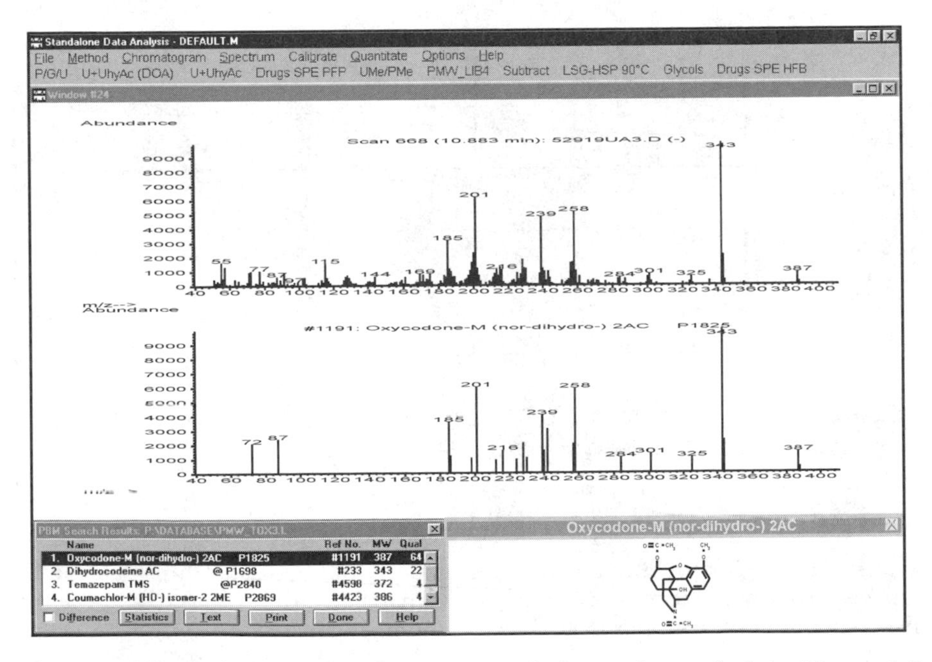

Figure 1.12 Unknown mass spectrum underlying the peak 9 in Figure 1.7 (upper part), the reference spectrum (middle part), and the structure and the hit list found by library search in Reference 92 (lower part).

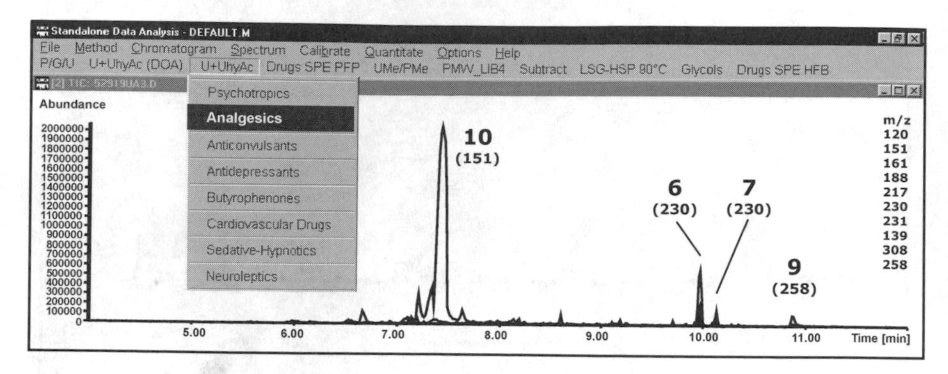

Figure 1.13 Mass chromatograms corresponding to fragment ions typical for analgesics generated from the same data file as used for Figure1.7.

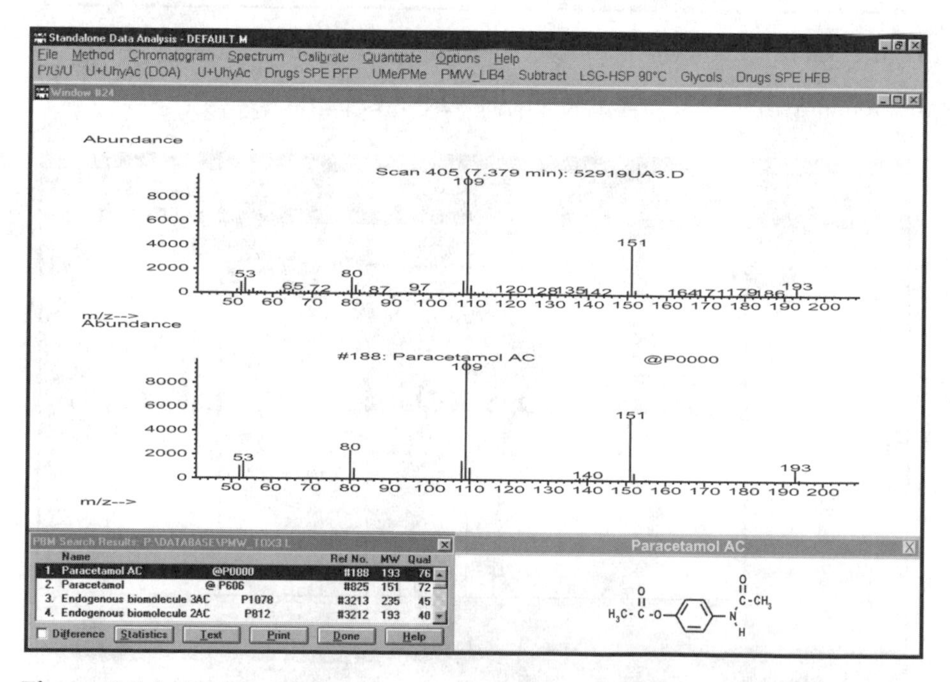

Figure 1.14 Unknown mass spectrum underlying the peak 10 in Figure 1.13 (upper part), the reference spectrum (middle part), and the structure and the hit list found by library search in Reference 92 (lower part).

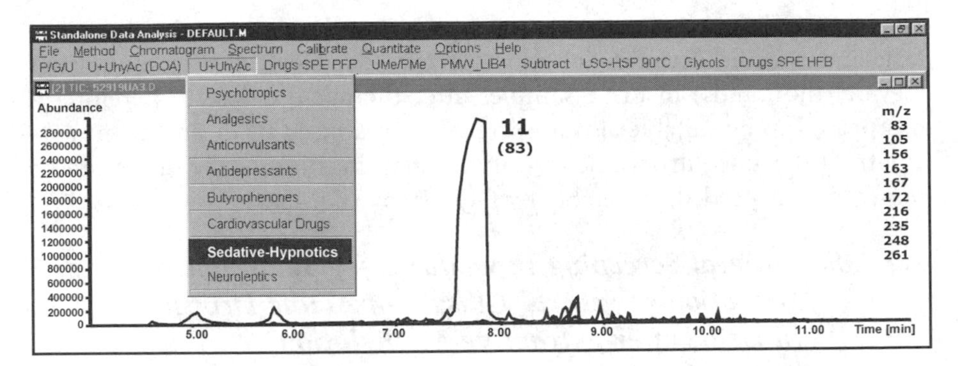

Figure 1.15 Mass chromatograms corresponding to fragment ions typical for sedative–hypnotics generated from the same data file as used for Figure 1.7.

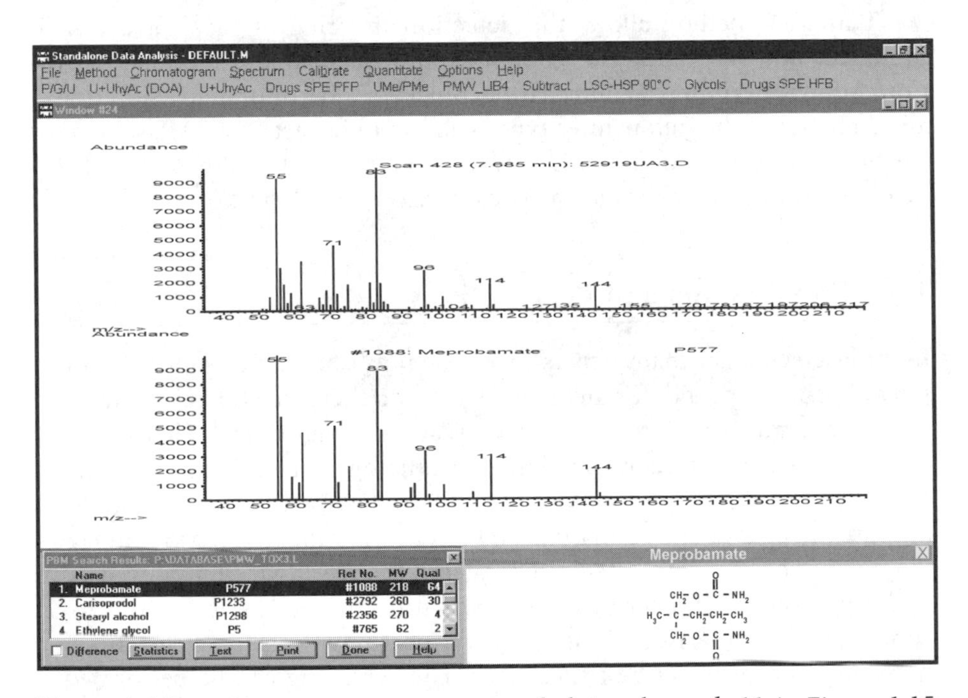

Figure 1.16 Unknown mass spectrum underlying the peak 11 in Figure 1.15 (upper part), the reference spectrum (middle part),and the structure and the hit list found by library search in Reference 92 (lower part).

in plasma. The described and exemplified procedure allows simultaneous, fast, and specific detection of most of the toxicologically relevant drugs (several thousands) in urine samples after therapeutic doses. Therefore, it has proved to be suitable also for screening of abused medicaments in psychiatry. After unequivocal identification, quantification of the drugs can be performed, if needed, preferably by GC/MS or LC/MS.

1.4.2.3 *General Screening Procedures for Simultaneous Detection of Several Classes of Acidic Drugs in Urine after Extractive Methylation*

A comprehensive GC/MS screening procedure for the detection of acidic drugs, poisons, and/or their metabolites in urine after extractive methylation was developed.[65–69] The analytes were separated by capillary GC and identified by computerized MS in the full scan mode. As already described above, the possible presence of acidic drugs and/or their metabolites could be indicated using mass chromatography with selective ions followed by peak identification using library search.[92,94]

This STA method allows the detection in urine of most of the ACE inhibitors and AT_1 blockers,[66] coumarin anticoagulants of the first generation,[65] dihyropyridine calcium channel blockers,[67] barbiturates,[68] diuretics,[317] antidiabetics of the sulfonylurea type (sulfonamide part),[32] NSAIDs,[69] and of various other acidic compounds.[32] At least the higher dosed drugs could also be detected in plasma samples after extractive methylation.

1.5 Conclusions and Perspectives

In the last few years, many papers have been published concerning GC/MS screening in body fluids for unknown drugs and their metabolites relevant to clinical toxicology, forensic toxicology, and doping control. They either describe procedures for confirmation of chromatographic or immunological results or for STA. Confirmation is usually performed in the SIM mode because only particular compounds have to be identified. GC/MS is today the method of choice for STA in clinical and forensic toxicology as well as in doping control. If the drug is unknown, full scan mode is the method of choice, since comparison of the full mass spectra with reference spectra is necessary and provides best specificity. The screening can be performed using mass chromatography followed by library search. Besides common toxicological mass spectral libraries [92] very large mass spectra collections with more than 200,000 entries [95,96] can also be used. Quantification in the SIM mode provides very good precision especially using stable isotopes as internal standards. However, they are commercially available for only a few drugs.

Such quantification procedures must be fully validated according to international guidelines. [47]

Finally, LC/MS has shown to be an ideal supplement, especially for more polar, unstable, or low-dosed drugs.

Acknowledgments

The author would like to thank his coworkers Thomas Kraemer, Carsten Kratzsch, Frank T. Peters, Dietmar Springer, Roland F. Staack, and Armin A. Weber for their support.

References

1. Maurer, H.H., Systematic toxicological analysis of drugs and their metabolites by gas chromatography-mass spectrometry [review], *J. Chromatogr.*, 580, 3, 1992.
2. Solans, A., Carnicero, M., de-la-Torre, R., and Segura, J., Comprehensive screening procedure for detection of stimulants, narcotics, adrenergic drugs, and their metabolites in human urine, *J. Anal. Toxicol.*, 19, 104, 1995.
3. Simpson, D., Braithwaite, R.A., Jarvie, D.R., Stewart, M.J., Walker, S., Watson, I.W., and Widdop, B., Screening for drugs of abuse (II): Cannabinoids, lysergic acid diethylamide, buprenorphine, methadone, barbiturates, benzodiazepines and other drugs, *Ann. Clin. Biochem.*, 34 (Pt 5), 460, 1997.
4. Lehrer, M., The role of gas chromatography/mass spectrometry. Instrumental techniques in forensic urine drug testing. Looking all around: honeybees use different cues in different eye regions, *Clin. Lab. Med.*, 18, 631, 1998.
5. Kraemer, T. and Maurer, H.H., Determination of amphetamine, methamphetamine and amphetamine-derived designer drugs or medicaments in blood and urine [review], *J. Chromatogr. B*, 713, 163, 1998.
6. Moeller, M.R., Steinmeyer, S., and Kraemer, T., Determination of drugs of abuse in blood [review], *J. Chromatogr. B*, 713, 91, 1998.
7. Kidwell, D.A., Holland, J.C., and Athanaselis, S., Testing for drugs of abuse in saliva and sweat [review], *J. Chromatogr. B*, 713, 111, 1998.
8. Sachs, H. and Kintz, P., Testing for drugs in hair. Critical review of chromatographic procedures since 1992 [review], *J. Chromatogr. B*, 713, 147, 1998.
9. Polettini, A., Groppi, A., Vignali, C., and Montagna, M., Fully-automated systematic toxicological analysis of drugs, poisons, and metabolites in whole blood, urine, and plasma by gas chromatography-full scan mass spectrometry, *J. Chromatogr. B*, 713, 265, 1998.
10. Kintz, P. and Samyn, N., Determination of "Ecstasy" components in alternative biological specimens [review], *J. Chromatogr. B*, 733, 137, 1999.

11. Polettini, A., Systematic toxicological analysis of drugs and poisons in biosamples by hyphenated chromatographic and spectroscopic techniques [review], *J. Chromatogr. B*, 733, 47, 1999.
12. Huestis, M., Oyler, J.M., Cone, E., Wstadik, A.T., Schoendorfer, D., and Joseph, R.E., Sweat testing for cocaine, codeine and metabolites by gas chromatography-mass spectrometry [review], *J. Chromatogr. B*, 733, 247, 1999.
13. Gaillard, Y. and Pepin, G., Testing hair for pharmaceuticals [review], *J. Chromatogr. B*, 733, 231, 1999.
14. Staub, C., Chromatographic procedures for determination of cannabinoids in biological samples, with special attention to blood and alternative matrices like hair, saliva, sweat and meconium [review], *J. Chromatogr. B*, 733, 119, 1999.
15. Nakahara, Y., Hair analysis for abused and therapeutic drugs [review], *J. Chromatogr. B*, 733, 161, 1999.
16. Maurer, H.H., Systematic toxicological analysis procedures for acidic drugs and/or metabolites relevant to clinical and forensic toxicology or doping control [review], *J. Chromatogr. B*, 733, 3, 1999.
17. Maurer, H.H., Screening procedures for simultaneous detection of several drug classes used in the high throughput toxicological analysis and doping control [review], *Comb. Chem. High Throughput Screen.*, 3, 461, 2000.
18. Maurer, H.H., Applications of GC-MS in Clinical and Forensic Toxicology and Doping Control, in *Current Practice of GC-MS*, Niessen, W.M.A., Ed., Marcel Dekker, New York, 2001, 355.
19. Kintz, P. and Samyn, N., Use of alternative specimens drugs of abuse in saliva and doping agents in hair [review], *Ther. Drug Monit.*, 24, 239, 2002.
20. Maurer, H.H., The role of gas chromatography-mass spectrometry with negative ion chemical ionization (GC-MS-NCI) in clinical and forensic toxicology, doping control and biomonitoring [review], *Ther. Drug Monit.*, 24, 247, 2002.
21. Moeller, M.R. and Kraemer, T., Drugs of abuse monitoring in blood for control of driving under the influence of drugs [review], *Ther. Drug Monit.*, 24, 210, 2002.
22. Kratzsch, C., Weber, A.A., Peters, F.T., Kraemer, T., and Maurer, H.H., Screening, library-assisted identification and validated quantification of fifteen neuroleptics and three of their metabolites in plasma by liquid chromatography/mass spectrometry with atmospheric pressure chemical ionization, *J. Mass Spectrom.*, 38, 283, 2003.
23. Maurer, H.H., Kraemer, T., Kratzsch, C., Paul, L.D., Peters, F.T., Springer, D., Staack, R.F., and Weber, A.A., Systematic toxicological analysis by GC-MS and LC-MS, in *Proceedings of the Symposium Toxicological Analysis and Certainty of Results, Leipzig, October 1 and 2, 2001*, Kleemann, W.J. and Teske, J., Eds., Schmidt-Roemhild Verlag, Leipzig, 2002.

24. Maurer, H.H., Kraemer, T., Kratzsch, C., Peters, F.T., and Weber, A.A., Negative ion chemical ionization gas chromatography-mass spectrometry (NICI-GC-MS) and atmospheric pressure chemical ionization liquid chromatography-mass spectrometry (APCI-LC-MS) of low-dosed and/or polar drugs in plasma, *Ther. Drug Monit.*, 24, 117, 2002.

25. Maurer, H.H., Kratzsch, C., Kraemer, T., Peters, F.T., and Weber, A.A., Screening, library-assisted identification and validated quantification of oral antidiabetics of the sulfonylurea-type in plasma by atmospheric pressure chemical ionization liquid chromatography-mass spectrometry (APCI-LC-MS), *J. Chromatogr. B*, 773, 63, 2002.

26. Saint-Marcoux, F., Lachatre, G., and Marquet, P., Evaluation of an improved general unknown screening procedure using liquid chromatography-electrospray-mass spectrometry by comparison with gas chromatography and high-performance liquid-chromatography — diode array detection, *J. Am. Soc. Mass Spectrom.*, 14, 14, 2003.

27. Marquet, P., Is LC-MS suitable for a comprehensive screening of drugs and poisons in clinical toxicology?, *Ther. Drug Monit.*, 24, 125, 2002.

28. Marquet, P., Progress of LC-MS in clinical and forensic toxicology [review], *Ther. Drug Monit.*, 24, 255, 2002.

29. Rittner, M., Pragst, F., Bork, W.R., and Neumann, J., Screening method for seventy psychoactive drugs or drug metabolites in serum based on high-performance liquid chromatography-electrospray ionization mass spectrometry, *J. Anal. Toxicol.*, 25, 115, 2001.

30. Gergov, M., Robson, J.N., Ojanpera, I., Heinonen, O.P., and Vuori, E., Simultaneous screening and quantitation of 18 antihistamine drugs in blood by liquid chromatography ionspray tandem mass spectrometry, *Forensic Sci. Int.*, 121, 108, 2001.

31. Bogusz, M.J., LC/MS in forensic toxicology, in *Advances in Forensic Applications of Mass Spectrometry*, Yinon, J., Ed., Chapter 2, this volume.

32. Pfleger, K., Maurer, H.H., and Weber, A., *Mass Spectral and GC Data of Drugs, Poisons, Pesticides, Pollutants and their Metabolites, part 4*, Wiley-VCH, Weinheim, 2000.

33. Drummer, O.H., Kotsos, A., and McIntyre, I.M., A class-independent drug screen in forensic toxicology using a photodiode array detector, *J. Anal. Toxicol.*, 17, 225, 1993.

34. Balikova, M., Application of HPLC with photodiode array detection for systematic toxicological analyses of drug groups, *Sb. Lek.*, 95, 339, 1994.

35. Bogusz, M., Hyphenated liquid chromatographic techniques in forensic toxicology [review], *J. Chromatogr. B*, 733, 65, 1999.

36. Gaillard, Y. and Pepin, G., Screening and identification of drugs in human hair by high-performance liquid chromatography-photodiode-array UV detection and gas chromatography-mass spectrometry after solid-phase extraction. A powerful tool in forensic medicine, *J. Chromatogr. A*, 762, 251, 1997.
37. Gaillard, Y. and Pepin, G., Use of high-performance liquid chromatography with photodiode-array UV detection for the creation of a 600-compound library. Application to forensic toxicology, *J. Chromatogr. A*, 763, 149, 1997.
38. Elliott, S.P. and Hale, K.A., Applications of an HPLC-DAD drug-screening system based on retention indices and UV spectra, *J. Anal. Toxicol.*, 22, 279, 1998.
39. Drummer, O.H., Methods for the measurement of benzodiazepines in biological samples [review], *J. Chromatogr. B*, 713, 201, 1998.
40. Valli, A., Polettini, A., Papa, P., and Montagna, M., Comprehensive drug screening by integrated use of gas chromatography/mass spectrometry and Remedi HS, *Ther. Drug Monit.*, 23, 287, 2001.
41. Maurer, H.H., Liquid chromatography-mass spectrometry in forensic and clinical toxicology [review], *J. Chromatogr. B*, 713, 3, 1998.
42. Van Bocxlaer, J.F., Clauwaert, K.M., Lambert, W.E., Deforce, D.L., Van den Eeckhout, E.G., and de-Leenheer, A.P., Liquid chromatography-mass spectrometry in forensic toxicology [review], *Mass Spectrom. Rev.*, 19, 165, 2000.
43. Bogusz, M.J., Liquid chromatography-mass spectrometry as a routine method in forensic sciences: a proof of maturity, *J. Chromatogr. B Biomed. Sci. Appl.*, 748, 3, 2000.
44. Peters, F.T., Kraemer, T., and Maurer, H.H., Drug testing in blood: validated negative-ion chemical ionization gas chromatographic-mass spectrometric assay for determination of amphetamine and methamphetamine enantiomers and its application to toxicology cases, *Clin. Chem.*, 48, 1472, 2002.
45. Peters, F.T., Schaefer, S., Staack, R.F., Kraemer, T., and Maurer, H.H., Screening for and validated quantification of amphetamines as well as of amphetamine- and piperazine-derived designer drugs in human blood plasma by gas chromatography/mass spectrometry, *J. Mass Spectrom.*, 38, 659, 2003.
46. Maurer, H.H., Kratzsch, C., Weber, A.A., Peters, F.T., and Kraemer, T., Validated assay for quantification of oxcarbazepine and its active dihydro metabolite 10-hydroxy carbazepine in plasma by atmospheric pressure chemical ionization liquid chromatography/mass spectrometry, *J. Mass Spectrom.*, 37, 687, 2002.
47. Peters, F.T. and Maurer, H.H., Bioanalytical method validation and its implications for forensic and clinical toxicology — A review [review], *Accred. Qual. Assur.*, 7, 441, 2002.

48. Polettini, A., A simple automated procedure for the detection and identification of peaks in gas chromatography — continuous scan mass spectrometry. Application to systematic toxicological analysis of drugs in whole human blood, *J. Anal. Toxicol.*, 20, 579, 1996.
49. Pfleger, K., Maurer, H.H., and Weber, A., *Mass Spectral and GC Data of Drugs, Poisons, Pesticides, Pollutants and their Metabolites*, VCH, Weinheim, Germany, 1992.
50. Takeda, A., Tanaka, H., Shinohara, T., and Ohtake, I., Systematic analysis of acid, neutral and basic drugs in horse plasma by combination of solid-phase extraction, non-aqueous partitioning and gas chromatography-mass spectrometry, *J. Chromatogr. B Biomed. Sci. Appl.*, 758, 235, 2001.
51. Drummer, O.H. and Gerostamoulos, J., Postmortem drug analysis: analytical and toxicological aspects, *Ther. Drug Monit.*, 24, 199, 2002.
52. Villain, M., Cirimele, V., and Kintz, P., Substance abuse in sports: detection of doping agents in hair by mass spectrometry, in *Advances in Forensic Applications of Mass Spectrometry*, Yinon, J., Ed., Chapter 3, this volume.
53. Moore, C., Negrusz, A., and Lewis, D., Determination of drugs of abuse in meconium [review], *J. Chromatogr. B*, 713, 137, 1998.
54. Engelhart, D.A., Lavins, E.S., and Sutheimer, C.A., Detection of drugs of abuse in nails, *J. Anal. Toxicol.*, 22, 314, 1998.
55. Gonzalez, G., Ventura, R., Smith, A.K., de-la-Torre, R., and Segura, J., Detection of non-steroidal anti-inflammatory drugs in equine plasma and urine by gas chromatography-mass spectrometry, *J. Chromatogr. A*, 719, 251, 1996.
56. Spahn, L.H. and Benet, L.Z., Acyl glucuronides revisited: is the glucuronidation process a toxification as well as a detoxification mechanism?, *Drug Metab. Rev.*, 24, 5, 1992.
57. Maurer, H.H., Methods for GC-MS, in *Mass Spectral and GC Data of Drugs, Poisons, Pesticides, Pollutants and their Metabolites, Part 4*, Pfleger, K., Maurer, H.H., and Weber, A., Eds., Wiley-VCH, Weinheim, Germany, 2000, 3.
58. Bickeboeller-Friedrich, J. and Maurer, H.H., Screening for detection of new antidepressants, neuroleptics, hypnotics, and their metabolites in urine by GC-MS developed using rat liver microsomes, *Ther. Drug Monit.*, 23, 61, 2001.
59. Maurer, H.H. and Bickeboeller-Friedrich, J., Screening procedure for detection of antidepressants of the selective serotonin reuptake inhibitor type and their metabolites in urine as part of a modified systematic toxicological analysis procedure using gas chromatography-mass spectrometry, *J. Anal. Toxicol.*, 24, 340, 2000.
60. Staack, R.F., Fritschi, G., and Maurer, H.H., Studies on the metabolism and the toxicological analysis of the new piperazine-like designer drug *N*-benzylpiperazine (BZP, A2) using gas chromatography-mass spectrometry (GC-MS), *J. Chromatogr. B*, 773, 35, 2002.

61. Staack, R.F., Fehn, J., and Maurer, H.H., New designer drug para-methoxymethamphetamine (PMMA): studies on its metabolism and toxicological detection in urine using gas chromatography-mass spectrometry, *J. Chromatogr. B*, 789, 27, 2003.
62. Paul, L.D. and Maurer, H.H., Studies on the metabolism and toxicological detection of the *Eschscholtzia californica* alkaloids californine and protopine in urine using gas chromatography-mass spectrometry, *J. Chromatogr. B*, 789, 43, 2003.
63. Maurer, H.H., Methods for GC-MS, in *Mass Spectral and GC Data of Drugs, Poisons, Pesticides, Pollutants and their Metabolites*, Pfleger, K., Maurer, H.H., and Weber, A., Eds., VCH, Weinheim, Germany, 1992, 3.
64. Toennes, S.W. and Maurer, H.H., Efficient cleavage of urinary conjugates of drugs or poisons in analytical toxicology using purified and immobilized β-glucuronidase and arylsulfatase packed in columns, *Clin. Chem.*, 45, 2173, 1999.
65. Maurer, H.H. and Arlt, J.W., Detection of 4-hydroxycoumarin anticoagulants and their metabolites in urine as part of a systematic toxicological analysis procedure for acidic drugs and poisons by gas chromatography-mass spectrometry after extractive methylation., *J. Chromatogr. B*, 714, 181, 1998.
66. Maurer, H.H., Kraemer, T., and Arlt, J.W., Screening for the detection of angiotensin-converting enzyme inhibitors, their metabolites, and AT II receptor antagonists, *Ther. Drug Monit.*, 20, 706, 1998.
67. Maurer, H.H. and Arlt, J.W., Screening procedure for detection of dihydropyridine calcium channel blocker metabolites in urine as part of a systematic toxicological analysis procedure for acidics by gas chromatography-mass spectrometry (GC-MS) after extractive methylation, *J. Anal. Toxicol.*, 23, 73, 1999.
68. Maurer, H.H., Tauvel, F.X., and Kraemer, T., Detection of non-steroidal anti-inflammatory drugs (NSAIDs), barbiturates and their metabolites in urine as part of a systematic toxicological analysis (STA) procedure for acidic drugs and poisons by GC-MS, in *Proceedings of the 38th International TIAFT Meeting in Helsinki*, Rasanen, I., Ed., TIAFT, Helsinki, 2001, 316.
69. Maurer, H.H., Tauvel, F.X., and Kraemer, T., Screening procedure for detection of non-steroidal antiinflammatory drugs (NSAIDs) and their metabolites in urine as part of a systematic toxicological analysis (STA) procedure for acidic drugs and poisons by gas chromatography-mass spectrometry (GC-MS) after extractive methylation, *J. Anal. Toxicol.*, 25, 237, 2001.
70. Bogusz, M.J., Maier, R.D., Schiwy, B.K., and Kohls, U., Applicability of various brands of mixed-phase extraction columns for opiate extraction from blood and serum., *J. Chromatogr. B*, 683, 177, 1996.

71. Reubsaet, K.J., Ragnar, N.H., Hemmersbach, P., and Rasmussen, K.E., Determination of benzodiazepines in human urine and plasma with solvent modified solid phase micro extraction and gas chromatography; rationalisation of method development using experimental design strategies, *J. Pharm. Biomed. Anal.*, 18, 667, 1998.

72. Hall, B.J. and Brodbelt, J.S., Determination of barbiturates by solid-phase microextraction (SPME) and ion trap gas chromatography-mass spectrometry., *J. Chromatogr. A*, 777, 275, 1997.

73. Ulrich, S., Kruggel, S., Weigmann, H., and Hiemke, C., Fishing for a drug: solid-phase microextraction for the assay of clozapine in human plasma, *J. Chromatogr. B Biomed. Sci. Appl.*, 731, 231, 1999.

74. Koester, C.J., Andresen, B.D., and Grant, P.M., Optimum methamphetamine profiling with sample preparation by solid-phase microextraction, *J. Forensic Sci.*, 47, 1002, 2002.

75. Musshoff, F., Junker, H.P., Lachenmeier, D.W., Kroener, L., and Madea, B., Fully automated determination of amphetamines and synthetic designer drugs in hair samples using headspace solid-phase microextraction and gas chromatography-mass spectrometry, *J. Chromatogr. Sci.*, 40, 359, 2002.

76. Musshoff, F., Junker, H.P., Lachenmeier, D.W., Kroener, L., and Madea, B., Fully automated determination of cannabinoids in hair samples using headspace solid-phase microextraction and gas chromatography-mass spectrometry, *J. Anal. Toxicol.*, 26, 554, 2002.

77. Fucci, N., De Giovanni, N., Chiarotti, M., and Scarlata, S., SPME-GC analysis of THC in saliva samples collected with "EPITOPE" device, *Forensic Sci. Int.*, 119, 318, 2001.

78. Liu, J., Hara, K., Kashimura, S., Kashiwagi, M., and Kageura, M., New method of derivatization and headspace solid-phase microextraction for gas chromatographic-mass spectrometric analysis of amphetamines in hair, *J. Chromatogr. B Biomed. Sci. Appl.*, 758, 95, 2001.

79. Okajima, K., Namera, A., Yashiki, M., Tsukue, I., and Kojima, T., Highly sensitive analysis of methamphetamine and amphetamine in human whole blood using headspace solid-phase microextraction and gas chromatography-mass spectrometry, *Forensic Sci. Int.*, 116, 15, 2001.

80. Jurado, C., Gimenez, M.P., Soriano, T., Menendez, M., and Repetto, M., Rapid analysis of amphetamine, methamphetamine, MDA, and MDMA in urine using solid-phase microextraction, direct on-fiber derivatization, and analysis by GC-MS, *J. Anal. Toxicol.*, 24, 11, 2000.

81. Sporkert, F. and Pragst, F., Use of headspace solid-phase microextraction (HS-SPME) in hair analysis for organic compounds, *Forensic Sci. Int.*, 107, 129, 2000.

82. Battu, C., Marquet, P., Fauconnet, A.L., Lacassie, E., and Lachatre, G., Screening procedure for 21 amphetamine-related compounds in urine using solid-phase microextraction and gas chromatography-mass spectrometry, *J. Chromatogr. Sci.*, 36, 1, 1998.

83. Lisi, A.M., Kazlauskas, R., and Trout, G.J., Gas chromatographic-mass spectrometric quantitation of urinary buprenorphine and norbuprenorphine after derivatization by direct extractive alkylation., *J. Chromatogr. B*, 692, 67, 1997.

84. Lisi, A.M., Kazlauskas, R., and Trout, G.J., Diuretic screening in human urine by gas chromatography-mass spectrometry: use of a macroreticular acrylic copolymer for the efficient removal of the coextracted phase-transfer reagent after derivatization by direct extractive alkylation., *J. Chromatogr.*, 581, 57, 1992.

85. Segura, J., Ventura, R., and Jurado, C., Derivatization procedures for gas chromatographic-mass spectrometric determination of xenobiotics in biological samples, with special attention to drugs of abuse and doping agents [review], *J. Chromatogr. B*, 713, 61, 1998.

86. Schutz, H., Gotta, J.C., Erdmann, F., Risse, M., and Weiler, G., Simultaneous screening and detection of drugs in small blood samples and bloodstains, *Forensic Sci. Int.*, 126, 191, 2002.

87. Weinmann, W., Renz, M., Vogt, S., and Pollak, S., Automated solid-phase extraction and two-step derivatisation for simultaneous analysis of basic illicit drugs in serum by GC/MS, *Int. J. Legal Med.*, 113, 229, 2000.

88. Staerk, U. and Kulpmann, W.R., High-temperature solid-phase microextraction procedure for the detection of drugs by gas chromatography-mass spectrometry, *J. Chromatogr. B Biomed. Sci. Appl.*, 745, 399, 2000.

89. Inoue, H., Maeno, Y., Iwasa, M., Matoba, R., and Nagao, M., Screening and determination of benzodiazepines in whole blood using solid-phase extraction and gas chromatography/mass spectrometry, *Forensic Sci. Int.*, 113, 367, 2000.

90. Lillsunde, P., Michelson, L., Forsstrom, T., Korte, T., Schultz, E., Ariniemi, K., Portman, M., Sihvonen, M.L., and Seppala, T., Comprehensive drug screening in blood for detecting abused drugs or drugs potentially hazardous for traffic safety., *Forensic Sci. Int.*, 77, 191, 1996.

91. Maurer, H.H., Toxicological analysis of drugs and poisons by GC-MS, *Spectroscopy Europe*, 6, 21, 1994.

92. Pfleger, K., Maurer, H.H., and Weber, A., *Mass Spectral Library of Drugs, Poisons, Pesticides, Pollutants and their Metabolites*, Agilent Technologies, Palo Alto, CA, 2000.

93. Maurer, H.H., Kraemer, T., Kratzsch, C., Paul, L.D., Peters, F.T., Springer, D., Staack, R.F., and Weber, A.A., What is the appropriate analytical strategy for effective management of intoxicated patients?, in *Proceedings of the 39th International TIAFT Meeting in Prague, 2001*, Balikova, M. and Navakova, E., Eds., Charles University, Prague, 2002, 61.

94. Pfleger, K., Maurer, H.H., and Weber, A., *Mass Spectral Library of Drugs, Poisons, Pesticides, Pollutants and their Metabolites*, Agilent Technologies, Palo Alto, CA, 2003, in preparation.
95. NIST, EPA, and NIH, *Mass Spectral Database*, National Institute of Standards and Technology, Gaithersburg, MD, 1998.
96. McLafferty, F.W. and Stauffer, D.B., *Registry of Mass Spectral Data, 7th ed.*, John Wiley & Sons, New York, 2001.
97. Pragst, F., Herzler, M., Herre, S., Erxleben, B.T., and Rothe, M., *UV Spektra of Toxic Compounds*, Helm-Verlag, Heppenheim, Germany, 2001.
98. Drummer, O.H., Chromatographic screening techniques in systematic toxicological analysis [review], *J. Chromatogr. B*, 733, 27, 1999.
99. Lai, C.K., Lee, T., Au, K.M., and Chan, A.Y., Uniform solid-phase extraction procedure for toxicological drug screening in serum and urine by HPLC with photodiode-array detection, *Clin. Chem.*, 43, 312, 1997.
100. Lambert, W.E., Van Bocxlaer, J.F., and de-Leenheer, A.P., Potential of high-performance liquid chromatography with photodiode array detection in forensic toxicology, *J. Chromatogr. B*, 689, 45, 1997.
101. Koves, E.M., Use of high-performance liquid chromatography-diode array detection in forensic toxicology, *J. Chromatogr. A*, 692, 103, 1995.
102. Lambert, W.E., Meyer, E., and de-Leenheer, A.P., Systematic toxicological analysis of basic drugs by gradient elution of an alumina-based HPLC packing material under alkaline conditions, *J. Anal. Toxicol.*, 19, 73, 1995.
103. Maier, R.D. and Bogusz, M., Identification power of a standardized HPLC-DAD system for systematic toxicological analysis, *J. Anal. Toxicol.*, 19, 79, 1995.
104. Tracqui, A., Kintz, P., and Mangin, P., Systematic toxicological analysis using HPLC/DAD, *J. Forensic Sci.*, 40, 254, 1995.
105. Kratzsch, C., Weber, A.A., Kraemer, T., and Maurer, H.H., Validated high-throughput assay for the determination of risperidone and its 9-hydroxy metabolite in plasma by atmospheric pressure chemical ionization liquid chromatography-mass spectrometry (APCI-LC-MS), in *Proceedings of the XIIth GTFCh Symposium in Mosbach*, Pragst, F. and Aderjan, R., Eds., Helm-Verlag, Heppenheim, Germany, 2001, 76.
106. Rivier, L., Robustness evaluation for automatic identification of general unknown by reference ESI mass spectra comparison in LC-MS, in *Abstract Book of the 39th International TIAFT Meeting in Prague, 2001*, Balikova, M. and Navakova, E., Eds., Charles University, Prague, 2002, 71.
107. Weinmann, W., Stoertzel, M., Vogt, S., and Wendt, J., Tune compounds for electrospray ionisation/in-source collision-induced dissociation with mass spectral library searching, *J Chromatogr. A*, 926, 199, 2001.

108. Bogusz, M.J., Maier, R.D., Kruger, K.D., Webb, K.S., Romeril, J., and Miller, M.L., Poor reproducibility of in-source collisional atmospheric pressure ionization mass spectra of toxicologically relevant drugs, *J. Chromatogr. A*, 844, 409, 1999.

109. Maurer, H.H., Schmitt, C.J., Weber, A.A., and Kraemer, T., Validated electrospray LC-MS assay for determination of the mushroom toxins alpha- and beta-amanitin in urine after immunoaffinity extraction, *J. Chromatogr. B*, 748, 125, 2000.

110. Schramm, W., Smith, R.H., Craig, P.A., and Kidwell, D.A., Drugs of abuse in saliva: a review, *J. Anal. Toxicol.*, 16, 1, 1992.

111. Cone, E.J., Presley, L., Lehrer, M., Seiter, W., Smith, M., Kardos, K.W., Fritch, D., Salamone, S., and Niedbala, R.S., Oral fluid testing for drugs of abuse: positive prevalence rates by Intercept immunoassay screening and GC-MS-MS confirmation and suggested cutoff concentrations, *J. Anal. Toxicol.*, 26, 541, 2002.

112. Jones, J., Tomlinson, K., and Moore, C., The simultaneous determination of codeine, morphine, hydrocodone, hydromorphone, 6-acetylmorphine, and oxycodone in hair and oral fluid, *J. Anal. Toxicol.*, 26, 171, 2002.

113. Katikaneni, L.D., Salle, F.R., and Hulsey, T.C., Neonatal hair analysis for benzoylecgonine: a sensitive and semiquantitative biological marker for chronic gestational cocaine exposure, *Biol. Neonate*, 81, 29, 2002.

114. Samyn, N., de Boeck, G., Cirimele, V., Verstraete, A., and Kintz, P., Detection of flunitrazepam and 7-aminoflunitrazepam in oral fluid after controlled administration of rohypnol, *J. Anal. Toxicol.*, 26, 211, 2002.

115. Samyn, N., de Boeck, G., and Verstraete, A.G., The use of oral fluid and sweat wipes for the detection of drugs of abuse in drivers, *J. Forensic Sci.*, 47, 1380, 2002.

116. Auwarter, V., Sporkert, F., Hartwig, S., Pragst, F., Vater, H., and Diefenbacher, A., Fatty acid ethyl esters in hair as markers of alcohol consumption. Segmental hair analysis of alcoholics, social drinkers, and teetotalers, *Clin. Chem.*, 47, 2114, 2001.

117. Kronstrand, R., Andersson, M.C., Ahlner, J., and Larson, G., Incorporation of selegiline metabolites into hair after oral selegiline intake, *J. Anal. Toxicol.*, 25, 594, 2001.

118. Lester, B.M., ElSohly, M., Wright, L.L., Smeriglio, V.L., Verter, J., Bauer, C.R., Shankaran, S., Bada, H.S., Walls, H.H., Huestis, M.A., Finnegan, L.P., and Maza, P.L., The maternal lifestyle study: drug use by meconium toxicology and maternal self-report, *Pediatrics*, 107, 309, 2001.

119. Paterson, S., McLachlan-Troup, N., Cordero, R., Dohnal, M., and Carman, S., Qualitative screening for drugs of abuse in hair using GC-MS, *J. Anal. Toxicol.*, 25, 203, 2001.

120. Pragst, F., Auwaerter, V., Sporkert, F., and Spiegel, K., Analysis of fatty acid ethyl esters in hair as possible markers of chronically elevated alcohol consumption by headspace solid-phase microextraction (HS-SPME) and gas chromatography-mass spectrometry (GC-MS), *Forensic Sci. Int.*, 121, 76, 2001.

121. Vandevenne, M., Vandenbussche, H., and Verstraete, A., Detection time of drugs of abuse in urine, *Acta Clin. Belg.*, 55, 323, 2000.

122. Jimenez, C., Ventura, R., and Segura, J., Validation of qualitative chromatographic methods: strategy in antidoping control laboratories, *J. Chromatogr. B Analyt. Technol. Biomed. Life Sci.*, 767, 341, 2002.

123. Goldberger, B.A. and Cone, E.J., Confirmatory tests for drugs in the workplace by gas chromatography-mass spectrometry, *J. Chromatogr. A*, 674, 73, 1994.

124. Casari, C. and Andrews, A.R., Application of solvent microextraction to the analysis of amphetamines and phencyclidine in urine, *Forensic Sci. Int.*, 120, 165, 2001.

125. Kraemer, T., Roditis, S.K., Peters, F.T., and Maurer, H.H., Amphetamine concentrations in human urine following single-dose administration of the calcium antagonist prenylamine — studies using FPIA and GC-MS, *J. Anal. Toxicol.*, 27, 68, 2003.

126. Kraemer, T., Wennig, R., and Maurer, H.H., The antispasmodic mebeverine leads to positive amphetamine results with the fluorescence polarization immuno assay (FPIA) — studies on the toxicological detection in urine by GC-MS and FPIA, *J. Anal. Toxicol.*, 25, 1, 2001.

127. Kraemer, T., Theis, G.A., Weber, A.A., and Maurer, H.H., Studies on the metabolism and toxicological detection of the amphetamine-like anorectic fenproporex in human urine by gas chromatography-mass spectrometry and fluorescence polarization immunoassay (FPIA), *J. Chromatogr. B*, 738, 107, 2000.

128. Kraemer, T., Bickeboeller-Friedrich, J., and Maurer, H.H., On the metabolism of the amphetamine-derived antispasmodic drug mebeverine: gas chromatography-mass spectrometry studies on rat liver microsomes and on human urine, *Drug Metab. Dispos.*, 28, 339, 2000.

129. Musshoff, F. and Kraemer, T., Identification of famprofazone ingestion, *Int. J. Legal Med.*, 111, 305, 1998.

130. Kraemer, T., Vernaleken, I., and Maurer, H.H., Studies on the metabolism and toxicological detection of the amphetamine-like anorectic mefenorex in human urine by gas chromatography-mass spectrometry and fluorescence polarization immunoassay, *J. Chromatogr. B*, 702, 93, 1997.

131. Maurer, H.H., Kraemer, T., Ledvinka, O., Schmitt, C.J., and Weber, A.A., Gas chromatography-mass spectrometry (GC-MS) and liquid chromatography-mass spectrometry (LC-MS) in toxicological analysis. Studies on the detection of clobenzorex and its metabolites within a systematic toxicological analysis procedure by GC-MS and by immunoassay and studies on the detection of alpha- and beta-amanitin in urine by atmospheric pressure ionization electrospray LC-MS, *J. Chromatogr. B*, 689, 81, 1997.

132. Maurer, H.H. and Kraemer, T., Toxicological detection of selegiline and its metabolites in urine using fluorescence polarization immunoassay (FPIA) and gas chromatography-mass spectrometry (GC-MS) and differentiation by enantioselective GC-MS of the intake of selegiline from abuse of methamphetamine or amphetamine, *Arch. Toxicol.*, 66, 675, 1992.
133. Kraemer, T. and Maurer, H.H., Toxicokinetics of amphetamines: metabolism and toxicokinetic data of designer drugs, of amphetamine, methamphetamine and their *N*-alkyl derivatives [review], *Ther. Drug Monit.*, 24, 277, 2002.
134. Musshoff, F., Illegal or legitimate use? Precursor compounds to amphetamine and methamphetamine [review], *Drug Metab. Rev.*, 32, 15, 2000.
135. Hegadoren, K.M., Baker, G.B., and Bourin, M., 3,4-methylenedioxy analogues of amphetamine: defining the risks to humans, *Neurosci. Biobehav. Rev.*, 23, 539, 1999.
136. Nichols, D.E., Differences between the mechanism of action of MDMA, MBDB, and the classic hallucinogens. Identification of a new therapeutic class: entactogens, *J. Psychoactive Drugs*, 18, 305, 1986.
137. Jones, A.L. and Simpson, K.J., Review article: mechanisms and management of hepatotoxicity in ecstasy (MDMA) and amphetamine intoxications, *Aliment. Pharmacol. Ther.*, 13, 129, 1999.
138. Walubo, A. and Seger, D., Fatal multi-organ failure after suicidal overdose with MDMA, "ecstasy": case report and review of the literature, *Hum. Exp. Toxicol.*, 18, 119, 1999.
139. Maurer, H.H., Bickeboeller-Friedrich, J., Kraemer, T., and Peters, F.T., Toxicokinetics and analytical toxicology of amphetamine-derived designer drugs ("Ecstasy"), *Toxicol. Lett.*, 112, 133, 2000.
140. Carvalho, M., Carvalho, F., and Bastos, M.L., Is hyperthermia the triggering factor for hepatotoxicity induced by 3,4-methylenedioxymethamphetamine (ecstasy)? An *in vitro* study using freshly isolated mouse hepatocytes, *Arch. Toxicol.*, 74, 789, 2001.
141. Garbino, J., Henry, J.A., Mentha, G., and Romand, J.A., Ecstasy ingestion and fulminant hepatic failure: liver transplantation to be considered as a last therapeutic option, *Vet. Hum. Toxicol.*, 43, 99, 2001.
142. Ricaurte, G.A., McCann, U.D., Szabo, Z., and Scheffel, U., Toxicodynamics and long-term toxicity of the recreational drug, 3,4-methylenedioxymethamphetamine (MDMA, "Ecstasy"), *Toxicol. Lett.*, 112, 143, 2000.
143. Rochester, J.A. and Kirchner, J.T., Ecstasy (3,4-methylenedioxymethamphetamine): history, neurochemistry, and toxicology, *J. Am. Board. Fam. Pract.*, 12, 137, 1999.
144. Bai, F., Lau, S.S., and Monks, T.J., Glutathione and *N*-acetylcysteine conjugates of alpha-methyldopamine produce serotonergic neurotoxicity: possible role in methylenedioxyamphetamine-mediated neurotoxicity, *Chem. Res. Toxicol.*, 12, 1150, 1999.

145. Colado, M.I., Granados, R., O'Shea, E., Esteban, B., and Green, A.R., The acute effect in rats of 3,4-methylenedioxyethamphetamine (MDEA, "eve") on body temperature and long term degeneration of 5-HT neurones in brain: a comparison with MDMA ("ecstasy"), *Pharmacol. Toxicol.*, 84, 261, 1999.

146. Morland, J., Toxicity of drug abuse — amphetamine designer drugs (ecstasy): mental effects and consequences of single dose use, *Toxicol. Lett.*, 112–113, 147, 2000.

147. Curran, H.V., Is MDMA ("Ecstasy") neurotoxic in humans? An overview of evidence and of methodological problems in research, *Neuropsychobiology*, 42, 34, 2000.

148. Chang, L., Grob, C.S., Ernst, T., Itti, L., Mishkin, F.S., Jose-Melchor, R., and Poland, R.E., Effect of ecstasy [3,4-methylenedioxymethamphetamine (MDMA)] on cerebral blood flow: a co-registered SPECT and MRI study, *Psychiatry Res.*, 98, 15, 2000.

149. Ernst, T., Chang, L., Leonido-Yee, M., and Speck, O., Evidence for long-term neurotoxicity associated with methamphetamine abuse: A 1H MRS study, *Neurology*, 54, 1344, 2000.

150. Hervias, I., Lasheras, B., and Aguirre, N., 2-Deoxy-D-glucose prevents and nicotinamide potentiates 3,4-methylenedioxymethamphetamine-induced serotonin neurotoxicity, *J. Neurochem.*, 75, 982, 2000.

151. Kish, S.J., Furukawa, Y., Ang, L., Vorce, S.P., and Kalasinsky, K.S., Striatal serotonin is depleted in brain of a human MDMA (Ecstasy) user, *Neurology*, 55, 294, 2000.

152. Ricaurte, G.A., Yuan, J., and McCann, U.D., (+/-)3,4-methylenedioxymethamphetamine ("Ecstasy")-induced serotonin neurotoxicity: studies in animals, *Neuropsychobiology*, 42, 5, 2000.

153. Bai, F., Jones, D.C., Lau, S.S., and Monks, T.J., Serotonergic neurotoxicity of 3,4-(+/-)-methylenedioxyamphetamine and 3,4-(+/-)-methylendioxymethamphetamine ("Ecstasy") is potentiated by inhibition of gamma-glutamyl transpeptidase, *Chem. Res. Toxicol.*, 14, 863, 2001.

154. Buchert, R., Obrocki, J., Thomasius, R., Vaterlein, O., Petersen, K., Jenicke, L., Bohuslavizki, K.H., and Clausen, M., Long-term effects of "ecstasy" abuse on the human brain studied by FDG PET, *Nucl. Med. Commun.*, 22, 889, 2001.

155. Cadet, J.L., Thiriet, N., and Subramanian, J., Involvement of free radicals in MDMA-induced neurotoxicity in mice, *Ann. Med. Interne (Paris)*, 152 Suppl. 3, 57, 2001.

156. Nixdorf, W.L., Burrows, K.B., Gudelsky, G.A., and Yamamoto, B.K., Enhancement of 3,4-methylenedioxymethamphetamine neurotoxicity by the energy inhibitor malonate, *J. Neurochem.*, 77, 647, 2001.

157. Moeller, M.R. and Hartung, M., Ecstasy and related substances — serum levels in impaired drivers [letter; comment], *J. Anal. Toxicol.*, 21, 591, 1997.

158. Taylor, E.H., Oertli, E.H., Wolfgang, J.W., and Mueller, E., Accuracy of five on-site immunoassay drugs-of-abuse testing devices, *J. Anal. Toxicol.*, 23, 119, 1999.

159. Iwersen-Bergmann, S. and Schmoldt, A., Direct semiquantitative screening of drugs of abuse in serum and whole blood by means of CEDIA DAU urine immunoassays, *J. Anal. Toxicol.*, 23, 247, 1999.

160. Scholer, A., Nicht-instrumentelle Immunoassays in der Suchtmittelanalytik (Drogenanalytik), *Toxichem. Krimtech.*, 66, 27, 1999.

161. Lekskulchai, V. and Mokkhavesa, C., Evaluation of Roche Abuscreen ONLINE amphetamine immunoassay for screening of new amphetamine analogues, *J. Anal. Toxicol.*, 25, 471, 2001.

162. Zhao, H., Brenneisen, R., Scholer, A., McNally, A.J., ElSohly, M.A., Murphy, T.P., and Salamone, S.J., Profiles of urine samples taken from Ecstasy users at Rave parties: analysis by immunoassays, HPLC, and GC-MS, *J. Anal. Toxicol.*, 25, 258, 2001.

163. Wikstroem, M., Holmgren, P., and Ahlner, J., *N*-Benzylpiperazine, a "new" drug of abuse in Sweden, in Rasanen, I., Ed., Abstract book, 38th Int. TIAFT Meeting, Helsinki, 2000, 88.

164. Balmelli, C., Kupferschmidt, H., Rentsch, K., and Schneemann, M., Fatal brain edema after ingestion of ecstasy and benzylpiperazine, *Dtsch. Med. Wochenschr*, 126, 809, 2001.

165. de Boer, D., Bosman, I.J., Hidvegi, E., Manzoni, C., Benko, A.A., dos, R.L., and Maes, R.A., Piperazine-like compounds: a new group of designer drugs-of-abuse on the European market, *Forensic Sci. Int.*, 121, 47, 2001.

166. Staack, R.F., Fritschi, G., and Maurer, H.H., GC-MS studies on the metabolism and on the toxicological analysis of the new piperazine-like designer drugs BZP, MDBP, TFMPP, mCPP, MeOPP, in *Proceedings of the 39th International TIAFT Meeting in Prague, 2001*, Balikova, M. and Navakova, E., Eds., Charles University, Prague, 2002, 115.

167. Staack, R.F., Fritschi, G., and Maurer, H.H., Studies on the metabolism and the toxicological analysis of the new piperazine-like designer drugs BZP and TFMPP using GC-MS, in *Proceedings of the XIIth GTFCh Symposium in Mosbach*, Pragst, F. and Aderjan, R., Eds., Helm-Verlag, Heppenheim, Germany, 2001, 149.

168. Shulgin, A., #142 PEA; Phenethylamine, in *Pihkal, A Chemical Love Story*, Dan Joy, Ed., Transform Press, Berkley, CA, 1991, 815.

169. Bye, C., Munro, F.A., Peck, A.W., and Young, P.A., A comparison of the effects of 1-benzylpiperazine and dexamphetamine on human performance tests, *Eur. J. Clin. Pharmacol.*, 6, 163, 1973.

170. Tekes, K., Tothfalusi, L., Malomvolgyi, B., Herman, F., and Magyar, K., Studies on the biochemical mode of action of EGYT-475, a new antidepressant, *Pol. J. Pharmacol. Pharm.*, 39, 203, 1987.

171. Maurer, H.H. and Kraemer, T., Amphetamine-derived designer drugs: metabolism and screening procedures [review], *J. Lab. Med.*, 26, 37, 2002.

172. Maurer, H.H., On the metabolism and the toxicological analysis of methylenedioxyphenylalkylamine designer drugs by gas chromatography-mass spectrometry, *Ther. Drug Monit.*, 18, 465, 1996.

173. Hensley, D. and Cody, J.T., Simultaneous determination of amphetamine, methamphetamine, methylenedioxyamphetamine (MDA), methylenedioxymethamphetamine (MDMA), and methylenedioxyethylamphetamine (MDEA) enantiomers by GC-MS, *J. Anal. Toxicol.*, 23, 518, 1999.

174. Ensslin, H.K., Kovar, K.A., and Maurer, H.H., Toxicological detection of the designer drug 3,4-methylenedioxyethylamphetamine (MDE, "Eve") and its metabolites in urine by gas chromatography-mass spectrometry and fluorescence polarization immunoassay, *J. Chromatogr. B*, 683, 189, 1996.

175. Helmlin, H.J., Bracher, K., Bourquin, D., Vonlanthen, D., and Brenneisen, R., Analysis of 3,4-methylenedioxymethamphetamine (MDMA) and its metabolites in plasma and urine by HPLC-DAD and GC-MS [see comments], *J. Anal. Toxicol.*, 20, 432, 1996.

176. Centini, F., Masti, A., and Comparini, I.B., Quantitative and qualitative analysis of MDMA, MDEA, MA and amphetamine in urine by headspace/solid phase micro-extraction and GC/MS, *Forensic Sci. Int.*, 83, 161, 1996.

177. Roesner, P., Junge, T., Fritschi, G., Klein, B., Thielert, K., and Kozlowski, M., Neue synthetische Drogen: Piperazin-, Procyclidin- und alpha-Aminopropiophenonderivate, *Toxichem. Krimtech.*, 66, 81, 1999.

178. Springer, D., Peters, F.T., Fritschi, G., and Maurer, H.H., Studies on the metabolism of the new pyrrolidino-propiophenone designer drugs PPP and MOPPP, in *Proceedings of the XIIth GTFCh Symposium in Mosbach*, Pragst, F. and Aderjan, R., Eds., Helm-Verlag, Heppenheim, Germany, 2001, 156.

179. Springer, D., Peters, F.T., Fritschi, G., and Maurer, H.H., Studies on the metabolism of the new pyrrolidino-propiophenone designer drugs MPPP and MDPPP, in *Proceedings of the 39th International TIAFT Meeting in Prague, 2001*, Balikova, M. and Navakova, E., Eds., Charles University, Prague, 2002, 122.

180. Springer, D., Peters, F.T., Fritschi, G., and Maurer, H.H., Studies on the metabolism and toxicological detection of the new designer drug 4′-methyl-alpha-pyrrolidinopropiophenone (MPPP) in urine using gas chromatography-mass spectrometry, *J. Chromatogr. B*, 773, 25, 2002.

181. Springer, D., Peters, F.T., Fritschi, G., and Maurer, H.H., New designer drug 4′-methyl-alpha-pyrrolidinohexanophenone (MPHP): Studies on its metabolism and toxicological detection in urine using gas chromatography-mass spectrometry, *J. Chromatogr. B*, 789, 79, 2003.

182. Martinez, M., Mercado, O., Santamaria, A., Galvan, S., Vazquez, M., Bucio, V., Hall, C., Hernandez, R., Hurtazo, A., Pego, E., Rodriguez, F., Salvatierra, R., Sosa, A., and Rios, C., The action of amfepramone on neurochemical and behavioral markers in rats, *Proc. West. Pharmacol. Soc.*, 41, 125, 1998.

183. Glennon, R.A., Yousif, M., Naiman, N., and Kalix, P., Methcathinone: a new and potent amphetamine-like agent, *Pharmacol. Biochem. Behav.*, 26, 547, 1987.

184. Kalix, P. and Glennon, R.A., Further evidence for an amphetamine-like mechanism of action of the alkaloid cathinone, *Biochem. Pharmacol.*, 35, 3015, 1986.

185. Bryant, S.G., Guernsey, B.G., and Ingrim, N.B., Review of bupropion, *Clin. Pharm.*, 2, 525, 1983.

186. Springer, D., Peters, F.T., Fritschi, G., and Maurer, H.H., GC-MS studies on the metabolism and toxicological analysis of the new pyrrolidino-hexanophenone designer drug 4′-methyl-alpha-pyrrolidinohexanophenone (MPHP), in *Proceedings of the 40th International TIAFT Meeting in Paris*, Marquet, P., Ed., TIAFT, Strasbourg, France, 2003.

187. Springer, D., Paul, L.D., Staack, R.F., Kraemer, T., and Maurer, H.H., Identification of the cytochrome P450 enzymes involved in the metabolism of 4′-methyl-pyrrolidinopropiophenone (MPPP), a novel scheduled designer drug, in human liver microsomes, *Drug Metab. Dispos.*, 31, 979, 2003.

188. Springer, D., Fritschi, G., and Maurer, H.H., Metabolism and toxicological detection of the new designer drug 4′-methoxy-alpha-pyrrolidinopropiophenone (MOPPP) studied in urine using gas chromatography-mass spectrometry, *J. Chromatogr. B*, submitted, 2003.

189. de Kanel, J., Vickery, W.E., Waldner, B., Monahan, R.M., and Diamond, F.X., Automated extraction of lysergic acid diethylamide (LSD) and *N*-demethyl-LSD from blood, serum, plasma, and urine samples using the Zymark RapidTrace with LC/MS/MS confirmation, *J. Forensic Sci.*, 43, 622, 1998.

190. Hoja, H., Marquet, P., Verneuil, B., Lotfi, H., Dupuy, J.L., and Lachatre, G., Determination of LSD and *N*-demethyl-LSD in urine by liquid chromatography coupled to electrospray ionization mass spectrometry, *J. Chromatogr. B*, 692, 329, 1997.

191. Cai, J. and Henion, J., Elucidation of LSD *in vitro* metabolism by liquid chromatography and capillary electrophoresis coupled with tandem mass spectrometry, *J. Anal. Toxicol.*, 20, 27, 1996.

192. Stout, P.R., Horn, C.K., and Klette, K.-L., Solid-phase extraction and GC-MS analysis of THC-COOH method optimized for a high-throughput forensic drug-testing laboratory, *J. Anal. Toxicol.*, 25, 550, 2001.

193. ElSohly, M.A., deWit, H., Wachtel, S.R., Feng, S., and Murphy, T.P., Delta 9-tetrahydrocannabivarin as a marker for the ingestion of marijuana versus Marinol: results of a clinical study, *J. Anal. Toxicol.*, 25, 565, 2001.

194. Broussard, L.A., Presley, L.C., Tanous, M., and Queen, C., Improved gas chromatography-mass spectrometry method for simultaneous identification and quantification of opiates in urine as propionyl and oxime derivatives, *Clin. Chem*, 47, 127, 2001.
195. O'Neal, C.L. and Poklis, A., The detection of acetylcodeine and 6-acetylmorphine in opiate positive urines, *Forensic Sci. Int.*, 95, 1, 1998.
196. Galloway, J.H., Ashford, M., Marsh, I.D., Holden, M., and Forrest, A.R., A method for the confirmation and identification of drugs of misuse in urine using solid phase extraction and gas-liquid chromatography with mass spectrometry, *J. Clin. Pathol.*, 51, 326, 1998.
197. Smith, M.L., Hughes, R.O., Levine, B., Dickerson, S., Darwin, W.D., and Cone, E.J., Forensic drug testing for opiates. VI. Urine testing for hydromorphone, hydrocodone, oxymorphone, and oxycodone with commercial opiate immunoassays and gas chromatography-mass spectrometry, *J. Anal. Toxicol.*, 19, 18, 1995.
198. Maurer, H.H. and Fritz, C.F., Metabolism of pholcodine in man, *Arzneim. - Forsch.*, 40, 564, 1990.
199. Maurer, H.H. and Fritz, C.F., Toxicological detection of pholcodine and its metabolites in urine and hair using radio immunoassay, fluorescence polarisation immunoassay, enzyme immunoassay and gas chromatography-mass spectrometry, *Int. J. Legal Med.*, 104, 43, 1990.
200. Meadway, C., George, S., and Braithwaite, R., Interpretation of GC-MS opiate results in the presence of pholcodine, *Forensic Sci. Int.*, 127, 131, 2002.
201. Cone, E.J. and Preston, K.J., Toxicological aspects of heroin substitution treatment [review], *Ther. Drug Monit.*, 24, 193, 2002.
202. Kuhlman, J.J., Magluilo, J., Cone, E.J., and Levine, B., Simultaneous assay of buprenorphine and norbuprenorphine by negative chemical ionization tandem mass spectrometry, *J. Anal. Toxicol.*, 20, 229, 1996.
203. Vincent, F., Bessard, J., Vacheron, J., Mallaret, M., and Bessard, G., Determination of buprenorphine and norbuprenorphine in urine and hair by gas chromatography-mass spectrometry, *J. Anal. Toxicol.*, 23, 270, 1999.
204. Tracqui, A., Kintz, P., and Mangin, P., HPLC/MS determination of buprenorphine and norbuprenorphine in biological fluids and hair samples, *J. Forensic Sci.*, 42, 111, 1997.
205. Hoja, H., Marquet, P., Verneuil, B., Lotfi, H., Dupuy, J.L., and Lachatre, G., Determination of buprenorphine and norbuprenorphine in whole blood by liquid chromatography-mass spectrometry, *J. Anal. Toxicol.*, 21, 160, 1997.
206. Polettini, A. and Huestis, M.A., Simultaneous determination of buprenorphine, norbuprenorphine, and buprenorphine-glucuronide in plasma by liquid chromatography-tandem mass spectrometry, *J. Chromatogr. B Biomed. Sci. Appl.*, 754, 447, 2001.

207. Maurer, H. and Pfleger, K., Screening procedure for the detection of opioids, other potent analgesics and their metabolites in urine using a computerized gas chromatographic-mass spectrometric technique, *Fresenius Z. Anal. Chem.*, 317, 42, 1984.

208. Smith, M.L., Shimomura, E.T., Summers, J., Paul, B.D., Jenkins, A.J., Darwin, W.D., and Cone, E.J., Urinary excretion profiles for total morphine, free morphine, and 6-acetylmorphine following smoked and intravenous heroin, *J. Anal. Toxicol.*, 25, 504, 2001.

209. Broussard, L.A., Presley, L.C., Pittman, T., Clouette, R., and Wimbish, G.H., Simultaneous identification and quantitation of codeine, morphine, hydrocodone, and hydromorphone in urine as trimethylsilyl and oxime derivatives by gas chromatography-mass spectrometry, *Clin. Chem.*, 43, 1029, 1997.

210. Smith, M.L., Hughes, R.O., Levine, B., Dickerson, S., Darwin, W.D., and Cone, E.J., Forensic drug testing for opiates. VI. Urine testing for hydromorphone, hydrocodone, oxymorphone, and oxycodone with commercial opiate immunoassays and gas chromatography-mass spectrometry, *J. Anal. Toxicol.*, 19, 18, 1995.

211. Dawson, M., Fryirs, B., Kelly, T., Keegan, J., and Mather, L.E., A rapid and sensitive high-performance liquid chromatography-electrospray ionization-triple quadrupole mass spectrometry method for the quantitation of oxycodone in human plasma, *J. Chromatogr. Sci.*, 40, 40, 2002.

212. Kraemer, T. and Maurer, H.H., Forensic toxicology: non-opioid analgesics, in *Handbook of Analytical Separation Sciences: Forensic Sciences*, Bogusz, M., Ed., Elsevier Science, Amsterdam, 2000, 259.

213. Maurer, H. and Pfleger, K., Screening procedure for detecting anti-inflammatory analgesics and their metabolites in urine, *Fresenius Z. Anal. Chem.*, 314, 586, 1983.

214. Maurer, H.H., Kraemer, T., and Weber, A., Toxicological detection of ibuprofen and its metabolites in urine using gas chromatography-mass spectrometry (GC-MS), *Pharmazie*, 49, 148, 1994.

215. Tomson, T. and Johannessen, S.I., Therapeutic monitoring of the new antiepileptic drugs [Review], *Eur. J. Clin. Pharmacol.*, 55, 697, 2000.

216. Behnke, C.E. and Reddy, M.N., Determination of felbamate concentration in pediatric samples by high-performance liquid chromatography, *Ther. Drug Monit.*, 19, 301, 1997.

217. Dasgupta, A. and Hart, A.P., Lamotrigine analysis in plasma by gas chromatography-mass spectrometry after conversion to a *tert*-butyldimethylsilyl derivative., *J. Chromatogr. B*, 693, 101, 1997.

218. Maurer, H.H., Detection of anticonvulsants and their metabolites in urine within a "general unknown" analysis procedure using computerized gas chromatography-mass spectrometry, *Arch. Toxicol.*, 64, 554, 1990.

219. Maurer, H. and Pfleger, K., Identification and differentiation of benzodiazepines and their metabolites in urine by computerized gas chromatography-mass spectrometry, *J. Chromatogr.*, 422, 85, 1987.

220. Augsburger, M., Rivier, L., and Mangin, P., Comparison of different immunoassays and GC-MS screening of benzodiazepines, *J. Pharm. Biomed. Anal.*, 7, 681, 1998.

221. Cardenas, S., Gallego, M., and Valcarcel, M., Gas chromatographic-mass spectrometric confirmation of selected benzophenones from benzodiazepines in human urine following automatic screening, *J. Chromatogr. A*, 823, 389, 1998.

222. Black, D.A., Clark, G.D., Haver, V.M., Garbin, J.A., and Saxon, A.J., Analysis of urinary benzodiazepines using solid-phase extraction and gas chromatography-mass spectrometry [see comments], *J. Anal. Toxicol.*, 18, 185, 1994.

223. Maurer, H.H., Identification and differentiation of barbiturates, other sedative–hypnotics and their metabolites in urine integrated in a general screening procedure using computerized gas chromatography–mass spectrometry, *J. Chromatogr.*, 530, 307, 1990.

224. Maurer, H. and Pfleger, K., Screening procedure for the detection of alkanolamine antihistamines and their metabolites in urine using computerized gas chromatography-mass spectrometry, *J. Chromatogr.*, 428, 43, 1988.

225. Geldmacher-von Malinckrodt, M., Heijst, A.V., and Koeppel, C., Qualitativer Nachweis von Giften — Farbreaktionen — Chlorierte Kohlenwasserstoffe, in *Einfache toxikologische Laboratoriumsuntersuchungen bei akuten Vergiftungen*, Gibitz, H.J. and Schütz, H., Eds., VCH, Weinheim, Germany, 1995, 168.

226. Gaulier, J.M., Merle, G., Lacassie, E., Courtiade, B., Haglund, P., Marquet, P., and Lachatre, G., Fatal intoxications with chloral hydrate, *J. Forensic Sci.*, 46, 1507, 2001.

227. Deinl, I., Mahr, G., and von Meyer, L., Determination of flunitrazepam and its main metabolites in serum and urine by HPLC after mixed-mode solid-phase extraction., *J. Anal. Toxicol.*, 22, 197, 1998.

228. Koppen, B., Dalgaard, L., and Christensen, J.M., Determination of trichloroethylene metabolites in rat liver homogenate using headspace gas chromatography., *J. Chromatogr.*, 442, 325, 1988.

229. Kraemer, T. and Maurer, H.H., Forensic toxicology: sedatives and hypnotics, in *Handbook of Analytical Separation Sciences: Forensic Sciences*, Bogusz, M., Ed., Elsevier Science, Amsterdam, 2000, 197.

230. Baumann, P., Pharmacokinetic-pharmacodynamic relationship of the selective serotonin reuptake inhibitors, *Clin. Pharmacokinet.*, 31, 444, 1996.

231. Richelson, E., Pharmacokinetic interactions of antidepressants, *J. Clin. Psychiatry*, 59 Suppl 10, 22, 1998.

232. Mitchell, P.B., Therapeutic drug monitoring of psychotropic medications, *Br. J. Clin. Pharmacol.*, 49, 303, 2000.

233. Eap, C.B., Bouchoux, G., Amey, M., Cochard, N., Savary, L., and Baumann, P., Simultaneous determination of human plasma levels of citalopram, paroxetine, sertraline, and their metabolites by gas chromatography-mass spectrometry, *J. Chromatogr. Sci.*, 36, 365, 1998.

234. Eap, C.B. and Baumann, P., Analytical methods for the quantitative determination of selective serotonin reuptake inhibitors for therapeutic drug monitoring purposes in patients, *J. Chromatogr. B*, 686, 51, 1996.

235. Maurer, H. and Pfleger, K., Screening procedure for detection of antidepressants and their metabolites in urine using a computerized gas chromatographic-mass spectrometric technique, *J. Chromatogr.*, 305, 309, 1984.

236. Maurer, H. and Pfleger, K., Screening procedure for detection of phenothiazine and analogous neuroleptics and their metabolites in urine using a computerized gas chromatographic-mass spectrometric technique, *J. Chromatogr.*, 306, 125, 1984.

237. Maurer, H. and Pfleger, K., Identification of phenothiazine antihistamines and their metabolites in urine, *Arch. Toxicol.*, 62, 185, 1988.

238. Maurer, H. and Pfleger, K., Toxicological detection of ethylenediamine and piperazine antihistamines and their metabolites in urine by computerized gas chromatography-mass spectrometry, *Fresenius Z. Anal. Chem.*, 331, 744, 1988.

239. Maurer, H. and Pfleger, K., Identification and differentiation of alkylamine antihistamines and their metabolites in urine by computerized gas chromatography-mass spectrometry., *J. Chromatogr.*, 430, 31, 1988.

240. Maurer, H. and Pfleger, K., Screening procedure for the detection of antiparkinsonian drugs and their metabolites in urine using a computerized gas chromatographic-mass spectrometric technique, *Fresenius Z. Anal. Chem.*, 321, 363, 1985.

241. Romberg, R.W., Needleman, S.B., Snyder, J.J., and Greedan, A., Methamphetamine and amphetamine derived from the metabolism of selegiline, *J. Forensic Sci.*, 40, 1100, 1995.

242. Hemmersbach, P. and de-la-Torre, R., Stimulants, narcotics and beta-blockers: 25 years of development in analytical techniques for doping control., *J. Chromatogr. B*, 687, 221, 1996.

243. Siren, H., Saarinen, M., Hainari, S., Lukkari, P., and Riekkola, M.L., Screening of beta-blockers in human serum by ion-pair chromatography and their identification as methyl or acetyl derivatives by gas chromatography-mass spectrometry, *J. Chromatogr.*, 632, 215, 1993.

244. Maurer, H. and Pfleger, K., Identification and differentiation of beta-blockers and their metabolites in urine by computerized gas chromatography-mass spectrometry, *J. Chromatogr.*, 382, 147, 1986.

245. Chi, H., Fang, H.J., Xu, Y.Q., Duan, H.J., Zhou, T.H., and Wu, Y., Study on the doping control method for the detection of narcotic analgesics and betablockers., *Yao Hsueh Hsueh Pao.*, 26, 688, 1991.

246. Lho, D.S., Hong, J.K., Paek, H.K., Lee, J.A., and Park, J., Determination of phenolalkylamines, narcotic analgesics, and beta-blockers by gas chromatography/mass spectrometry., *J. Anal. Toxicol.*, 14, 77, 1990.

247. Delbeke, F.T., Debackere, M., Desmet, N., and Maertens, F., Qualitative gas chromatographic and gas chromatographic-mass spectrometric screening for beta-blockers in urine after solid-phase extraction using Extrelut-1 columns, *J. Chromatogr.*, 426, 194, 1988.

248. Leloux, M.S., de Jong, E.G., and Maes, R.A., Improved screening method for beta-blockers in urine using solid-phase extraction and capillary gas chromatography-mass spectrometry, *J. Chromatogr.*, 488, 357, 1989.

249. Black, S.B., Stenhouse, A.M., and Hansson, R.C., Solid-phase extraction and derivatisation methods for beta-blockers in human post mortem whole blood, urine and equine urine, *J. Chromatogr. B Biomed. Appl.*, 685, 67, 1996.

250. Amendola, L., Molaioni, F., and Botre, F., Detection of beta-blockers in human urine by GC-MS-MS-EI: perspectives for the antidoping control, *J. Pharm. Biomed. Anal.*, 23, 211, 2000.

251. Maurer, H.H., Identification of antiarrhythmic drugs and their metabolites in urine, *Arch. Toxicol.*, 64, 218, 1990.

252. Baker, E.H. and Sandle, G.I., Complications of laxative abuse, *Annu. Rev. Med.*, 47, 127, 1996.

253. Turner, J., Batik, M., Palmer, L.J., Forbes, D., and McDermott, B.M., Detection and importance of laxative use in adolescents with anorexia nervosa, *J. Am. Acad. Child Adolesc. Psychiatry*, 39, 378, 2000.

254. Maurer, H.H., Toxicological detection of the laxatives bisacodyl, picosulfate, phenolphthalein and their metabolites in urine integrated in a "general-unknown" analysis procedure using gas chromatography-mass spectrometry, *Fresenius J. Anal. Chem.*, 337, 144, 1990.

255. Kudo, K., Miyazaki, C., Kadoya, R., Imamura, T., Jitsufuchi, N., and Ikeda, N., Laxative poisoning: toxicological analysis of bisacodyl and its metabolite in urine, serum, and stool, *J. Anal. Toxicol.*, 22, 274, 1998.

256. Maurer, H.H. and Kraemer, T., Rapid detection of anthraquinone laxatives in urine using high-performance thin-layer chromatography, *Bull. Soc. Sci. Med. Grand-Duche Lux.*, 127 Suppl, 476, 1990.

257. Perkins, S.L. and Livesey, J.F., A rapid high-performance thin-layer chromatographic urine screen for laxative abuse, *Clin. Biochem.*, 26, 179, 1993.

258. Stolk, L.M. and Hoogtanders, K., Detection of laxative abuse by urine analysis with HPLC and diode array detection, *Pharm. World Sci.*, 21, 40, 1999.

259. Liu, R.H., McKeehan, A.M., Edwards, C., Foster, G., Bensley, W.D., Langner, J.G., and Walia, A.S., Improved gas chromatography/mass spectrometry analysis of barbiturates in urine using centrifuge-based solid-phase extraction, methylation, with d5-pentobarbital as internal standard., *J. Forensic Sci.*, 39, 1504, 1994.

260. Meatherall, R., GC/MS confirmation of barbiturates in blood and urine, *J. Forensic Sci.*, 42, 1160, 1997.

261. Pocci, R., Dixit, V., and Dixit, V.M., Solid-phase extraction and GC/MS confirmation of barbiturates from human urine, *J. Anal. Toxicol.*, 16, 45, 1992.

262. Namera, A., Yashiki, M., Okada, K., Iwasaki, Y., Ohtani, M., and Kojima, T., Automated preparation and analysis of barbiturates in human urine using the combined system of PrepStation and gas chromatography-mass spectrometry, *J. Chromatogr. B*, 706, 253, 1998.

263. Kojima, T., Taniguchi, T., Yashiki, M., Miyazaki, T., Iwasaki, Y., Mikami, T., and Ohtani, M., A rapid method for detecting barbiturates in serum using EI-SIM., *Int. J. Legal Med.*, 107, 21, 1994.

264. Joern, W.A., Unexpected volatility of barbiturate derivatives: an extractive alkylation procedure for barbiturates and benzoylecgonine [letter], *J. Anal. Toxicol.*, 18, 423, 1994.

265. Laakkonen, U.M., Leinonen, A., and Savonen, L., Screening of non-steroidal anti-inflammatory drugs, barbiturates and methyl xanthines in equine urine by gas chromatography-mass spectrometry, *Analyst*, 119, 2695, 1994.

266. Rivier, L., Grunauer, C., and Saugy, M., Direct extractive methylation and automatic GC-MS screening and identification of drugs of abuse and other compounds of potential abuse in sport, in Mueller, R.K., Ed., Proc. 31st Int. TIAFT Meeting, Molinapress, Leipzig, 1994, 419.

267. Kim, K.R., Shim, W.H., Shin, Y.J., Park, J., Myung, S., and Hong, J., Capillary gas chromatography of acidic non-steroidal antiinflammatory drugs as *tert*-butyldimethylsilyl derivatives, *J. Chromatogr.*, 641, 319, 1993.

268. Kim, K.R. and Yoon, H.R., Rapid screening for acidic non-steroidal anti-inflammatory drugs in urine by gas chromatography–mass spectrometry in the selected-ion monitoring mode, *J. Chromatogr. B*, 682, 55, 1996.

269. Cardenas, S., Gallego, M., Valcarcel, M., Ventura, R., and Segura, J., A partially automated pretreatment module for routine analyses for seventeen non-steroid antiinflammatory drugs in race horses using gas chromatography/mass spectrometry, *Anal. Chem.*, 68, 118, 1996.

270. Leis, H.J., Fauler, G., Raspotnig, G., and Windischhofer, W., An improved method for the measurement of the angiotensin-converting enzyme inhibitor lisinopril in human plasma by stable isotope dilution gas chromatography/negative ion chemical ionization mass spectrometry, *Rapid. Commun. Mass. Spectrom.*, 13, 650, 1999.

271. Franklin, M.E., Addison, R.S., Baker, P.V., and Hooper, W.D., Improved analytical procedure for the measurement of captopril in human plasma by gas chromatography–mass spectrometry and its application to pharmacokinetic studies, *J. Chromatogr. B*, 705, 47, 1998.

272. Hammes, W., Hammes, B., Buchsler, U., Paar, F., and Bokens, H., Simultaneous determination of moexipril and moexiprilat, its active metabolite, in human plasma by gas chromatography-negative-ion chemical ionization mass spectrometry, *J. Chromatogr. B*, 670, 81, 1995.

273. Shioya, H., Shimojo, M., and Kawahara, Y., Determination of enalapril and its active metabolite enalaprilat in plasma and urine by gas chromatography–mass spectrometry, *Biomed. Chromatogr.*, 6, 59, 1992.

274. Goto, N., Kamata, T., and Ikegami, K., Trace analysis of quinapril and its active metabolite, quinaprilat, in human plasma and urine by gas chromatography-negative-ion chemical ionization mass spectrometry, *J. Chromatogr.*, 578, 195, 1992.

275. Leis, H.J., Fauler, G., Raspotnig, G., and Windischhofer, W., Quantitative determination of the angiotensin-converting enzyme inhibitor lisinopril in human plasma by stable isotope dilution gas chromatography/negative ion chemical ionization mass spectrometry, *Rapid. Commun. Mass. Spectrom.*, 12, 1591, 1998.

276. Lotfi, H., Dreyfuss, M.F., Marquet, P., Debord, J., Merle, L., and Lachatre, G., A screening procedure for the determination of 13 oral anticoagulants and rodenticides, *J. Anal. Toxicol.*, 20, 93, 1996.

277. Chalermchaikit, T., Felice, L.J., and Murphy, M.J., Simultaneous determination of eight anticoagulant rodenticides in blood serum and liver, *J. Anal. Toxicol.*, 17, 56, 1993.

278. Rengel, I. and Friedrich, A., Detection of anticoagulant rodenticides (4-hydroxycoumarins) by thin-layer chromatography and reversed-phase high-performance liquid chromatography with fluorescence detection, *Vet. Res. Commun.*, 17, 421, 1993.

279. Jones, A., HPLC determination of anticoagulant rodenticide residues in animal livers, *Bull. Environ. Contam. Toxicol.*, 56, 8, 1996.

280. Kuijpers, E.A., den Hartigh, J., Savelkoul, T.J., and de Wolff, F.A., A method for the simultaneous identification and quantitation of five superwarfarin rodenticides in human serum, *J. Anal. Toxicol.*, 19, 557, 1995.

281. Berny, P.J., Buronfosse, T., and Lorgue, G., Anticoagulant poisoning in animals: a simple new high- performance thin-layer chromatographic (HPTLC) method for the simultaneous determination of eight anticoagulant rodenticides in liver samples, *J. Anal. Toxicol.*, 19, 576, 1995.

282. Meyer, H., Scherling, D., and Karl, W., Nitrendipine: identification and synthesis of main metabolites, *Arzneim. -Forsch.*, 33, 1528, 1983.

283. Meyer, H., Wehinger, E., Bossert, F., and Scherling, D., Nimodipine: synthesis and metabolic pathway, *Arzneim. -Forsch.*, 33, 106, 1983.

284. Scherling, D., Buhner, K., Krause, H.P., Karl, W., and Wunsche, C., Biotransformation of nimodipine in rat, dog, and monkey, *Arzneim. -Forsch.*, 41, 392, 1991.

285. Scherling, D., Karl, W., Ahr, G., Ahr, H.J., and Wehinger, E., Pharmacokinetics of nisoldipine. III. Biotransformation of nisoldipine in rat, dog, monkey, and man, *Arzneim. -Forsch.*, 38, 1105, 1988.

286. Scherling, D., Karl, W., Ahr, H.J., Kern, A., and Siefert, H.M., Biotransformation of nitrendipine in rat, dog, and mouse, *Arzneim. -Forsch.*, 41, 1009, 1991.

287. Scherling, D., Karl, W., Radtke, M., Ahr, H.J., and Siefert, H.M., Biotransformation of nifedipine in rat and dog, *Arzneim. -Forsch.*, 42, 1292, 1992.

288. Kann, J., Krol, G.J., Raemsch, K.D., Burkholder, D.E., and Levitt, M.J., Bioequivalence and metabolism of nitrendipine administered orally to healthy volunteers, *J. Cardiovasc. Pharmacol.*, 6 Suppl 7, S968, 1984.

289. Beresford, A.P., Macrae, P.V., Alker, D., and Kobylecki, R.J., Biotransformation of amlodipine. Identification and synthesis of metabolites found in rat, dog and human urine/confirmation of structures by gas chromatography–mass spectrometry and liquid chromatography–mass spectrometry, *Arzneim. -Forsch.*, 39, 201, 1989.

290. Beresford, A.P., Macrae, P.V., and Stopher, D.A., Metabolism of amlodipine in the rat and the dog: a species difference, *Xenobiotica*, 18, 169, 1988.

291. Beresford, A.P., Macrae, P.V., Stopher, D.A., and Wood, B.A., Analysis of amlodipine in human plasma by gas chromatography, *J. Chromatogr.*, 420, 178, 1987.

292. Beresford, A.P., McGibney, D., Humphrey, M.J., Macrae, P.V., and Stopher, D.A., Metabolism and kinetics of amlodipine in man, *Xenobiotica*, 18, 245, 1988.

293. Stopher, D.A., Beresford, A.P., Macrae, P.V., and Humphrey, M.J., The metabolism and pharmacokinetics of amlodipine in humans and animals, *J. Cardiovasc. Pharmacol.*, 12 Suppl 7, S55, 1988.

294. Baarnhielm, C., Backman, A., Hoffmann, K.J., and Weidolf, L., Biotransformation of felodipine in liver microsomes from rat, dog, and man, *Drug Metab. Dispos.*, 14, 613, 1986.

295. Raemsch, K.D. and Sommer, J., Pharmacokinetics and metabolism of nifedipine, *Hypertension*, 5, II18, 1983.

296. Abernethy, D.R. and Schwartz, J.B., Pharmacokinetics of calcium antagonists under development, *Clinical Pharmacokinetics*, 15, 1, 1988.

297. Jean, C. and Laplanche, R., Assay of isradipine and of its major metabolites in biological fluids by capillary gas chromatography and chemical ionization mass spectrometry, *J. Chromatogr.*, 428, 61, 1988.

298. Tokuma, Y., Fujiwara, T., and Noguchi, H., Determination of nilvadipine in human plasma by capillary column gas chromatography-negative-ion chemical-ionization mass spectrometry, *J. Chromatogr.*, 345, 51, 1985.

299. Dru, J.D., Hsieh, J.Y., Matuszewski, B.K., and Dobrinska, M.R., Determination of felodipine, its enantiomers, and a pyridine metabolite in human plasma by capillary gas chromatography with mass spectrometric detection, *J. Chromatogr. B*, 666, 259, 1995.

300. Tokuma, Y., Fujiwara, T., and Noguchi, H., Combined capillary column gas chromatography/electron capture negative ion chemical ionization mass spectrometry of dihydropyridine calcium antagonists, *Biomed. Environ. Mass Spectrom.*, 13, 251, 1986.

301. Rosseel, M.T., Bogaert, M.G., and Huyghens, L., Determination of the calcium antagonist nimodipine in plasma by capillary gas chromatography and nitrogen detection, *J. Chromatogr.*, 533, 224, 1990.

302. Van Harten, J., Lodewijks, M.T., Guyt Scholten, J.W., Van Brummelen, P., and Breimer, D.D., Gas chromatographic determination of nisoldipine and one of its metabolites in plasma, *J. Chromatogr.*, 423, 327, 1987.

303. Soons, P.A. and Breimer, D.D., Gas chromatographic analysis of nitrendipine and its pyridine metabolite in human plasma, *J. Chromatogr.*, 428, 362, 1988.

304. Krol, G.J., Noe, A.J., Yeh, S.C., and Raemsch, K.D., Gas and liquid chromatographic analyses of nimodipine calcium antagonist in blood plasma and cerebrospinal fluid, *J. Chromatogr.*, 305, 105, 1984.

305. Josefsson, M., Zackrisson, A.L., and Norlander, B., Sensitive high-performance liquid chromatographic analysis of amlodipine in human plasma with amperometric detection and a single-step solid-phase sample preparation, *J. Chromatogr. B*, 672, 310, 1995.

306. Shimooka, K., Sawada, Y., and Tatematsu, H., Analysis of amlodipine in serum by a sensitive high-performance liquid chromatographic method with amperometric detection, *J. Pharm. Biomed. Anal.*, 7, 1267, 1989.

307. Boutagy, J., Rumble, F., and Dunagan, F., Determination of isradipine and the oxidative pyridine metabolite in human plasma by high-performance liquid chromatography, *J. Chromatogr.*, 487, 483, 1989.

308. Uno, T., Ohkubo, T., and Sugawara, K., Enantioselective high-performance liquid chromatographic determination of nicardipine in human plasma, *J. Chromatogr. B*, 698, 181, 1997.

309. Muck, W.M., Enantiospecific determination of nimodipine in human plasma by liquid chromatography-tandem mass spectrometry, *J. Chromatogr. A*, 712, 45, 1995.

310. Yasuda, T., Tanaka, M., and Iba, K., Quantitative determination of amlodipine in serum by liquid chromatography with atmospheric pressure chemical ionization tandem mass spectrometry, *J. Mass Spectrom.*, 31, 879, 1996.

311. Tokuma, Y. and Noguchi, H., Stereoselective pharmacokinetics of dihydropyridine calcium antagonists, *J. Chromatogr. A*, 694, 181, 1995.

312. Ventura, R. and Segura, J., Detection of diuretic agents in doping control, *J. Chromatogr. B*, 687, 127, 1996.

313. Deventer, K., Delbeke, F.T., Roels, K., and Van Eenoo, P., Screening for 18 diuretics and probenecid in doping analysis by liquid chromatography-tandem mass spectrometry, *Biomed. Chromatogr.*, 16, 529, 2002.

314. Thieme, D., Grosse, J., Lang, R., Mueller, R.K., and Wahl, A., Screening, confirmation and quantification of diuretics in urine for doping control analysis by high-performance liquid chromatography–atmospheric pressure ionisation tandem mass spectrometry, *J. Chromatogr. B Biomed. Sci. Appl.*, 757, 49, 2001.

315. Garbis, S.D., Hanley, L., and Kalita, S., Detection of thiazide-based diuretics in equine urine by liquid chromatography–mass spectrometry, *J. AOAC. Int.*, 81, 948, 1998.

316. Ventura, R., Fraisse, D., Becchi, M., Paisse, O., and Segura, J., Approach to the analysis of diuretics and masking agents by high-performance liquid chromatography-mass spectrometry in doping control, *J. Chromatogr.*, 562, 723, 1991.

317. Lisi, A.M., Trout, G.J., and Kazlauskas, R., Screening for diuretics in human urine by gas chromatography-mass spectrometry with derivatisation by direct extractive alkylation., *J. Chromatogr.*, 563, 257, 1991.

318. Carreras, D., Imaz, C., Navajas, R., Garcia, M.A., Rodriguez, C., Rodriguez, A.F., and Cortes, R., Comparison of derivatization procedures for the determination of diuretics in urine by gas chromatography-mass spectrometry, *J. Chromatogr. A*, 683, 195, 1994.

319. Maurer, H.H., Arlt, J.W., Kraemer, T., Schmitt, C.J., and Weber, A.A., Analytical development for low molecular weight xenobiotic compounds, *Arch. Toxicol.*, 19 Suppl., 189, 1997.

320. Lehr, K.H. and Damm, P., Simultaneous determination of the sulphonylurea glimepiride and its metabolites in human serum and urine by high-performance liquid chromatography after pre-column derivatization, *J. Chromatogr.*, 526, 497, 1990.

321. Huupponen, R., Ojala, K.P., Rouru, J., and Koulu, M., Determination of metformin in plasma by high-performance liquid chromatography, *J. Chromatogr.*, 583, 270, 1992.

322. Valdes Santurio, J.R. and Gonzalez Porto, E., Determination of glibenclamide in human plasma by solid-phase extraction and high-performance liquid chromatography, *J. Chromatogr. B*, 682, 364, 1996.

323. Caldwell, J., Hutt, A.J., and Fournel-Gigleux, S., The metabolic chiral inversion and dispositional enantioselectivity of the 2-arylpropionic acids and their biological consequences., *Biochem. Pharmacol.*, 37, 105, 1988.

324. Careri, M., Manini, P., and Mori, G., Fast separation of phenolic acids in normal-phase partition chromatography, *Anal. Proc.*, 32, 129, 1995.

325. Davies, N.M., Methods of analysis of chiral non-steroidal anti-inflammatory drugs, *J. Chromatogr. B*, 691, 229, 1997.

326. Bhushan, R. and Joshi, S., TLC separation of antihistamines on silica gel plates impregnated with transition metal ions, *J. Planar Chromatogr. — Mod. TLC*, 9, 70, 1996.

327. De-Kanel, J., Vickery, W.E., and Diamond, F.X., Simultaneous analysis of 14 non-steroidal anti-inflammatory drugs in human serum by electrospray ionization-tandem mass spectrometry without chromatography, *J. Am. Soc. Mass Spectrom.*, 9, 255, 1998.

328. Maurer, H.H., Formation of Artifacts, in *Mass Spectral and GC Data of Drugs, Poisons, Pesticides, Pollutants and Their Metabolites*, Pfleger, K., Maurer, H.H., and Weber, A., Eds., VCH Verlag, Weinheim, Germany, 1992, 21.

329. Kraemer, T., Weber, A.A., and Maurer, H.H., Improvement of sample preparation for the STA — Acceleration of acid hydrolysis and derivatization procedures by microwave irradiation, in *Proceedings of the Xth GTFCh Symposium in Mosbach*, Pragst, F., Ed., Helm-Verlag, Heppenheim, Germany, 1997, 200.

330. Maurer, H. and Pfleger, K., Screening procedure for detecting butyrophenone and bisfluorophenyl neuroleptics in urine using a computerized gas chromatographic-mass spectrometric technique, *J. Chromatogr.*, 272, 75, 1983.

331. Maurer, H. and Pfleger, K., Toxicological detection of ajmaline, prajmaline and their metabolites in urine integrated in a "general-unknown" analysis procedure using gas chromatography-mass spectrometry, *Fresenius Z. Anal. Chem.*, 330, 459, 1988.

Liquid Chromatography/Mass Spectrometry in Forensic Toxicology

2

MACIEJ J. BOGUSZ

Contents

0-8493-1522-0/04/$0.00+$1.50

2.1 Introduction

Toxicology may be generally defined as a scientific search for a causal relationship between exposure to a given compound and observed biological effect. Analytical toxicology provides all the technically refined tools indispensable for detection of toxic compounds or — in other words — for the assessment of the exposure. These tools, as well as the whole analytical strategy, may be different in various toxicological disciplines. In forensic toxicology, only those methods may be successfully applied which combine efficient separation of the relevant compound from the biological matrix with the most specific detection. LC/MS fulfils the requirements of forensic toxicological analysis due to high selectivity, possibility of detection of active metabolites, and efficient screening in unclear death cases. For this reason, also, due to gradual introduction of low-cost LC/MS instruments, this technique is now finding an increasingly important place in forensic toxicological laboratories.

LC/MS underwent major evolution in the last decade. From an expensive, difficult, and not always reliable hyphenated technique, it changed to a robust analytical tool, applicable in almost all analytical situations. This was caused by the introduction of atmospheric pressure ionization (API) sources. The advent of API/LC/MS converted all earlier LC/MS interfaces like thermospray and particle beam ionization of fast atom bombardment to obsolete techniques.

The statement of Krull and Cohen may illustrate the dimension of this change:

> There are at least two fundamental views on the future role of MS in biotechnology HPLC. Is the mass spectrometer an expensive sophisticated LC detector? Or, is the chromatograph an expensive sample-preparation device for a mass spectrometer? One of the newer developments in LC/MS is that this distinction will cease to be important in the future.[1]

The change of the status of LC/MS from hyphenated method to routine analytical tool attracted a new generation of users, who are taking the hyphenation for granted and treat a LC/MS instrument as a single unit. Similar evolution was observed for GC/MS in the past. Technical and methodical progress in LC/MS in the last few years was reviewed in several books and publications.[2–4] Special reviews were devoted to application of LC/MS in forensic toxicology.[5–9]

2.2 Methodical Considerations Relevant to Forensic Analysis

Recent years have seen some publications on the general aspects of LC/MS that are of relevance for forensic toxicological analysis. These studies concerned the quality of the sample and its influence on the chromatographic signal, optimization of chromatographic analysis, application of various ionization sources, and various mass analyzers in toxicology.

2.2.1 Separation Issues

2.2.1.1 Sample Pretreatment Methods and Matrix Suppression

It is common knowledge that poor sample preparation procedure may ruin any detection, even under the application of most modern and sophisticated techniques. This is valid also for LC/MS. It should be noted that in the earlier phase of application some authors tried to apply LC/MS/MS in a flow-injection mode, i.e., the extracts of serum or urine were directly injected into API/MS without chromatographic separation. This procedure was checked only for a limited number of drugs (morphine, codeine, amphetamine, and benzoylecgonine) and potentially interfering compounds. Additionally, the matrix effects were not recognized.[10]

Müller et al.[11] studied the effect of coextracted serum matrix on the signal of test substances — codeine (as a positive ion) and glafenine (as a negative ion). Severe ion suppression was observed for both ionization methods after the injection of serum matrix originating from protein precipitation and SPE. Less suppression was observed in the case of solvent extraction, while the combination of protein precipitation with SPE caused no suppression. The authors stated that the suppression effects were caused by polar, nonretained matrix components appearing on the beginning of the chromatogram.

Zhou et al.[12] carried out a systematic study on the relation of matrix suppression to the extraction methods, chromatographic conditions, and concentration of analytes. In this study, blank serum matrix samples were injected into the HPLC column with the postcolumn infusion of four test compounds at three concentration levels. The areas of suppression were located along the whole chromatogram. On the solvent front, salt and other polar unretained species were present. Other endogenous compounds were eluted later, sometimes in very high concentrations, causing severe ion suppression which was independent of the analyte concentration.

Avery compared the ion suppression effects caused by extracts of human and animal (rat, dog, monkey, rabbit, and guinea pig) plasma.[13] The responses for analyte and internal standard were measured in isocratic and gradient elution, using APCI and ionspray ionization sources. It was stated that each species showed different suppression. Therefore, the validation should be performed with samples originating from the same species.

Tang et al.[14] studied matrix effects in postcolumn-infusion experiments. Extracted blanks were injected while the ion transitions of the infused analytes were monitored. Both suppression and enhancement of ionization was observed. These phenomena were compensated by changing the ionization energy, ionization source (ACPI instead of ESI), sample pretreatment method, or by including matrix ions in acquisition methods.

Bansal and Liang introduced two novel terms to assess the matrix effect in an LC/MS/MS assay. Matrix factor (MF) is calculated from the ratio of peak area of pure analyte to the peak area of the analyte injected together with blank matrix. Extraction uniformity (EU) is the ratio of the extraction efficiency of analyte to that of internal standard. If the EU was close to 1, it was possible to perform the validation for the same matrix from different species.[15] Some authors applied column switching as a method to minimize the influence of matrix compounds. Sun et al.[16] applied protein precipitation in a 96-well format as a sample pretreatment method for the determination of nicotine and cotinine in plasma with LC/MS/MS. In order to avoid matrix ion suppression from late-eluting components, a dual-column switching system was applied and the analysis time was shortened to 2.5 min. Column switching after homogenization and protein precipitation was applied for determination of various drugs (benzodiazepines, felodipine, nitrendipine, nicardipine) in animal tissues to study drug distribution. The drugs were separated on short (2.1 × 10 mm) columns and analyzed with ESI or ACPI/MS/MS.[17]

From all above-mentioned studies, two general conclusions may be drawn: the extracts should be of high quality, and the chromatographic separation should not be neglected or sacrificed. LC/MS is not the panacea that will replace the optimal sample pretreatment and separation.

2.2.1.2 Composition of the Mobile Phase and Type of Column

The composition of the mobile phase may greatly affect the sensitivity of LC/MS analysis. Temesi and Law[18] studied the influence of mobile phase on electrospray response in ESI/MS, using 35 various acidic, neutral, and basic drugs as test substances. Methanol and ACN were used as organic modifiers, and ammonium acetate, ammonium formate, and TFA were used as electrolytes. Generally, in the positive ionization mode, methanol gave stronger signals than ACN. Electrospray response decreased with increasing concentration of ammonium acetate/formate. Very large quantitative individual differences between drugs were observed. The authors concluded that a thorough optimization of all eluent parameters is essential for single analyte analysis.

According to Naidong et al.,[19] reversed-phase chromatography, although widely used, is less compatible with MS/MS detection than normal-phase chromatography. This is particularly true for highly polar compounds, which are hardly retained on reversed-phase columns, even on the mobile phase containing mostly water. A low percentage of organic phase modifier affects the sensitivity due to nonoptimal spraying conditions. Conversely, normal-phase columns are used with a mobile phase containing high percentage of organic solvent, assuring better dispersing and evaporation of electrospray droplets. That was demonstrated in series of comparative experiments with nicotine, cotinine, and albuterol, analyzed with LC/ESI/MS/MS after separation on C18 and silica columns. The mobile phase for a reversed-phase column contained 10% acetonitrile, and for a normal-phase column, 70% acetonitrile. The sensitivity observed after normal-phase separation was distinctly higher due to the longer retention times of polar analytes and the separation from matrix compounds causing ion suppression. A similar approach was used by the same group of authors for the analysis of morphine and its glucuronides in serum,[20,21] for fentanyl,[22] and for hydrocodone/hydromorphone.[23] The combination of straight-phase short (50 × 3 mm) column and high flow rate (4 ml/min) allowed achieving ultrafast analysis (below 1 min) of nonpolar and polar analytes like morphine and its glucuronides and midazolam and its hydroxylated metabolites with LC/MS/MS.[24] Recently, Naidong et al.[25] eliminated evaporation and reconstitution steps in 96-well SPE. This was possible due to application of straight-phase LC/MS/MS, where the eluting solvent was compatible for injection into mobile phase containing high percentage of organic solvent. Such an approach allowed processing a batch of 96 samples (containing, e.g., fentanyl, omeprazole, or pseudoephedrine) in 1 h.

Zweigenbaum et al.[26] demonstrated the feasibility of rapid sample preparation and fast analysis by LC/ESI/MS/MS. As model drugs, six benzodiazepine derivatives were used. The drugs were extracted from plasma using 96-well plates and separated on C18 column (15 mm × 2.1 mm, particle size 3 μm) in less than 30 sec. As a result, 1152 samples could be analyzed in less than 12 h. Cheng et al.[27] studied the possibility of the rapid LC/MS/MS analysis of a mixture of various compounds. Amitriptyline, ibuprofen, lidocaine, naproxen, and prednisolone were used as model drugs. From theoretical considerations, a column capable of ultrafast liquid chromatography should provide a good peak capacity at a short gradient run time and high flow rate. The study was done on XTerra MS C18 columns 20 mm × 2.1 mm, 2.5 μm, 30 mm × 2.1 mm, 3.5 μm, and 50 mm × 2.1 mm, 5.0 μm. The drugs were eluted in acetonitrile–acid or acetonitrile–base gradient. The gradient duration time (5 to 100% ACN) was 0.7 to 1.7 min, and the flow rate was 1.5 ml/min. MS/MS and UV detection was applied. In these conditions the drug mixture was separated in less than 1 min. The study of electrospray response at different pH values showed that one of the drugs examined (lidocaine) did not follow the rules of electrospray theory. The authors stated that the optimization for particular drugs should be done checking different solvents, buffer types, and pH values since the best results are sometimes achieved outside the expected range.

2.2.2 Detection Issues

2.2.2.1 Choice of Ionization Source

In the scientific literature, a general trend could be observed regarding the use of electrospray ionization in toxicological analysis. Most probably, this is caused by an overwhelming use of this source in the LC/MS of large molecules. However, the choice of ESI or APCI may be of relevance for particular substances. As a rule, the response of ESI for very polar substances is higher than that of APCI. On the other hand, less polar compounds, including a majority of parent drugs, need an active ionization mode for better detectability. It is therefore a good practice to try both ionization sources in the

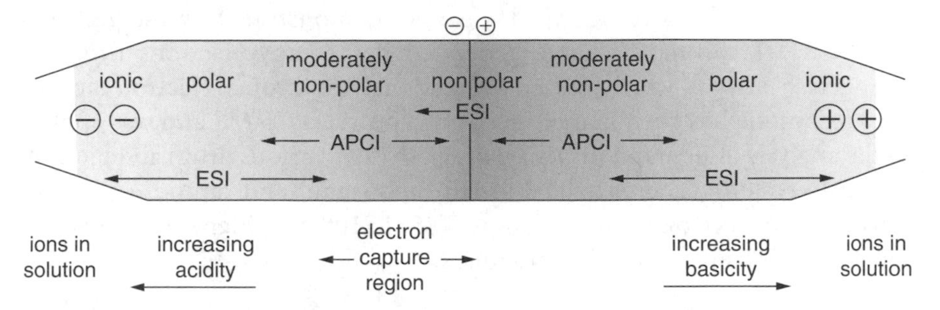

Figure 2.1 Ionization–continuum diagram showing the regions of effective usefulness of the various interfaces for LC–MS, including atmospheric pressure chemical ionization (APCI), electrospray ionization (ESI), and electron impact (EI) ionization. (Reprinted with permission from *Anal. Chem.* 73, 5441, 2001. Copyright 2001 American Chemical Society.)

developing stage of the method. Unfortunately, this approach is not often followed or not reported in the literature.

Nevertheless, some studies were specifically devoted to the usefulness of various ion sources in toxicological analysis. Dams et al. carried out a comparative performance study of the three ion sources for the ion-trap–based mass spectrometer (pneumatically assisted ESI, sonic spray, and APCI), using morphine as a model compound analyzed under different elution conditions. The influence of solvent modifier, buffer, pH, and volatile acids was studied. The composition of the mobile phase had a serious impact on the ionization efficiency. Strong similarities were observed between the performance characteristics of ESI and sonic spray, whereas APCI showed completely different behavior. The authors concluded that APCI source showed the greatest potential, due to robustness, applicability to higher flow rates, and positive response to acids or buffers. [28] Thurman et al. assessed the sensitivity of APCI and ESI sources used for detection of 75 pesticides belonging to different classes. The experiments were performed in chromatographic conditions optimized for particular groups of pesticides in full-scan mode and in positive and negative ionization modes. It was stated that neutral and basic pesticides were better detectable with APCI, while ESI gave better signal for cationic and anionic herbicides. Generally, the optimization of mobile phase composition and use of proper ion source was of paramount importance. [29]

Yang and Henion compared the performance of APCI and recently introduced atmospheric pressure photoionization (APPI) for the determination of idoxifene and its major metabolite. APPI assured six to eight times higher sensitivity than APCI. Both sources showed similar response to matrix suppression. The authors concluded that APPI might be an additional tool in LC/MS.[30]

New possibilities of LC/APCI/MS were demonstrated by the group of Blair.[31] APCI corona discharge generated low-energy electrons from the nitrogen sheath gas, which ionized suitable analytes through electron capture. This technique has been named electron capture APCI-MS and was applied to the analysis of pentafluorobenzyl derivatives of various drugs and biomolecules like steroids, prostaglandins, thromboxane, and amino acids. The sensitivity of electron capture APCI was 25- to 100-fold higher than negative ion APCI for underivatized compounds.

2.2.2.2 *Choice of Mass Analyzer (Q vs. QQQ vs. IT vs. qTOF)*

It goes without saying that the use of tandem mass spectrometry instead of single stage quadrupole gives obvious advantages concerning specificity and sensitivity. The main obstacle of LC/MS/MS is the price. However, in the last few years some manufacturers launched lower-cost bench-top tandem instruments that may be widely applicable in forensic toxicological laboratories.

Some research groups performed comparative studies concerning various mass analyzers applied in LC/MS. Clauwaert et al.[32] applied the LC/ESI/qTOF/MS for quantitative measurements of MDMA and its metabolite MDA in body fluids. The LOQ was 1 μg/l and the linear dynamic range extended over four decades. It was concluded that this technique achieves the linear dynamic range for LC/ESI/MS/MS. Zhang and Henion[33] compared the applicability of LC/ESI/qTOF/MS and LC/ESI/MS/MS for quantitative analysis of idoxifene in human plasma. The drug and its deuterated analog as an internal standard was isolated from plasma using a semiautomated 96-well hexane extraction and separated on a 30 mm × 2 mm CN column. This column could separate target compound from the matrix, which was not possible with the ODS column. For the TOF/MS instrument, an exact mass of the protonated quasi-molecular ion was measured (*m/z* 524.1441), whereas a MS/MS instrument was applied in SRM mode, using *m/z* 524.2 as a parent ion and *m/z* 97.9 as a product ion. The comparison showed that the LC/MS/MS technique was about ten times more sensitive than LC/qTOF/MS (the LOQs for idoxifene were 0.5 μg/l and 5 μg/l, respectively). Both methods showed satisfactory dynamic range. In conclusion, the authors stated that LC/qTOF/MS might be used to complement LC/MS/MS in certain cases. Comparison between an ion trap and single-stage quadrupole mass spectrometer was done for GC/MS by Vorce et al.[34] The authors compared the sensitivity, precision, and stability of ion ratios obtained for both mass analyzers, using urine extracts spiked with amphetamine, methamphetamine, THC-carboxylic acid, phencyclidine, morphine, codeine, 6-monoacetylmorphine, and benzoylecgonine on five concentration levels for each drug. The sensitivity of the ion trap used in the full-scale mode was comparable to quadrupole, working in the selected ion monitoring mode. The advantage of the ion trap was that it provided a full spectra of substances. The ion ratios were slightly more precise for quadrupole. These data may be also useful for the assessment of ion trap LC/MS.

2.3 LC/MS Analysis of Illicit Drugs

2.3.1 Opiate Agonists

2.3.1.1 Analysis of Street Drugs

Dams et al.[35] applied sonic spray LC/MS (ion trap) for profiling of street heroin samples. Chromatographic separation was performed on monolithic silica column (chromolith performance 100 × 4.6 mm) in gradient elution in acetonitrile–water at a flow of 5 ml/min. A postcolumn split of 1/20 was applied; the analysis time was 5 min. The protonated molecular ions of seven constituents of street heroin (morphine, codeine, 6-MAM, heroin, acetylcodeine, papaverine, noscapine, and levallorphan, used as internal standard) were monitored. The limits of SSI/MS detection ranged from 0.25 to 1 ng on-column.

2.3.1.2 Analysis of Biological Fluids

The advent of LC/MS brought very important progress in determination of opiates and its metabolites in biological fluids. The most important opiate agonist — heroin — is very rapidly deacetylated to 6-monoacetylmorphine and consecutively to morphine. Morphine in turn is demethylated to normorphine. Both active heroin metabolites — morphine and normorphine — are then glucuronidated to 3- and 6-glucuronides. A similar process applies to all natural and semi-synthetic opiates. [36] LC/MS is the only analytical technique which allows specific detection of parent opiates and all polar metabolites without derivatization and without acidic or enzymatic cleavage.

Zuccaro et al.[37] developed an LC/ESI/MS method for the simultaneous determination of heroin, 6-MAM, morphine, M3G, and M6G in serum. The drugs were extracted with SPE C_2 cartridges and separated on a straight-phase silica column in a methanol–ACN–formic acid mobile phase. The authors used a silica column in order to separate all substances in one run under isocratic conditions. The LOD for heroin was 0.5 μg/l, for 6-MAM, 4 μg/l. The method was applied for a pharmacokinetic study on heroin-treated mice. Bogusz et al.[38] used atmospheric pressure chemical ionization LC/APCI/MS for a determination of heroin metabolites (6-MAM, morphine, M3G, and M6G) in blood, cerebrospinal fluid, vitreous humor, and urine of heroin victims. The drugs were extracted with C18 cartridges; the LOD for 6-MAM was 0.5 μg/l. Low molar ratios of M3G/morphine and M6G/mor-

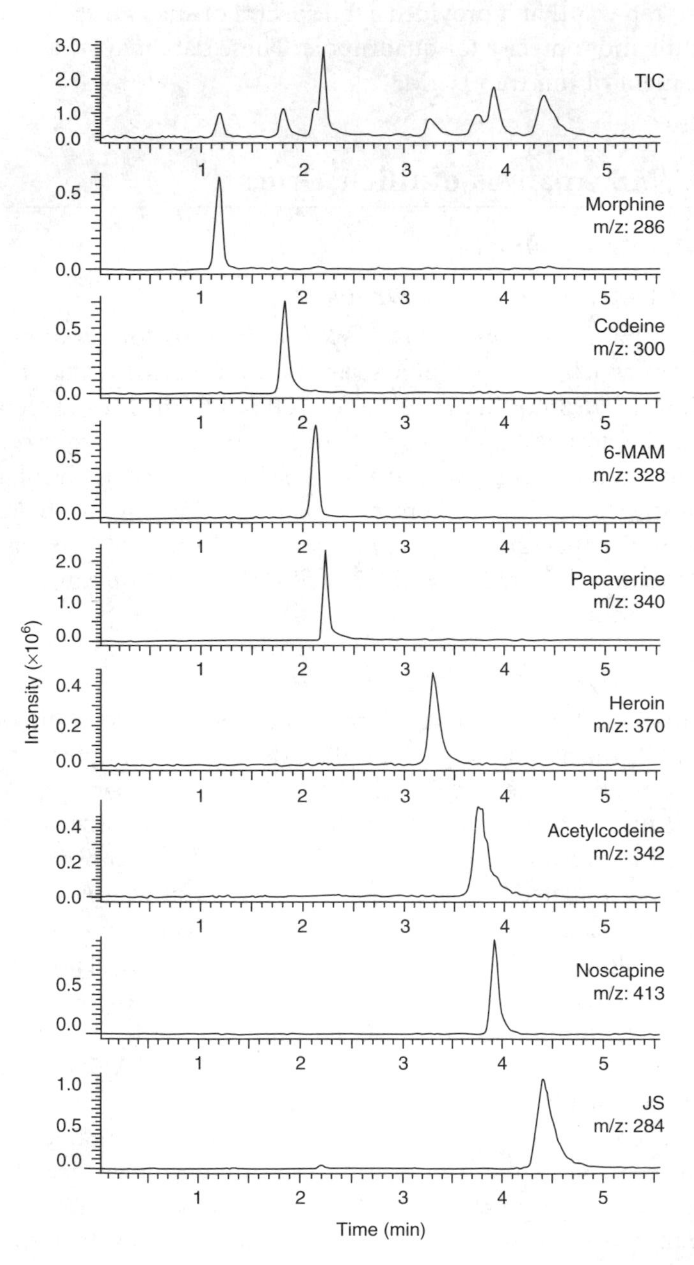

Figure 2.2 Total ion and selected ion chromatograms of a mixture (1 mg/l) of the opium alkaloids and the internal standard. (Reprinted with permission from *Anal. Chem.* 74, 3206, 2002. Copyright 2002 American Chemical Society.)

phine in blood indicated a short survival time after drug intake. In the next study, Bogusz et al.[39] extended the LC/APCI/MS method for the determination of 6-MAM, M3G, M6G, morphine, codeine, and C6G, using deuterated internal standards for each compound. The detection limits ranged from 0.5 to 10 μg/l. This method was applied for routine forensic examination of blood samples taken from suspected heroin abusers.[7]

In the last few years, several LC/MS methods for determination of morphine, codeine, and corresponding glucuronides were published. The methods were generally based on solid-phase extraction and ESI/MS or ESI/MS/MS detection and are summarized in Table 2.1. Determination of free morphine and both its glucuronides in body fluids may be helpful in the interpretation of a given case from a forensic and clinical point of view. High free-morphine fraction generally indicates acute poisoning in a very early stage, particularly in a person who has not taken heroin or morphine chronically. Also, the differentiation between pharmacologically active M6G and inactive M3G, as well as the free morphine:morphine glucuronides ratio, is of practical importance for interpretation of the severity of poisoning.[45] The Naidong group[24] applied ultrafast LC/MS/MS of morphine and its glucuronides, using the combination of short (50 × 3 mm), straight-phase column and high flow rate (4 ml/min).

It must be stressed that heroin after deacetylation follows metabolic routes which are common with morphine and — to some extent — also with codeine. Therefore, only 6-MAM may be regarded as a heroin-specific metabolite and as a marker of heroin use, and is detectable in blood and urine samples of heroin addicts.[7,38,45] Strictly speaking, the presence of 6-MAM in urine, blood, or other body fluids evidences the intake of pure diacetylmorphine (DAM). The differentiation between the intake of pure DAM and illicit heroin became relevant after the introduction heroin prescription programs in some countries like Switzerland, Great Britain, Germany, and the Netherlands. One of the basic requirements of these programs is that the participants must not use any illicit drugs, particularly illicit heroin. In illicit heroin not only DAM is present but also several other opiates like 6-MAM, acetylcodeine, codeine, papaverine and noscapine, as well as various adulterants. It must be stressed that only acetylcodeine (AC) may be regarded as a specific marker of illicit heroin. AC is produced from codeine during the acetylation of opium. Its content in illicit heroin ranges from 2 to 7%.[46] O'Neal and Poklis[47,48] developed a GC/MS method for detection of AC in urine and found this drug in over 30% of morphine-positive specimens in concentrations ranged from 1 to 4600 μg/l. 6-MAM was found in over 70% of these samples. Codeine — a possible metabolite of AC — was found in all urine samples. Staub et al.[49] used also GC/MS and detected AC in over 85% and 6-MAM in over 94% of 71 urine samples obtained from illegal heroin consumers.

Table 2.1 LC/MS Methods Used for Opioids

Drug	Sample	Isolation	Column, Elution Conditions	Detection	LOD (μg/l)	Ref.
M, M3G, M6G	Plasma	SPE	Silica, ACN-HCOOH isocr.	ESI-QQQ, MRM	0.5–1.0	20
M, M3G, M6G	Plasma	SPE 96	Silica, ACN-TFA isocr.	ESI-QQQ, MRM	0.5–10	21
M, 6-MAM, Cod, NorCod, Pholcod	Plasma, urine	L/l	C8, ACN-$HCOONH_4$ isocr.	ESI-Q, SIM	10.0	40
M, M3G, M6G, NorM	Serum, urine	SPE	C18, ACN-HCOOH grad.	ESI-QQQ, MRM	0.3–2.5	41
M, M3G, M6G, 6-MAM, Cod, C6G	Serum, urine	SPE	C18, ACN-$HCOONH_4$ isocr.	APCI-Q, SIM	0.1–10	39
M, M3G, M6G, 6-MAM, Cod, C6G	Serum	SPE	C18, ACN-$HCOONH_4$ grad.	ESI-Q, SIM	0.5–5.0	42
M, M3G, M6G	Serum	SPE	C18, ACN-$HCOONH_4$ isocr.	ESI-QQQ, MRM	1.0–5.0	43
M, M3G, M6G	Plasma	SPE	C18, ACN-HCOOH isocr.	ESI-QQQ, MRM	0.25–0.5	44
Fentanyl	Plasma	SPE 96	Silica, ACN-TFA isocr.	ESI-QQQ, MRM	0.05	22
HYM, HYC	Plasma	SPE 96	Silica, ACN-TFA isocr.	ESI-QQQ, MRM	0.1	23
HYM, DHM, H3G, other metabolites	Plasma (rat)	SPE	C18, ACN-$HCOONH_4$ isocr.	ESI-QQQ, MRM	2–5	62
Buprenorphine, BUG, NBUG	Plasma	SPE	C18, ACN-$HCOONH_4$ grad.	ESI-QQQ, MRM	0.1	52
Ketobemidone, Nor-K	Urine	SPE	C8, ACN-HCOOH grad.	ESI-Q, SIM	25.0	57

Notes: M = morphine, M3G = morphine-3-glucuronide, M6G = morphine-6-glucuronide, 6-MAM = 6-monoacetylmorphine, NorM = normorphine, Cod = codeine, C6G = codeine-6-glucuronide, Pholcod = pholcodine, BUG = buprenorphine glucuronide, HYC = hydrocodone, HYM = hydromorphone, DHM = dihydromorphine, H3G = hydromorphone-3-glucuronide, Nor-K = nor-ketobemidone, Q = single stage quadrupole, QQQ = triple stage quadrupole, SIM = selected ion monitoring, MRM = multiple reaction monitoring, ACN = acetonitrile, TFA = trifluoroacetic acid.

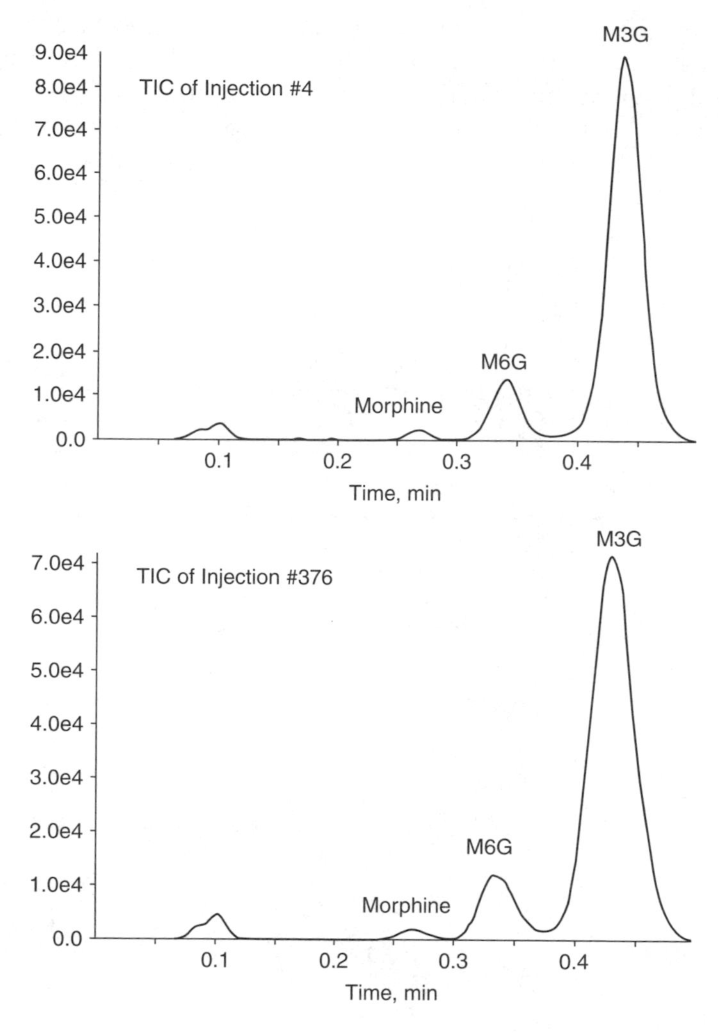

Figure 2.3 Total ion chromatograms of injections #4 and #376, both of a morphine/M6G/M3G calibration standard (5:10:100 μg/l) during the continuous analysis of 384 extracted morphine samples in human plasma. (From Shou W.Z., Chen Y.L., Eerkes A., Tang Y.Q., Magis L., Jiang X., and Naidong W., Ultrafast liquid chromatography/tandem mass spectrometry bioanalysis of polar analytes using packed silica columns. *Rapid Commun. Mass Spectrom.* 16, 1613, 2002. Reproduced by permission of John Wiley & Sons Ltd.)

Bogusz et al.[50] determined heroin markers in 25 morphine-positive urine samples with LC/APCI/MS. Codeine-6-glucuronide was found in all samples, codeine in 24, noscapine in 22, 6-MAM in 16, papaverine in 14, DAM in 12, and AC in 4 samples. Katagi et al.[40] developed an automatic method for the determination of heroin and its metabolites monoacetylmorphine (6-MAM) and morphine, as well as acetylcodeine, codeine, and dihydrocodeine in urine. Urine samples were applied on cation-trapping exchange column, washed with ammonium acetate, and after column switching the drugs were eluted and separated on analytical cation exchange column in ACN–ammonium acetate (70:30). The detection was done with ESI-MS in full scan or SIM mode. Protonated quasi-molecular ions or acetonitrile adducts were

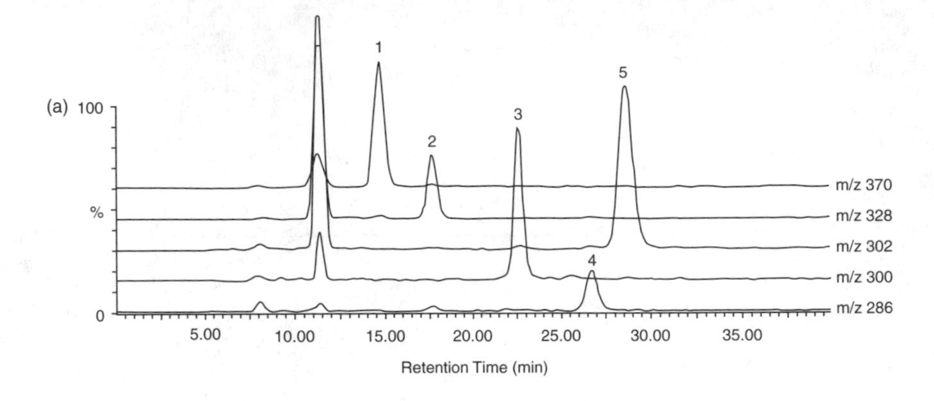

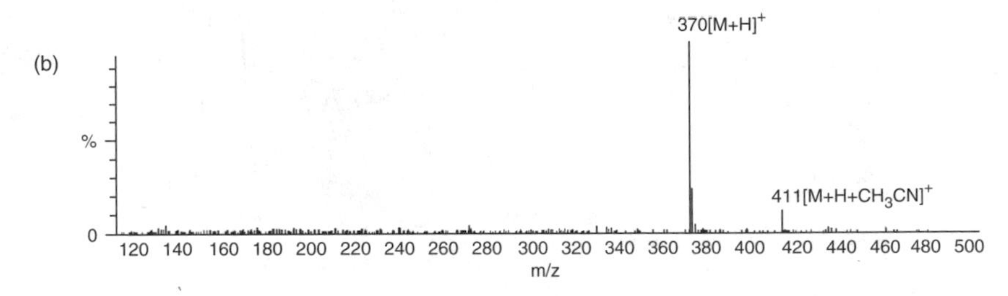

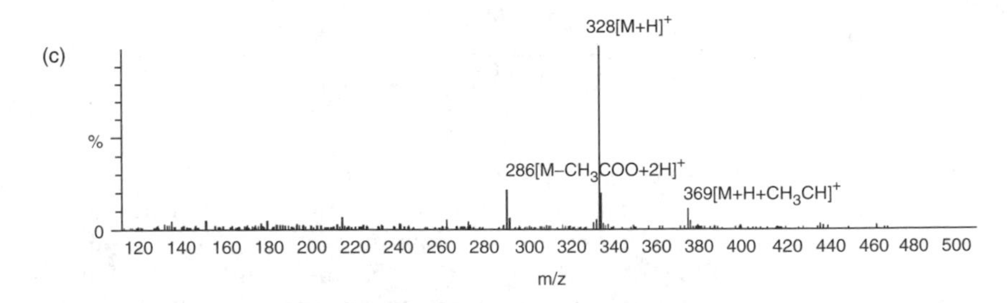

Figure 2.4 Mass chromatograms (a) obtained for a urine sample spiked with diacetylmorphine (1), monoacetylmorphine (2), codeine (3), morphine (4), and dihydrocodeine (5) to the concentration of 100 μg/l. Mass spectra produced for diacetylmorphine, monoacetylmorphine and morphine are presented as (a), (b), and (c), respectively. (Reprinted from Katagi M., Nishikawa M., Tatsuno M., Miki A., and Tsushihashi H., Column switching high-performance liquid chromatography-electrospray ionization mass spectrometry for identification of heroin metabolites in human urine. *J. Chromatogr. B* 751, 177, 2001. Copyright 2001, with permission from Elsevier Science.)

monitored. The LODs ranged from 2 to 30 μg/l in full scan mode and from 0.1 to 3 μg/l in SIM.

Recent publications also contained research concerning the determination of synthetic or semisynthetic opiates. Gaulier et al.[51] reported a suicidal poisoning of a 25-year-old male heroin addict with a high dose of buprenorphine. Buprenorphine (BU) and its active metabolite norbuprenorphine (NBU) were determined in body fluids and organs with LC/ESI/MS after deproteinization and SPE. In gastric content, only BU was found in concentration of 899 mg/l. The following concentrations were found in selected matrices: in blood BU 3.3 mg/l, NBU 0.4 mg/l; in bile BU 2035 mg/l, NBU 536 mg/l; and in brain BU 6.4 mg/l, NBU 3.9 mg/l. Besides BU and NBU, high concentrations of 7-aminoflunitrazepam were found in blood (1.2 mg/l), urine (4.9 mg/l), and gastric content (28.6 mg/l). Polettini and Huestis[52] developed a LC/ESI/MS/MS method for determination of buprenorphine and its metabolites — norbuprenorphine and buprenorphine glucuronide — in human plasma. SPE extraction with C18 cartridges and gradient elution was applied. For buprenorphine, and norbuprenorphine, as well as for deuterated analogs used as internal standards, the protonated quasi-molecular ions were monitored for the buprenorphine glucuronide protonated quasi-molecular ion and buprenorphine aglycone. The LOQ was 0.1 μg/l for all compounds. On the basis of the transition *m/z* 590→414, norbuprenorphine glucuronide was also tentatively detected. The reference standard of this compound was not available. The authors stated that useful fragmentation of the buprenorphine molecule was not possible; after the increase of fragmentation energy this compound dissipated to very small particles. On the other hand, Moody et al.,[53,54,] who also used ESI/LC/MS/MS and fragmentation, and achieved detection at the level of 0.1 μg/l in MRM mode. Additionally, Hoja et al.[55] and Bogusz et al.,[56] who applied ESI and APCI/MS, respectively, observed fragmentation of buprenorphine — which was particularly distinct in the case of the APCI/MS used by Bogusz. This confirms the need for comparison of both ionization sources (ESI and APCI) when the method for a particular compound is established.

Naidong et al.[23] published LC/ESI/MS/MS procedures for determination of hydrocodone and hydromorphone in plasma. The drugs and deuterated analogs were extracted with solvent and separated from glucuronides using a 50 mm × 2 mm silica column and a mobile phase consisting of ACN–water–formic acid (80:20:1). The LOQ was 0.1 μg/l. The same group determined fentanyl in plasma, using automated 96-well solid-phase extraction, straight-phase chromatography, and ESI/MS/MS.[22] The LOQ was 0.05 μg/l in plasma, based on 0.25 ml sample volume. It should be stressed that the Naidong group is consequently using straight-phase columns and a mobile phase containing high concentration of organic modifier for opiate agonists.

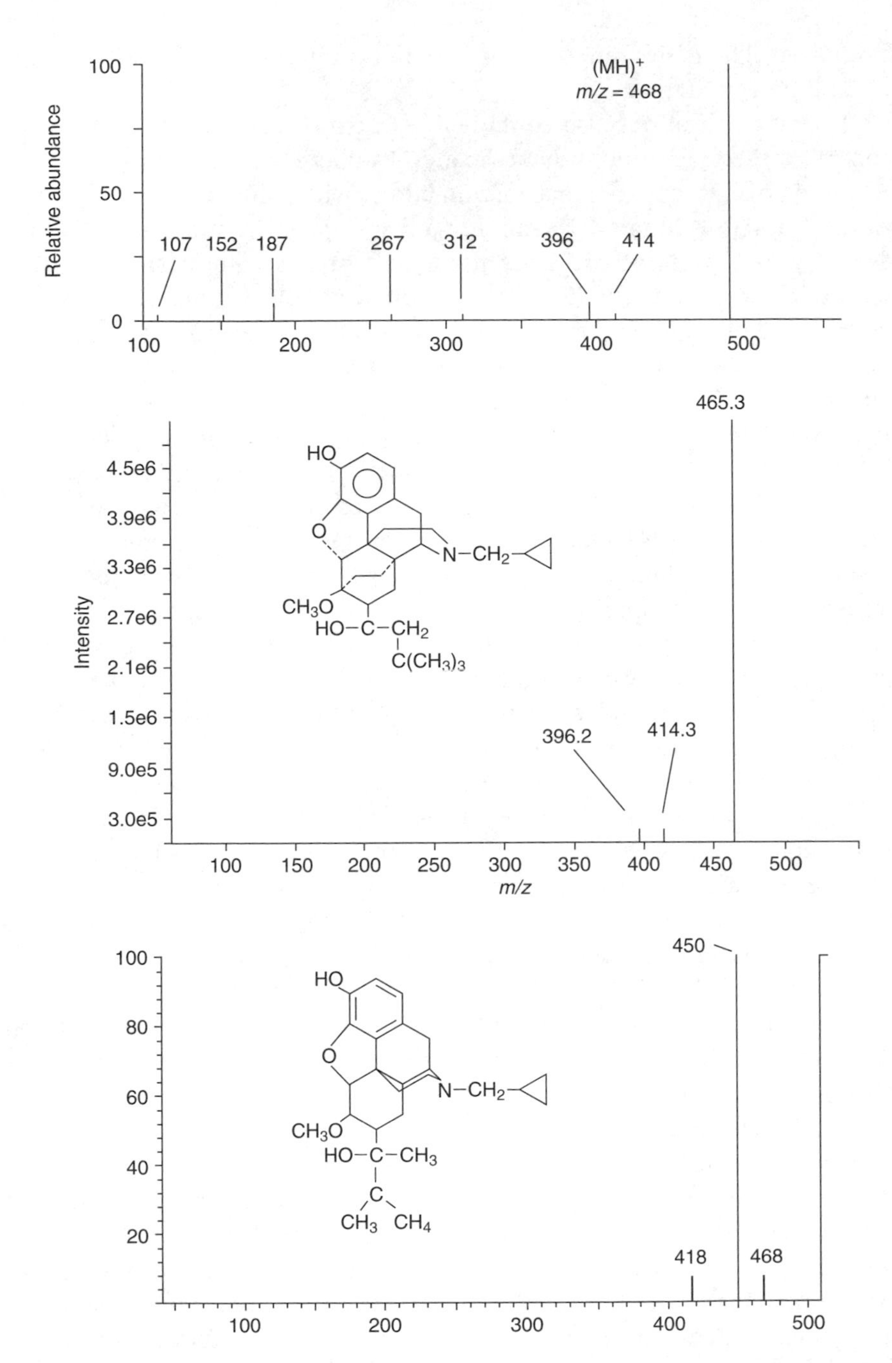

Figure 2.5 Mass spectra of buprenorphine obtained with LC/ESI/MS/MS (upper), LC/ESI/MS (middle), and LC/APCI/MS (lower). (Reprinted from Bogusz M.J., Opioid Agonists. In Bogusz M.J., Ed., *Forensic Science*, Vol. 2, *Handbook of Analytical Separations*. Elsevier, Amsterdam, 2000, 3. Copyright 2000, with permission from Elsevier Science.)

Ketobemidone is a synthetic opioid agonist and narcotic analgesic which is frequently abused, particularly in Scandinavian countries. Breindahl et al.[57] developed an LC/ESI/MS method for the determination of ketobemidone and its demethylated metabolite in urine. Mixed-bed SPE cartridges were used for isolation with a recovery of over 90%. Protonated quasi-molecular ions for both substances as well as three fragments for ketobemidone were monitored. The LOD was 25 μg/l. Sunstrom et al. [58] applied ESI/MS/MS for determination of ketobemidone, and its five phase I metabolites, as well as glucuronides of ketobemidone and norketobemidone in human urine. The same group used ESI/qTOF/MS besides LC/MS/MS for the determination of the glucuronides of ketobemidone, nor-, and hydroxymethoxyketobemidone in urine. [59] The accuracy of the mass measurement was better than 2 ppm.

The enantiomers of tramadol and its active metabolite *O*-desmethyltramadol were determined with LC/APCI/MS-MS.[60] The substances were isolated from plasma with automated SPE and separated on a Chiralpak AD column in isohexane–ethanol–diethylamine (97:3:0.1) mobile phase. The transitions of protonated quasi-molecular ions of drugs and internal standard (ethyltramadol) to the same ion *m/z* 58 were monitored. Also, other tramadol metabolites could be detected. Juzwin et al.[61] determined tramadol *N*-oxide and its major metabolites in the plasma of experimental animals, using LC/ESI/MS/MS. The LOQ was 6 μg/l and the method was applied for preclinical studies.

Methadone is a synthetic opioid of very long elimination half-life (15 to 55 h) which is used mainly as a heroin substitute in the treatment of heroin addicts. Methadone is metabolized to inactive 2-ethylidene-1,5-dimethyl-3,3-diphenylpyrrolidine (EDDP) and, to a lesser extent, to 2-ethyl-5-methyl-3,3-diphenyl-1-pyrroline (EMDP). All these compounds contain a chiral center, and the enantiomers could be separated. This is of pharmacological relevance since (*R*)-(–)-methadone (levomethadone) is 25 to 50 times more potent than the (*S*)-(+)-methadone. Commercial methadone preparations may contain the racemic form or levomethadone. Kintz et al.[63] published the first ESI/LC/MS method for enantioselective separation of methadone and EDDP in hair, using deuterated analogs of all compounds for quantitation. The results showed that the (*R*)-(–)-methadone is preferably deposited in human hair. Ortelli et al.[64] applied LC/MS for enantioselective determination of methadone in saliva and serum. The method was applied for the analysis of samples taken from heroin addicts participating in methadone maintenance program. Disposition of methadone after nasal, intravenous, and oral application to volunteers was measured by LC/MS.[65] The parent drug and EDDP were measured in blood, showing high bioavailability of methadone by the nasal route.

2.3.2 Amphetamine and Ecstasy

Amphetamine and methamphetamine are powerful stimulants used for over 100 years. Some of their methylenedioxy analogs like methylenedioxymethamphetamine (MDMA), methylenedioxyethylamphetamine (MDEA) or methylenedioxyamphetamine (MDA) were already synthesized in 1914 and tested in treatment of psychiatric disorders. In contrast to amphetamine, MDMA did not become widely abused until the late 1970s. This drug is known under the names "Ecstasy," "XTC," and "Adam." Paradoxically, the widespread abuse of psychoactive phenethylamines was propagated by eminent pharmacologist Alexander Shulgin who, in his book *Pihkal, A Chemical Love Story*,[66] published procedures for synthesizing 179 various drugs of this group. Also, the information concerning recommended dosages and expected symptoms was given. The book is freely available and certainly helped to proliferate psychoactive phenethylamines in the society. Amphetamine derivatives are usually manufactured as tablets and distributed illegally in discotheques. These drugs, besides stimulating action, may alter thermoregulation, and they caused a growing number of death cases, mainly due to heat stroke at rave parties.[67–69] It must be stressed that the recreational use of Ecstasy is very often associated with consumption of other psychotropic drugs or alcohol. A review of 81 Ecstasy-related death cases in England, as well as examination of urine samples taken from 64 attendees of rave parties, revealed that in the majority of cases, MDMA was present in combination with amphetamine, methamphetamine, other designer amphetamines, or opiates.[69,70] This suggests that the majority of the ravers are multidrug users. A similar observation was also made by Bogusz.[7]

2.3.2.1 Analysis of Biological Fluids

Bogusz et al. compared the sensitivity of UV/DAD and APCI/MS detection for amphetamines. Amphetamine, methamphetamine, MDMA, MDA, and MDEA, as well as eight other phenethylamines, were extracted with ether from serum and urine, derivatized with phenylisothiocyanate and subjected to HPLC with APCI/MS or UV/DAD detection. LC/APCI/MS assured about 10 times higher detection than UV with LODs ranged from 1 to 5 μg/l.[71] In the next study of the same group,[72] the derivatization, as well as diode array detection, was abandoned. Fourteen amphetamines and related compounds were isolated from biofluids with SPE cartridges and subjected to LC/APCI/MS examination in SIM mode. Again, the limit of detection ranged from 1 to 5 μg/l. This method was applied in routine casework.[47] Kataoka et al.[73] applied SPME for isolation of amphetamine, methamphetamine, MDMA, MDEA, and MDA from urine. The drugs were desorbed by mobile-phase flow and detected with ESI/MS (SIM), with a LOD below 1 μg/l urine.

Clauwaert et al.[32] applied the LC/ESI/TOF/MS for quantitative measurements of MDMA and its metabolite MDA in whole blood, serum, vitreous humor, and urine. The LOQ was 1 μg/l and the linear dynamic range extended over four decades. It was concluded that this technique achieves the linear dynamic range for LC/ESI/MS/MS. The same group determined MDMA, MDEA, and MDA in whole blood, urine, and vitreous humor of rabbits using HPLC with fluorometric and tandem mass spectrometric detection. The LOD for fluorometric detection was 0.8 μg/l in blood and 2 μg/l in urine. Very good correlation between both methods was found. The methods were used for a thanatochemical distribution study in rabbits.[74]

Paramethoxyamphetamine (PMA) is a new designer amphetamine of relatively high toxicity. Mortier et al.[75] described the method for determination of PMA together with amphetamine, MDMA, MDEA, and MDA in biological fluids. Solvent extraction followed by sonic spray LC/MS(IT) was applied. The matrix suppression was measured and was negligible for human urine and liver or kidney.

Several therapeutic drugs act as amphetamine/methamphetamine precursors, and it is important to differentiate between the administration of these drugs and illicit preparations. LC/MS has been applied for these purposes. Benzphetamine (BMA) and its metabolites benzylamphetamine, hydroxybenzphetamine, hydroxybenzylamphetamine, methamphetamine, and amphetamine were isolated from urine with SPE, separated on an alkaline mobile phase and detected with ESI/MS (SIM). The LODs were from 0.3 to 10 μg/l urine.[76] This method allowed discrimination of the ingestion of BMA, amphetamine, and methamphetamine. Dimethylamphetamine and its metabolites dimethylamine-*N*-oxide, methamphetamine, and amphetamine were determined in urine by ESI/LC/MS (SIM) after isolation with SPE. LODs were 5 to 50 μg/l urine.[77]

Selegiline, an inhibitor of monoamine oxidase-B used in the treatment of Parkinson's disease, is metabolized to methamphetamine and amphetamine. Katagi et al.[78] determined this drug as well as the specific metabolites selegiline-*N*-oxide, methamphetamine, and amphetamine in human urine. All compounds were isolated from urine using SPE C18 cartridges, separated on a cation exchange column and subjected to ESI/MS detection in full scan acquisition mode. In urine samples collected from selegiline patients, only metabolites were detected. The method may serve for differentiation between selegiline and methamphetamine intake. The excretion of selegiline-*N*-oxide after selegiline administration in volunteers was studied with ESI/LC/MS.[79]

2.3.2.2 Analysis of Alternative Samples

Wood et al.[80] applied LC/ESI/MS/MS for determination of amphetamines (amphetamine, methamphetamine, MDMA, MDA, MDEA, and ephedrine) in saliva. Collected saliva (50 μl) was diluted with a 200 μl methanolic solution of an internal standard mixture (deuterated analogs) and centrifuged. Supernatant was subjected to HPLC in an ACN–ammonium acetate mobile phase on a C18 column. The drugs were detected using transitions of protonated quasi-molecular ions in optimized conditions. The LOQs were 1 to 5 μg/l. This method was used further for determination of designer amphetamines in oral fluid, sweat wipe, and plasma after controlled administration of MDMA and in real life conditions (after ingestion of Ecstasy tablets at a party). Toxicokinetic data were gathered, indicating very high intra- and interindividual variability of saliva results. In sweat, very low concentrations of MDMA were detected in the first 5 h after drug intake.[81] Stanaszek et al.[82] applied LC/APCI/MS for determination of amphetamines, as well as other drugs of abuse, in hair.

2.3.3 Cocaine

Cocaine may be administered in different ways — by smoking, intranasal ("snorting"), oral, or intravenous route. This drug is converted in the human body into a multitude of active and nonactive metabolites, mostly of high polarity. These metabolites are amenable to GC/MS analysis only after derivatization.[83]

2.3.3.1 Analysis of Biological Fluids

The advent of LC/MS brought substantial progress in liquid chromatographic detection of cocaine metabolites, which may be found without any derivatization with this technique. Sosnoff et al. detected BZE in dried blood spots with tandem LC/MS.[84] A LOD of 2 μg/l was achieved from a 12-μl sample. This technique was used for epidemiological screening in a study involving newborns. Singh et al.[85] applied acetonitrile precipitation for isolation of cocaine and metabolites from plasma. The drugs were determined with LC/APCI/MS/MS. The LOQ was 2 μg/l for cocaine, benzoylecgonine, and norcocaine, and 5 μg/l for ecgonine methyl ester. The procedure has been applied for pharmacokinetic study in rats. In a study of Jeanville et al. a fast-gradient ESI/LC/MS/MS was applied for automated urinalysis. Urine samples were only filtered and transferred into a 96-well plate before injection. Total analysis time was 2 min. However, the issue of possible matrix interference was not addressed.[86] In the next paper of Jeanville et al.,[87] centrifuged urine samples were injected into LC/MS/MS system equipped with an on-line extraction unit. Total analysis time was below 4 min. The LODs for EME, BZE, and cocaine were 0.5, 2, and 0.5 μg/l, respectively. The same research group studied the applicability of LC/ESI/qTOF for direct determination of cocaine and ecgonine methyl ester in rat plasma.[88] Similar performance as for the triple-quadrupole instrument was observed.

Skopp et al. carried out the study on the stability of cocaine and metabolites (benzoylecgonine, ecgonine methyl ester, and ecgonine) at different temperatures using LC/ESI/MS/MS.[89,90] The drugs were isolated from human plasma by SPE and separated on a narrow-bore C8 column. For each compound, one product ion originating from the protonated quasi-molecular ion was monitored. Only ecgonine appeared to be stable at room temperature. The conversion of cocaine to final metabolite ecgonine was stoichiometric. A validated LC/APCI/MS/MS procedure for cocaine and benzoylecgonine in human plasma was published by Lin et al.[91] The transitions *m/z* 304 →182 and 290 → 168 were monitored for cocaine and benzoylecgonine, respectively. The LOQ was 2.5 μg/l for both compounds.

Table 2.2 presents liquid chromatographic methods published recently for amphetamines and cocaine analysis.

2.3.3.2 Analysis of Alternative Samples

Clauwaert et al.[92] determined cocaine, benzoylecgonine, and cocaethylene in hair samples. Hair samples were hydrolyzed with HCL and extracted with Confirm HCX cartridges. The extracts were subjected to LC/ESI/MS and HPLC with fluorescence detection. The procedure was applied in 29 forensic drug overdose cases and close agreement for results obtained from both detection methods was observed.

2.3.3.3 Prenatal Exposure to Cocaine

The analysis of cocaine and its metabolites in meconium, placenta, or amniotic fluid may give a valuable indication of prenatal exposure to this drug. Xia et al.[93] established a LC/ESI/MS/MS method for the determination of cocaine and 12 metabolites in meconium. The method was tested on 22 drug-positive meconium samples. In 21 cases the presence of at least 8 metabolites was stated. Ecgonine appeared to be the most sensitive marker of cocaine exposure. The same method was applied in the study on distribution of cocaine and metabolites in blood, amniotic fluid, placental, and fetal tissues of pregnant rats after intravenous cocaine administration.[94]

2.3.4 Cannabis

2.3.4.1 Analysis of Street Drugs

Particle beam ionization (PBI) LC/MS was applied for separation and identification of cannabinoids (THC, cannabinol, cannabidiol, acidic cannabinoids) in hashish samples.[95] Quantitative data were not given. Since PBI interface has been largely abandoned, this paper is rather of historical value.

Table 2.2 LC/MS Methods Used for Amphetamines and Cocaine

Drug	Sample	Isolation	Column, Elution Conditions	Detection	LOD (μg/l)	Ref.
14 Amphetamines and related compounds	Plasma	SPE	C18, ACN-$HCOONH_4$ isocr.	APCI-Q, SIM	1–5	72
A, MA, MDMA, MDEA, MDA	Urine	SPME	CN, ACN-CH_3COONH_4 isocr.	ESI-Q, SIM	0.4–0.8	73
MDMA, MDEA, MDA	Plasma	l/l	C18, ACN-CH_3COONH_4 isocr.	ESI-qTOF/SIM	1.0	32
A, MA, MDMA, MDEA, MDA, E	Saliva	SPE	C18, ACN-CH_3COONH_4 isocr.	ESI-QQQ, MRM	1.0	80
Cocaine, BZE, EME	Urine	SPE	C18, ACN-H_2O gradient	ESI-QQQ, MRM	0.5–2.0	87
Cocaine, BZE, EME, ECG	Plasma	SPE	C8, MeOH/ACN-CH_3COONH_4 isocr.	ESI-QQQ, MRM	2.8–4.4	90
Cocaine, BZE, cocaethylene	Hair	SPE	C18, ACN-CH_3COONH_4 gradient	ESI-QQQ, MRM	25 pg/mg	92

Notes: A = amphetamine, MA = methamphetamine, MDMA = methylenedioxymethamphetamine, MDEA = methylenedioxyethylamphetamine, MDA = methylenedioxyamphetamine, E = ephedrine, BZE = benzoylecgonine, EME = ecgonine methyl ester, ECG = ecgonine, Q = single stage quadrupole, QQQ = triple stage quadrupole, qTOF = quadrupole/time-of-flight, SIM = selected ion monitoring, MRM = multiple reaction monitoring, ACN = acetonitrile.

Bäckström et al.[96] described a combination of supercritical fluid chromatography with APCI/MS for determination of cannabinoids in *Cannabis* products. The LODs (on-column) were 0.69 ng for THC, 0.55 ng for cannabidiol, and 2.10 ng for cannabinol, respectively.

2.3.4.2 *Analysis of Biological Fluids*

In the study of Mireault,[97] blood or urine samples were extracted with SPE cartridge and analyzed on C8 column in methanol–ammonium acetate. An ion trap instrument equipped with APCI source was used in MS/MS mode (positive ions). Detection limit of 1 μg/l was achieved for THC, 11-OH-THC, and THC-COOH. In the second study of the same group,[98] an APCI/MS/MS was applied. The limit of quantitation of 0.25 μg/l was reported for all three

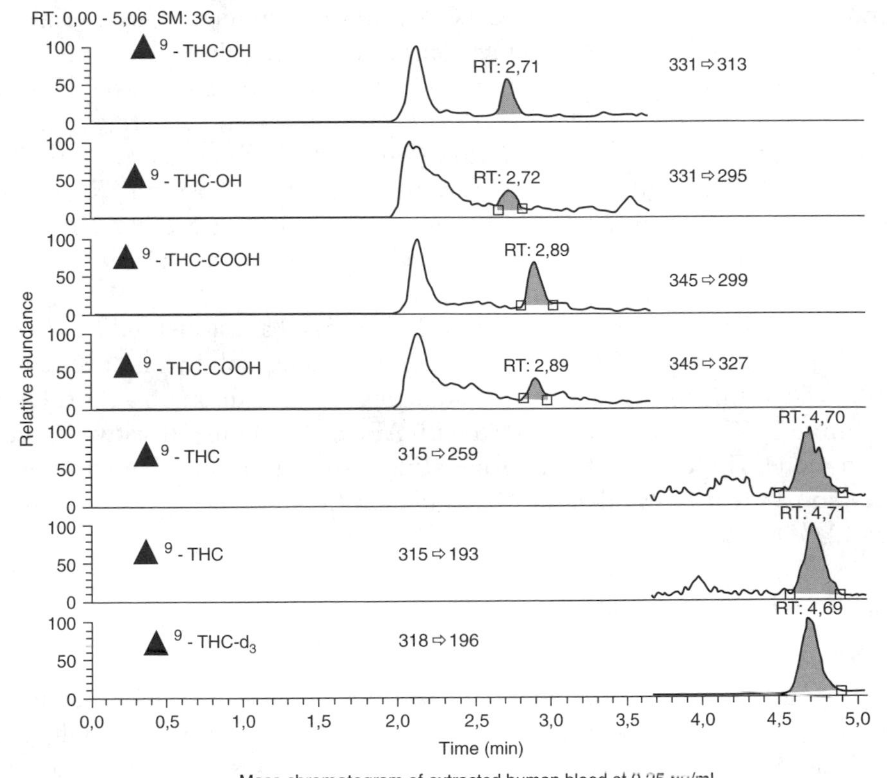

Figure 2.6 LC/APCI/MS/MS chromatogram of 11-OH-THC, THC-COOH, THC, and deuterated THC (internal standard) extracted from whole blood spiked to the concentration of 0.25 μg/l each. (From Picotte P., Mireault P., and Nolin G., A rapid and sensitive LC/APCI/MS/MS method for the determination of Δ^9-tetrahydrocannabinol and its metabolites in human matrices. Poster at the 48th ASMS conference, Long Beach, CA, 2000. With permission.)

compounds. In biological extracts, the limit of detection was 10 to 40 times lower for a quadrupole instrument than for an ion trap.

Breindahl and Andreasen applied ESI/LC/MS for determination of THC-COOH in urine.[99] Urine was subjected to basic hydrolysis and solid phase extraction. THC-COOH and its deuterated analog were analyzed with HPLC/ESI/MS, using a C8 reversed-phase column, gradient elution in acetonitrile–formic acid, and SIM detection (positive ions). In-source collision-induced dissociation was applied and protonated quasi-molecular ion as well as two fragment ions were monitored. LOD was 15 μg/l. Tai and Welch[100] determined THC-COOH with HPLC/ESI/MS (negative ions). The drug was extracted from urine with SPE and subjected to isocratic separation on an ODS column in the methanol–ammonium acetate mobile phase. Only deprotonated quasi-molecular ions of the drug and its deuterated analog were monitored. Weinmann et al. used LC/MS/MS for simultaneous determination of THC-COOH and its glucuronide in urine.[101] In this method, the cleavage of conjugates was omitted. THC-COOH and its glucuronide were extracted from urine with ethyl acetate–ether (1:1) and separated on C8 column in a gradient of ammonium formate buffer and acetonitrile. ESI/MS/MS was used for detection using protonated quasi-molecular ions as precursor ions and two fragment ions for each drug as product ions. The specificity of the method was checked using enzymatic hydrolysis of THC-COOH-glucuronide.

In another study, Weinmann et al.[102] developed a fast method for determination of THC-COOH in urine using automated solid phase extraction after alkaline hydrolysis. HPLC was done in gradient elution on a short ODS column. THC-COOH was detected with APCI/MS/MS in a negative ionization mode. Three product ions, originating from deprotonated quasi-molecular ion, were used for identification and quantitation. The LOD was 2.0 μg/l and LOQ was 5.1 μg/l.

2.3.5 Hallucinogens

2.3.5.1 LSD

LSD (lysergic acid diethylamide) is an extremely potent hallucinogenic drug. Its single dose ("trip") ranges from 30 to 100 μg. The drug is rapidly and extensively metabolized, and only a very small fraction is excreted unchanged with urine.[103] In the first phase of application of LC/MS for the analysis of LSD, this drug and its demethylated metabolite was determined in urine. Webb et al.[104] developed an immunoaffinity extraction of LSD from urine followed by LC/ESI/MS. Methysergide was used as internal standard for quantitation. Protonated molecular ion and two fragments were monitored; the LOD was 0.5 μg/l at 5 ml urine volume. In the next study of the same group, methysergide was replaced by a triple deuterated LSD analog used as internal

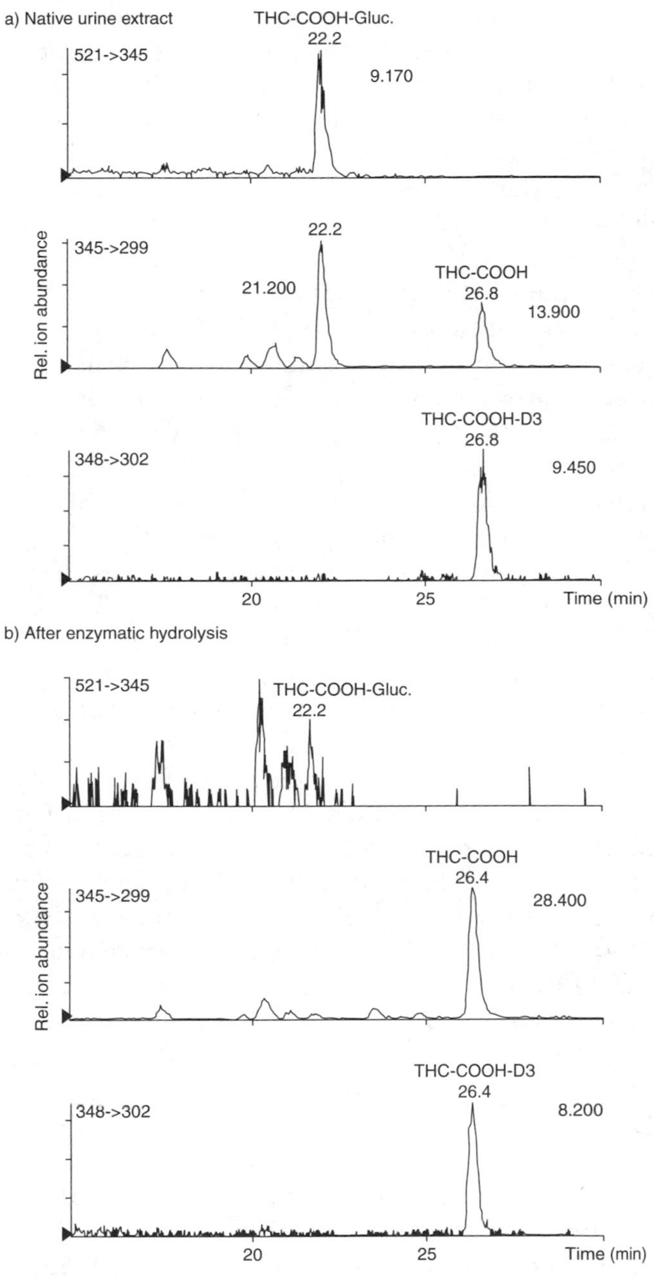

Figure 2.7 LC/ESI/MS/MS of THC-COOH and THC-COOH-glucuronide extracted from native urine (a) and after enzymatic hydrolysis. (Reprinted from Weinmann W., Vogt S., Goerke R., Muller C., and Bromberger A., Simultaneous determination of THC-COOH and THC-COOH-glucuronide in urine samples by LC/MS/MS. *Forensic Sci. Int.* 113, 381, 2000. Copyright 2000, with permission from Elsevier Science.)

standard.[105] This was associated with much better accuracy and precision. The LOD (0.5 μg/l) has not been improved. Hoja et al.[106] applied ESI-LC-MS for determination of LSD and *N*-demethyl-LSD in urine after Extrelut extraction. The LOQ was 0.05 and 0.1 μg/l for LSD and metabolite, respectively. De Kanel et al.[107] extracted LSD and its demethylated metabolite from 1 ml blood, serum, plasma, and urine with automated mixed-phase SPE. The drugs were detected with LC/ESI/MS/MS using phenyl column for separation. The LOD of 0.025 μg/l for both substances was reported. In the study of Bodin et al.[108] LSD was extracted from urine with organic solvent, back extracted to an acetate buffer and subjected to LC/ESI/MS. The LOD was 0.02 μg/l. Since only small fraction of LSD is eliminated with urine, its window of detection in urine following use is not longer than 12 to 22 h.[109] The turning point in the analysis of LSD came when Poch et al.[110] demonstrated the importance of determination of 2-oxo-3-hydroxy-LSD, a prevalent metabolite of LSD excreted with urine. This metabolite, as well as LSD, nor-LSD, and iso-LSD were determined in urine with LC/APCI/MS (ion trap). The concentrations of 2-oxo-3-hydroxy-LSD were distinctly higher than those of the parent drug and other metabolites. In the next study, Poch et al.[111] compared three detection methods for LSD and metabolites: LC/APCI/MS, LC/APCI/MS/MS (ion trap), and GC/MS. The latter method was used only for the parent drug. Very good agreement between both LC/MS methods was found. According to the authors, the detection window for LSD use may be significantly increased due to determination of the metabolite. Sklerov et al.[112] determined LSD and 2-oxo-3-hydroxy-LSD in blood and urine. The drugs were isolated from urine with alkaline solvent extraction; for blood a solvent extraction followed by SPE was used. LC/ESI/MS was applied with in-source collision-induced dissociation and monitoring of three ions for each compound. LOD was 0.1 μg/l for LSD and 0.4 μg/l for metabolite in urine. In authentic cases, 2-oxo-3-hydroxy-LSD was found in high concentrations in urine but was not present in blood samples. A LC/ESI/MS/MS technique was applied by Canezin et al.[113] for determination of LSD and iso-LSD in plasma with LOD of 0.02 μg/l for both compounds. In urine metabolites also were detected, such as 2-oxo-3-hydroxy-LSD, nor-LSD, nor-iso-LSD, 13- and 14-OH-LSD-glucoronides, lysergic acid ethylamide, trioxydated-LSD, and lysergic acid ethyl-2-hydroxyethylamide. The method was applied in two cases of drug abuse. Klette et al.[114] studied the specificity of LC/MS assay of 2-oxo-3-hydroxy-LSD. Fifteen compounds structurally related to LSD, as well as a wide range of unrelated compounds, were studied. None of the examined drugs interfered with the detection of the LSD metabolite. It was also found that 2-oxo-3-hydroxy-LSD is stable for 60 days in urine samples stored in a freezer or in a refrigerator at pH 4.6 to 8.4. Recent chromatographic methods for cannabinoids and LSD are depicted in Table 2.3.

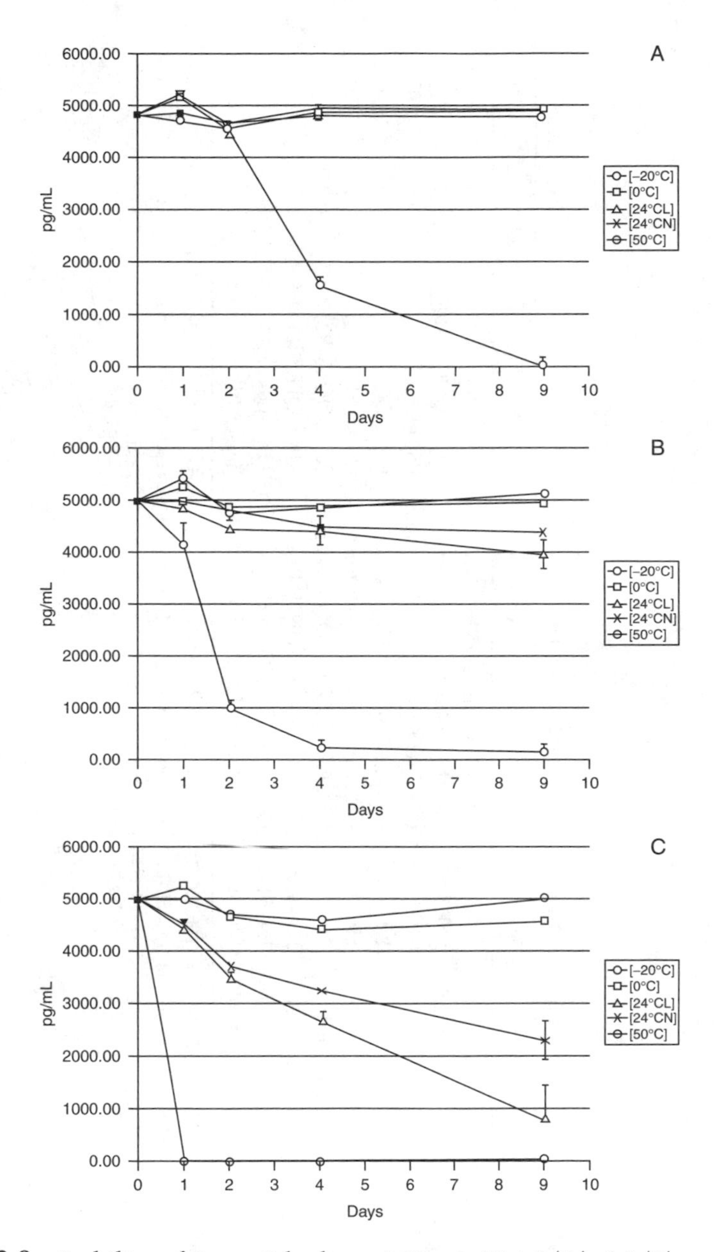

Figure 2.8 Stability of 2-oxo-3-hydroxy-LSD at pH 4.6 (A), 6.5 (B), and 8.4 (C) at different storage conditions. (From Klette K.L., Horn C.K., Stout P.R., and Anderson C.J., LC-MS analysis of human urine specimens for 2-oxo-3-hydroxy LSD: Method validation for potential interferants and stability study for 2-oxo-3-hydroxy LSD under various storage conditions. *J. Anal. Toxicol.* 26, 193, 2002. Reproduced from the *Journal of Analytical Toxicology* by permission of Preston Publications, a division of Preston Industries, Inc.)

Table 2.3 LC/MS Methods Used for Cannabinoids and Hallucinogens

Drug	Sample	Isolation	Column, Elution Conditions	Detection	LOD (μg/l)	Ref.
THC, OH-THC, THCCOOH	Plasma, urine	SPE	C8, MeOH-$HCOONH_4$ isocr.	APCI-QIT, SIM	1.0	97
THC, OH-THC, THCCOOH	Plasma, urine	SPE	C8, MeOH-$HCOONH_4$ isocr.	APCI-QQQ, MRM	0.25	98
THCCOOH	Urine	SPE	C18, MeOH-CH_3COONH_4 isocr.	ESI-Q, SIM	0.5	100
THCCOOH, THCCOOH-G	Urine	L/l	C8, ACN-$HCOONH_4$ grad.	ESI-QQQ	10.0	101
LSD, Nor-LSD	Urine	SPE	C18, ACN-CH_3COONH_4 isocr.	ESI-Q, SIM	0.5	105
LSD, Nor-LSD	Blood, urine	SPE	Phenyl, ACN-CH_3COONH_4	ESI-QQQ, MRM	0.025	107
LSD, O-H-LSD	Urine	L/l+SPE	C18, ACN-CH_3COONH_4 isocr.	APCI-QIT, SIM	0.2	110
LSD, O-H-LSD	Urine	L/l+SPE	C18, ACN-CH_3COONH_4 grad.	APCI-QIT, SIM	0.4	111
LSD, O-H-LSD	Blood, urine	L/l+SPE	C18, MeOH-$HCOONH_4$ grad.	ESI-Q, SIM	0.1–0.4	112

Notes: THC = tetrahydrocannabinol, THCCOOH-G = THCCOOH-glucuronide, O-H-LSD = 2-oxo-3-hydroxy-LSD, Q = single stage quadrupole, QQQ = triple stage quadrupole, QIT = quadrupole-ion trap, SIM = selected ion monitoring, MRM = multiple reaction monitoring, ACN = acetonitrile.

2.3.5.2 *Other Hallucinogens*

Ibogaine, an indole alkaloid occurring in the rain forest shrub *Tabernanthe iboga*, has been used for a long time by Western Africans in low doses to combat fatigue and hunger, and in higher doses as a hallucinogen in religious rituals. This drug, as well as its analog 18-methoxycoronaridine, promotes drug abstinence from addictive substances, and was used in experimental clinical trials.[115,116] Ibogaine has been determined, together with other drugs of abuse, by LC/APCI/MS after solid phase extraction by Bogusz et al.[56] The LOD of 0.2 μg/l in plasma was achieved. Zhang et al.[117] determined 18-methoxycoronaridine and its three metabolites with LC/ESI/MS/MS in human liver microsomes in a study devoted to the metabolic pathway of this drug.

Naturally occurring and synthetic indoleamines such as *N,N*-dimethyl-5-hydroxytryptamine (bufotenine), *N,N*-dimethyltryptamine (DMT), 5-methoxy-DMT, *N*-methyltryptamine (NMT), among others, show hallucinogenic properties and are sometimes abused. These drugs were determined in urine samples of psychiatric patients with LC/ESI/MS/MS[118] and with LC/APCI/MS/MS in rat brain after experimental administration.[119]

Mushrooms belonging to the genus *Psilocybe*, *Panaeolus*, *Conocybe*, and *Gymnopilus* contain hallucinogenic indole-derivative psilocybine. These mushrooms were consumed earlier by the Aztecs and are being widely abused in present times.[120] Bogusz et al. [121] analyzed psilocybine and its dephosphorylated active metabolite psilocine in honey mixed with "magic mushrooms." The drug was extracted with methanol and subjected to LC/APCI/MS analysis in SIM mode. Only psilocine was found in honey preparation.

2.3.6 LC/MS of Multiple Drugs of Abuse

The tendency to use multiple drugs is a common feature of drug addicts all over the world. This was found, in Canadian[122] and German[123] sources. Therefore, the analytical toxicologists face the challenge of determining a wide range of compounds in a very limited amount of sample. For this reason, Bogusz et al.[56] developed a LC/APCI/MS procedure for the determination of opiates, cocaine and metabolites, LSD, and ibogaine, using a single SPE isolation procedure and mobile phase consisting of the same components (acetonitrile and ammonium formate buffer, pH 3.0). This method also was used later on 14 psychoactive phenethylamines such as amphetamine and designer amphetamines.[72] For each compound, a protonated molecular ion and one to three fragments were monitored. Practical experience from this procedure was published by Bogusz.[7]

Cailleux et al.[124] extracted opiate agonists (morphine, 6-MAM, codeine, norcodeine, pholcodine, and codethyline) as well as nalorphine and cocaine and its metabolites (benzoylecgonine, ecgonine methyl ester, cocaethylene, and anhydromethylecgonine) from blood, plasma, or urine with

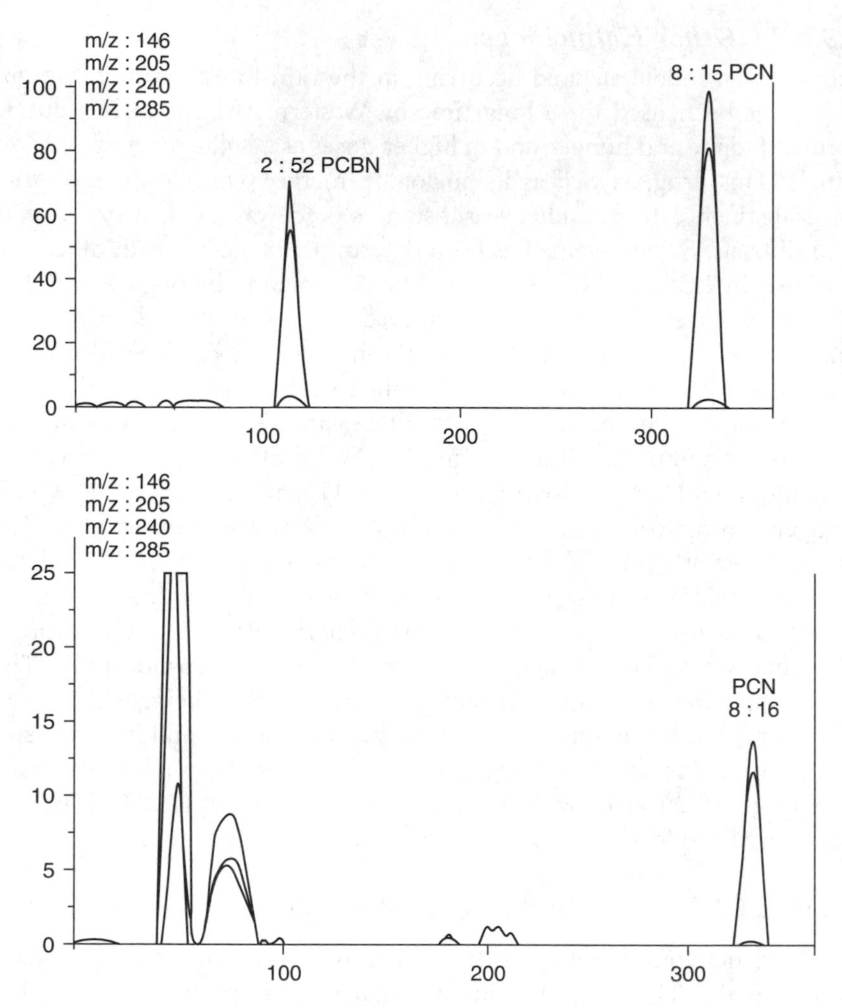

Figure 2.9 LC/APCI/MS chromatogram of psilocybine (PCBN) and psilocine (PCN), 50 ng each (upper picture) and chromatogram of the extract of "herbal honey" (lower picture). (From Bogusz M.J., Maier R.D., Schäfer A.T., and Erkens M., Honey with Psilocybe mushrooms: a revival of a very old preparation on the drug market? *Int. J. Legal Med.* 111, 147, 1998. © Springer-Verlag, with permission.)

chloroform–isopropanol (95:5) at pH 9. The drugs were separated on an octyl column in ACN–ammonium formate–formic acid. Protonated molecular ions and one fragment for each substance were monitored using ESI/MS/MS. The quantitation was done using deuterated internal standards. The limits of quantitation were 10 μg/l for opiates and 5 μg/l for cocaine and were higher than those reported after solid phase extraction.

Mortier et al.[125] stressed the importance of proper sample preparation for ESI/LC/MS/MS determination of drugs of abuse in saliva. Amphetamines (amphetamine, methamphetamine, MDMA, MDEA, and MDA), and opiates (morphine, codeine, 6-MAM), as well as cocaine and benzoylecgonine were isolated from spiked saliva with three protein precipitation procedures. In each case, serious ionization suppression was observed. The author warned that during the use of a highly selective method like LC/MS/MS in an MRM mode, the influence of the matrix may go unnoticed and he postulated the use of optimized sample preparation methods. The same Belgian group[126] established a LC/ESI/qTOF/MS method for the determination of various drugs of abuse: opiates (morphine and codeine), and amphetamines (amphetamine, methamphetamine, MDMA, MDEA, and MDA) as well as cocaine and benzoylecgonine in oral fluid. For a mixed-mode SPE, 200 μl of sample was used. The drugs were separated in a methanol–ammonium formate gradient on a narrow-bore phenyl column. The recoveries varied from 52 to 99%; the LOQ was 2 μg/l for all compounds. The method was applied for real samples obtained from car drivers suspected of drug use.

2.4 LC/MS Analysis of Therapeutic Drugs of Forensic Relevance

2.4.1 Incapacitating Drugs

Some drugs may be used as a chemical weapon to render a victim defenseless, usually in the cases of forced sexual abuse. These compounds are known as "date rape drugs." The drugs belonging to this category cause fast loss of consciousness after a low dose. The victims often develop amnesia and are not able to give a reliable report of the attacks. All groups of drugs used for incapacitating purposes were presented in a recent monograph.[127] The most important drugs used for incapacitation are flunitrazepam and other benzodiazepines, usually used together with alcoholic beverages, gamma-hydroxybutyrate (GHB), and related products; and hallucinogens and opioids.

2.4.1.1 Benzodiazepines

Flunitrazepam, a potent hypnotic drug, is of particular toxicological importance due to its high toxicity in combination with ethyl alcohol and subsequent misuse in drug-facilitated sexual assaults. Since flunitrazepam quickly disappears from the blood, simultaneous determination of its active metabolites, particularly of the prevalent 7-aminoflunitrazepam, is very important. Contrary to GC/MS, LC/MS allowed determination of polar metabolites of flunitrazepam without derivatization. Bogusz et al.[128] developed a LC/APCI/MS method for determination of flunitrazepam and its metabolites 7-aminoflunitrazepam, *N*-desmethylflunitrazepam, and 3-OH-flunitrazepam in blood or plasma. The comparison of ionization response showed that APCI gave a signal 7 to 30 times stronger than ESI for all compounds except 7-aminoflunitrazepam. After solid-phase extraction on C18 cartridges, the drugs were separated on a C18 column in ACN–ammonium formate buffer (pH 3.0) and detected with APCI/MS. For each compound, a protonated quasi-molecular ion was monitored. The LODs ranged from 0.2 to 1 μg/l. This method has been applied in routine toxicological casework.[7] LeBeau et al. determined flunitrazepam, 7-aminoflunitrazepam, and *N*-desmethylflunitrazepam in blood and urine after SPE in a mixed-phase cartridge.[129] A LC/ESI/MS/MS (ion trap) procedure was applied. Protonated quasi-molecular ions of drugs involved were monitored and the identity was confirmed by fragmentation. The LODs in blood were from 0.5 to 1 μg/l. An ion trap LC/ESI/MS/MS was also used for determination of flunitrazepam in serum by Darius et al.[130] The drug was extracted with tertiary butylmethyl ether and separated on C18 column in ACN–water mobile phase. Product ions of flunitrazepam and internal standard (clonazepam) were monitored. The LOD was 0.2 μg/l. Unfortunately, the metabolites, particularly 7-aminoflunitrazepam, were not determined. Yuan et al. described an automated in-tube solid-phase microextraction combined with LC/ESI/MS for determination of diazepam, nordiazepam, temazepam, oxazepam, 7-aminoflunitrazepam, and *N*-desmethylflunitrazepam in serum and urine.[131] Flunitrazepam itself was not included in this study. The isolation technique used allowed solvent-free, automatic extraction in 15 min for each consecutive sample. The extraction procedure was optimized, using six various extraction capillaries and various extraction conditions. The drugs were detected with LC/ESI/MS (full scan *m/z* 100 to 400) and SIM. The LODs of 0.02 to 2 μg/l were achieved. Two drawbacks of the procedure are relatively low recovery in serum (below 50%) and peak broadening caused by automatic desorption. A LC/APCI/MS/MS (ion trap) procedure was applied for determination of flunitrazepam, 7-aminoflunitrazepam and *N*-desmethylflunitrazepam in plasma after SPE on Oasis MCX cartridges by Kollroser et al.[132] The analytical recoveries were above 90% for all compounds; the limits of detection were 0.25 to 2 μg/l.

Midazolam is a short-acting benzodiazepine used for induction of anesthesia that may also be abused as an incapacitating agent. This drug and its active hydroxylated metabolite were extracted from serum with ether–isopropanol (98:2) at alkaline pH and separated on ODS column (Nucleosil C18, 150 × 1 mm). The drugs were determined with ESI/LC/MS in SIM mode. Protonated quasi-molecular ions and fragments of both compounds were monitored. The LOQ for both compounds was 0.5 μg/l.[133] In another study, midazolam and hydroxymidazolam, as well as triazolam and hydroxytriazolam used as internal standards, were extracted from plasma with Oasis HLB

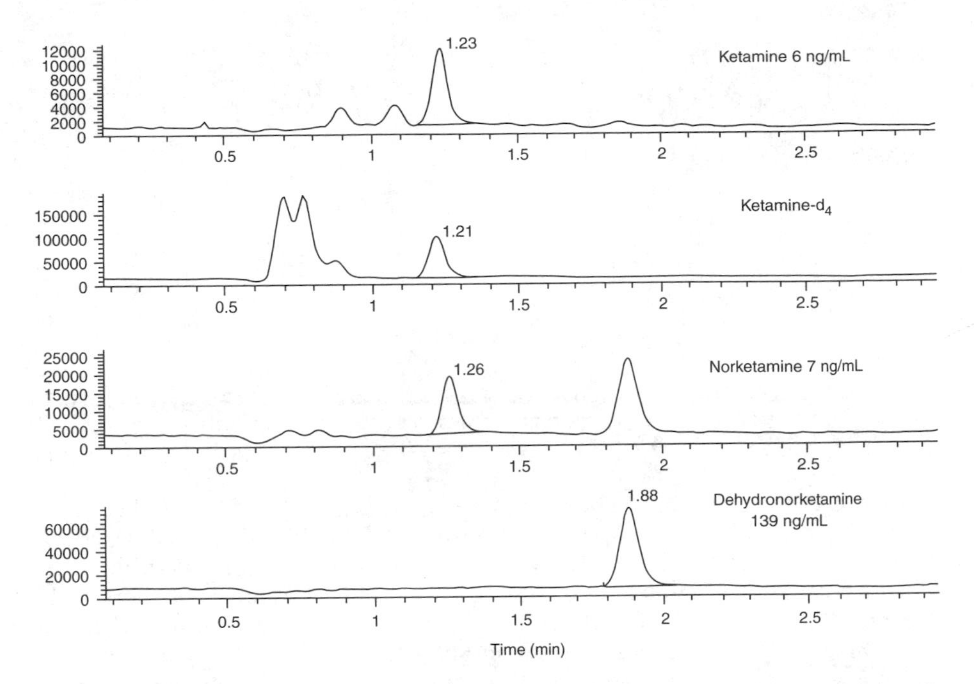

Figure 2.10 LC-ESI-MS (ion trap) of urine extract containing ketamine and metabolites. (From Moore K.A., Sklerov J., Levine B., and Jacobs A.J., Urine concentrations of ketamine and norketamine following illegal consumption. *J. Anal. Toxicol.* 25, 583, 2001. Reproduced from the *Journal of Analytical Toxicology* by permission of Preston Publications, a division of Preston Industries, Inc.)

cartridges and determined with LC/APCI/MS/MS.[134] All drugs were eluted within 2 min, the LOQ was 0.1 μg/l using multiple reaction monitoring.

2.4.1.2 *Ketamine*

Ketamine is an anesthetic drug, known in the drug scene as "Special-K" or "Kit-Kat." Ketamine may be abused at rave parties or to facilitate sexual assault.[135,136] Moore et al.[137] developed a LC/ESI/MS (ion trap) procedure for determination of ketamine and its metabolites norketamine and (presumptive) dehydronorketamine in urine after SPE on CLEAN SCREEN cartridges. Limits of detection were 3 μg/l for ketamine and norketamine. The method was applied in 33 forensic cases, and the concentrations 6 to 7744 μg/l and 7 to 7986 μg/l for ketamine and norketamine, respectively, were found.

2.4.2 Antidiabetic Drugs

Antidiabetic drugs are of particular forensic relevance since they are often misused by people who want to simulate an illness, as well as are used as a suicide poison or as a lethal weapon.

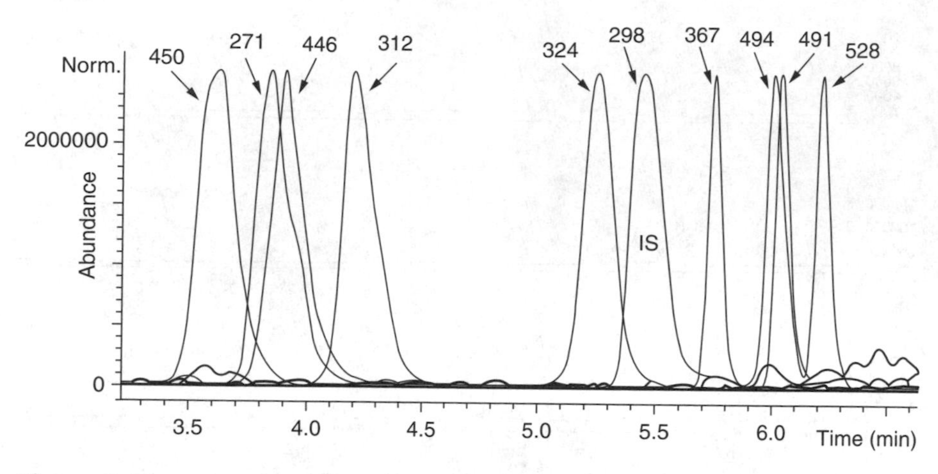

Figure 2.11 Merged mass chromatogram of glisoxepide (*m/z* 450), tolbutamide (*m/z* 271), glipizide (*m/z* 446), tolazamide (*m/z* 312), gliclazide (*m/z* 324), trimipramine (IS, *m/z* 298), glibornuride (*m/z* 367), glibenclamide (*m/z* 494), glimepiride (*m/z* 491), and gliquidone (*m/z* 528). (Reprinted from Maurer H.H., Kratzsch, Kraemer T., Peters F.T., and Weber A.W., Screening, library-assisted identification and validated quantification of oral antidiabetics of the sulfonylurea-type in plasma by atmospheric pressure chemical ionization liquid chromatography-mass spectrometry. *J. Chromatogr. B* 773, 63, 2002. Copyright 2002, with permission from Elsevier Science.)

2.4.2.1 Oral Antidiabetic Drugs

Ramos et al.[138] published a fast LC/APCI/MS/MS method for determination of glibenclamide in human plasma after acetonitrile precipitation, using deuterated analog as internal standard. The retention time of drug was 3 min, the LOQ 1 μg/l. Sulfonylurea antidiabetics (tolbutamide, chlorpropamide, glibenclamide, and glipizide) were detected and quantified in serum using LC/ESI/MS.[139] SPE method applied (on C18 cartridges) gave much cleaner extracts than acidic toluene extraction. The drugs were separated in methanol–acetic acid gradient and detected by ESI-MS in full scan mode (positive ions, *m/z* 265 to 510). For quantitation, protonated quasi-molecular ions were used. The LOQ was 10 μg/l for all drugs. Maurer et al.[140] published a LC/APCI/MS procedure for screening and library-assisted quantification of antidiabetics glibenclamide, glibornuride, gliclazide, glimepiride, glipizide, gliquidone, glisoxepide, tolazamide and tolbutamide. The drugs were extracted from plasma with ether–ethyl acetate and separated on Select B column in gradient elution conditions. Protonated quasi-molecular ions were used for screening and quantification, and full scan spectra for identification. Limits of detection varied from 2 to 3000 μg/l.

2.4.2.2 *Insulin*

Identification, detection, and quantitation of insulin in blood are very important in forensic toxicology. The presence of bovine or porcine insulin in the blood of a nondiabetic person may serve as evidence of poisoning. Since insulin and C-peptide are released from proinsulin, the determination of both products may be of relevance. Very high level of human insulin without elevation of the C-peptide indicates exogenous administration. When both insulin and C-peptide are present in high concentrations, the presence of pancreatic tumor or administration of insulin-releasing drugs, like sulfonylureas, is possible. The first LC/MS method for insulin was published by Stocklin et al.[141] Insulin and its deuterated analog were isolated by immunoaffinity chromatography and solid-phase extraction. The LOD of 3 pmol/l (about 17 ng/l) was reported. The method was used for hemolyzed postmortem forensic blood samples, for which immunoassays gave inconsistent results. A straightforward LC/ESI/MS method for determination of human, bovine, and porcine insulin, as well as C-peptide, was published by Darby et al.[142] The drugs were isolated from acidified plasma with C18 SPE cartridges, separated on a 150 × 2.1 mm C18 column and detected with ESI/MS (ion trap). Multiply charged molecular ions $(M + 3H)^{+3}$ and $(M + 4H)^{+4}$ were monitored. Both ions were selected for quantitation since the intensity ratio was variable between runs. The LOQ was 0.4 μg/l for insulin and 0.1 μg/l for C-peptide; the comparison with radioimmunoassay showed full agreement of values. The stability of insulin from blood stored in various conditions was studied. Zhu et al.[143] developed a LC/ESI/MS/MS assay for the quantitation of human insulin, its analog, and the catabolites in plasma. The compounds were isolated through precipitation, followed by SPE. The quantitation range was 1 to 500 μg/l.

2.4.3 Muscle Relaxants

Muscle relaxants usually belong to classes of quaternary ammonium compounds of high polarity and low thermal stability. Therefore, liquid chromatography is a method of choice for their separation.

Farenc et al. applied LC-ESI-MS for the determination of rocuronium (a neuromuscular blocking agent used widely during general anesthesia) in the plasma of patients. The drug and internal standard (verapamil) were extracted from plasma with dichloromethane and separated on an ODS column in ACN–0.1% TFA (gradient elution). Protonated quasi-molecular ions for both compounds were monitored. The LOQ was 25 μg/l.[144] Ballard et al.[145] developed a LC/qTOF/MS/MS method for determination of pancuronium, vecuronium, tubocurarine, rocuronium, and succinylcholine in postmortem tissues. These drugs were isolated using a combination of solvent extraction followed by an ion pairing SPE on C18 cartridges. Detection was based on an exact mass measurement of product ions. The method has been applied in cases of murder involving serial killers of medical profession. Kerskes et al.[146] performed a systematic study on detectability of pancuronium, rocuronium, vecuronium, gallamine, suxametonium, mivacurium, and atracurium in biological fluids using LC/ESI/MS (ion trap) after solid-phase extraction. The method was successfully applied in toxicological emergency case and after therapeutic administration of drugs.

2.4.4 Antidepressant and Antipsychotic Drugs

LC/MS has been frequently applied in the last few years for determination of antidepressants. Several authors determined multiple drugs in one LC/MS procedure. Kumazawa et al.[147] applied solid-phase microextraction followed by LC/ESI/MS/MS for the determination of eleven phenothiazine derivatives from whole blood and urine. The extraction efficiencies were studied and ranged from 0.0002 to 0.12% for blood and from 2.6 to 39.8% for urine, respectively. The sensitivity in selected reaction monitoring mode was described as satisfactory. The LC/ESI/qTOF technique was used by Zhang et al.[148] for the determination of doxcpin, desipramine, imipramine, amitriptyline, and trimipramine in human plasma. The separation time was 18 sec, using a C18 15 × 2.1 mm column and flow rate of 1.4 ml/min. The analyzer allowed an accurate mass determination; the LOD was 5 μg/l. Kollroser and Schober[149] developed a LC/APCI/MS/MS (ion trap) method for rapid determination of amitriptyline, nortriptyline, doxepin, dosulepin, dibenzepin, opipramol, and melitracen in human plasma. Plasma samples were only diluted with 0.1% formic acid and subjected to on-line extraction with an Oasis extraction column. The samples were then forwarded to a C18 analytical column by valve switching. LC/ESI/MS and LC/MS/MS were applied for the detection of antidepressants and neuroleptics in the hair of psychiatric patients. The drugs (maprotiline, pipamperone, and citalopram) were extracted from powdered hair with methanol and purified with SPE. Mass spectra library search was performed for ESI/CID and MS/MS spectra.[150] Also, McClean et al.[151] used LC/ESI/MS/MS (ion trap) for determination of antipsychotic drugs (chlorpromazine, trifluoperazine, flupenthixol, risperidone, and trimipramine in hair subjected to alkaline hydrolysis and solvent extraction.

LC/MS has been frequently applied for selected antipsychotic drugs and their metabolites.

Hoskins et al.[152] described a method for the determination of monoamine oxidase inhibitor moclobemide and its two metabolites in human plasma. The drugs were isolated with Bond Elut C18 SPE cartridges at alkaline pH and separated on a Nova-Pak phenyl column. Protonated quasi-molecular ions of all drugs and the internal standard were monitored by ESI/MS. The LODs were 1 to 5 μg/l, LOQ was 10 μg/l. LC/MS methods for fluoxetine and its metabolite were frequently published. A fast ESI/LC/MS/MS method for determination of fluoxetine and norfluoxetine was developed by Sutherland et al.[153] The drugs and internal standard (doxepin) were extracted from plasma with hexane–isoamylalcohol (98:2) and the aqueous phase was frozen and discarded. The organic phase was back-extracted with 2% formic acid; the aqueous phase was frozen and, after discarding of the hexane layer, thawed, and injected into LC/MS. The total chromatographic run time was 2.6 min. The transition of protonated quasi-molecular ions of drugs to product ions was monitored. The LOQ was 0.15 μg/l for both compounds. In a procedure published by Shen et al.,[154] fluoxetine was extracted with solvent in 96-well plates and enantiomers of the drug were separated on a vancomycin column. The detection was done with an APCI/MS/MS in MRM mode. The limit of quantitation was 2 μg/l. Li et al.[155] published a highly sensitive LC/ESI/MS/MS procedure for the determination of fluoxetine and norfluoxetine in human plasma. The drugs were extracted from plasma with solvent and separated within 5 min. The LOD of 0.1 μg/l was reported at a sample size of 0.2 ml.

Olanzapine, a thienobenzodiazepine, is an antipsychotic agent used broadly for the acute treatment and maintenance of schizophrenia. Due to the concentration-related response and toxicity, therapeutic monitoring of this drug is indicated. Berna et al.[156,157] published two procedures for LC-ESI/APCI-MS-MS determination of olanzapine in plasma. The drug was extracted with SPE cartridges or with organic solvent mixture in single or 96-well format. The LOQ was 0.25 μg/l. Bogusz et al.[158] applied LC/APCI/MS for determination of olanzapine in serum. SPE extraction on C18 cartridges was applied. The LOQ was 1 μg/l. In full-scan LC/MS a postulated olanzapine-10-*N*-glucuronide was found in urine. Aravagiri and Marder[159] applied LC/ESI/MS/MS for therapeutic drug monitoring of clozapine, clozapine-*N*-oxide, and norclozapine in the serum of schizophrenic patients. The drugs were isolated with solvent extraction and separated on ODS column. The transitions of protonated quasi-molecular ions to single fragments were monitored. An LOQ of 1 μg/l for all substances was reported. The same authors used LC/ESI/MS/MS for the determination of risperidone and 9-hydroxyrisperidone in plasma.[160] Solvent extraction and separation on a phenyl-hexyl column was applied. The LOQ was 0.1 μg/l. Venlafaxine, a phenethylamine antidepressant, as well as its *O*-desmethylated metabolite were determined in postmortem blood and tissues in 12 cases of fatal venlafaxine poisoning.[161] The drug was extracted with butyl chloride and determined by LC/ESI/MS in a positive ionization mode. In all cases, various other psychotropic drugs and alcohol were also detected.

2.4.5 Other Therapeutic Compounds of Forensic Relevance

Methylphenidate is a central stimulant with a history of abuse which is used for the treatment of narcolepsy. The drug has two chiral centers and is marketed as a racemic mixture, whereas *d-threo*-methylphenidate is pharmacologically more active than the *l*-isomer. Ramos et al.[162] described an LC/APCI/MS/MS procedure for determination of methylphenidate enantiomers in plasma. A vancomycin-based Chirobiotic V column was used for separation. The LOQ was 87 ng/l. In the next study, Ramos et al.[163] developed an LC/APCI/MS/MS method for the determination of racemic methylphenidate. Alkaline cyclohexane extraction in a 96-well plate format was utilized. The drug was separated on C18 column using switching valve, diverting first part of eluent to waste. Total run time was 3 min, a LOQ of 50 ng/l plasma was reported.

Scopolamine and the internal standard atropine were extracted from 0.2 ml serum with Oasis HLB cartridges in an automated procedure.[164] The drugs were separated on a C18 column in gradient elution on a mobile phase consisting of ACN, ammonium acetate, and formic acid. The detection was done with ESI/MS/MS (positive ions). Protonated quasi-molecular ions of scopolamine and atropine were used as parent ions and two product ions for each drug were monitored. The method was applied for pharmacokinetic studies with a LOQ of 20 ng/l serum.

Nicotine and its seven metabolites were determined in urine by LC/ESI/MS/MS after enzymatic hydrolysis and solvent extraction. For quantitation, deuterated or ^{13}C-labeled analogs were used for each compound. The LODs ranged from 5 to 38 ng/l. The most dominant metabolite was identified as 3-OH-cotinine.[165]

Sildenafil became an extremely popular drug and at least 100 million tablets have been prescribed worldwide. It may cause not only several adverse effects[166] but also life-threatening cardiac failure. Weinmann et al.[167] described postmortem detection and identification of sildenafil (Viagra) and its metabolites by LC/ESI/MS and LC/ESI/MS/MS. The drug and three metabolites were isolated from the urine and organs of a highly putrefied body of an 80-year-old man who died due to coronary sclerosis. Analytical results were compared with experimental data obtained from a volunteer who ingested 25 mg sildenafil. French authors[168] used LC/ESI/MS for the determination of sildenafil and verapamil in blood, organs, and hair of a 43-year-old man who was found dead after sexual relations with a prostitute. The deceased had a history of cardiovascular disease and erectile dysfunction. Sildenafil may be also used as a doping substance for racehorses due to its protective action against exercise-induced pulmonary hemorrhage. Rudy et al.[169] presented an LC/ESI/MS/MS method for identification of sildenafil and metabolites in equine plasma and urine. A liquid/liquid-alkaline extraction was applied, followed by separation on a CN-column and qTOF detection. Product ions in the range of *m/z* 100 to 500 from four parent ions, corresponding to sildenafil and its three metabolites, were monitored.

2.5 LC/MS Analysis of Natural Compounds of Forensic Relevance

Digitalis glycosides: digoxin, digitoxin, deslanoside, digoxigenin, and digitoxigenin were determined with ESI/LC/MS/MS.[170] The drugs were extracted from whole blood or urine by SPE using Oasis HLB cartridges at alkaline pH. Chromatographic separation was performed on ODS column in ACN–ammonium formate gradient. An abundant ammonia adduct $(M + NH_4)^+$ and much less abundant protonated quasi-molecular ion $(M + H)^+$ were observed in ESI. In APCI, the substances dissipated to many fragments. Time-scheduled selected reaction monitoring was applied, using $(M + NH_4)^+$ ions as precursor ions and one fragment for each compound as a product ion. The recovery ranged from 11 to 64%; the LODs were from 0.05 to 1 μg/l. The method was applied for therapeutic drug monitoring of cardiac glycosides.

Lacassie et al.[171] reported a case of a 36-year-old female who ingested a concoction of foxglove leaves (*Digitalis purpurea*) and was admitted to an emergency unit some hours later. She developed sinus bradycardia, abdominal pain, nausea, and vomiting. After five days of treatment she was discharged without symptoms. The cardiac glycosides acetyldigitoxin, convallatoxin, deslanoside, digitoxigenin, digitoxin, digoxin, gitaloxin, gitoxin, latanoside C, methyldigoxin, oleandrin, proscilardin, and strophantidin in blood and urine samples of the patient were determined with LC/ESI/MS. The drugs were extracted with an organic solvent mixture after acetonitrile precipitation and separated on an ODS column in a gradient of ACN–ammonium formate pH 3.0. For each analyte the protonated quasi-molecular ion $(M + H)^+$ and one or two fragment ions were monitored. The recovery was 67.8 to 98.6%, and the limit of detection 1 to 10 μg/l. This sensitivity was adequate for acute poisoning cases but not for therapeutic drug monitoring. Glycoside levels in blood were monitored from 8 to 100 h after intoxication. In a case reported by Gaillard and Pepin, veratridine and cevadine (toxins present in *Veratrum album*) were identified and quantitated in the blood of two persons found in a mountain lake. In their stomachs, seeds of *Veratrum* were identified. LC/ESI/MS was used. Measured blood concentrations were 0.17 and 0.40 μg/l for veratridine and 0.32 and 0.48 μg/l for cevadine.[172]

Oleander *(Nerium oleander* L.) is an evergreen, widely distributed shrub, which contains cardenolides — toxic cardiac glycosides such as oleandrin, oleandrigenin, and others. Preparations of oleander are used for cancer treatment. Oleandrin, odoroside, neritaloside, and the aglycone oleandrigenin, present in the extract of *Nerium oleander* (Anvirzel), were analyzed with qTOF/MS. CID mass spectra were obtained with mass accuracy greater than 5 ppm. LOD for oleandrin was 20 pg injected. The method was applied for the determination of oleander glycosides in human plasma after intramuscular injection of Anvirzel.[173] A 45-year-old female took *Nerium oleander* for suicidal purposes, showing nausea, vomiting, cardiovascular shock, and sinus bradycardia. Oleandrin was measured in plasma at 1.1 μg/l and was also identified in urine using ESI/LC/MS.[174] Arao et al.[175] applied SSI/LC/MS/MS (ion trap) for determination of oleandrin, oleandrigenin, and desacetyloleandrin in the heart, blood, and cerebrospinal fluid in the case of a fatal suicidal poisoning of a 49-year-old woman who learned about oleander toxicity on TV. Only oleandrin was found in concentrations of 9.8 and 10.1 μg/l in her heart, blood, and cerebrospinal fluid, respectively.

Capsaicinoids — capsaicin, dihydrocapsaicin, and nonivamide — occur naturally in red pepper (*Capsicum annuum*). Due to its extremely irritative properties, pepper spray is used in self-defense weapons, which are also applied in criminal acts. Reilly et al. developed a LC/ESI/MS/MS procedure for the determination of capsaicin, dihydrocapsaicin, and nonivamide in self-defense weapons[176] and in blood and tissue of experimental animals.[177] The LOQ in the latter procedure was 1 μg/l.

Ricin is a toxic protein (mol. wt. over 60,000 Da) present in the seeds of castor beans (*Ricinus communis L*). This compound has a history of murderous application and appeared in terrorist handbooks. Darby et al.[178] analyzed ricin and its marker ricinine in acetone extract of castor bean and in trypsin hydrolysates. LC/UV, ESI/LC/MS, and MALDI/TOF/MS were applied. The last technique gave the best results, both for intact ricin molecule and for digested peptide fragments.

Colchicine, an alkaloid present in flowers of the autumn crocus (*Colchicum autumnale*) has been used for centuries for the treatment of gout. Due to its toxicity, several fatal cases were reported.[179] Tracqui et al.[180] extracted colchicine from blood, plasma, or urine with dichlormethane at pH 8 and determined with LC/ESI/MS in full scan and SIM mode. The LOD in SIM mode was 0.6 μg/l. Jones et al.[181] detected colchicine in the biofluids and organs of a 73-year-old man who died 18 h after receiving a 1 mg intravenous dose of colchicine. He was treated with oral colchicine in previous days. Dichloromethane extraction, followed by LC/ESI/MS, was applied. The concentrations found were 50 μg/l in cardiac blood, 575 μg/l in liver, and 12,000 μg/l in bile.

Maurer et al.[182] determined α- and β-amanitine in the urine of patients intoxicated with *Amanita* mushrooms. The toxins were isolated with immunoaffinity extraction columns and determined with LC-ESI-MS, with a LOD of 2.5 μg/l.

2.6 Screening LC/MS Procedures in Forensic Toxicology

The application of HPLC as a tool of general unknown analysis is relatively new and is still in the developmental stage. Generally, two approaches were taken: the use of HPLC with diode array detection (HPLC/DAD) or LC/MS. HPLC/DAD has a high identification potential due to the combination of retention parameter and UV spectrum as identification parameters. Several databases comprising more than one thousand substances were established. The review of these databases was done by Bogusz.[5] The establishing of a universally applicable HPLC database was hindered by poor interlaboratory reproducibility of retention time values for the same drugs analyzed in nominally identical conditions in different laboratories. This problem was solved by introduction of retention index scale, using alkyl aryl ketones[183] or 1-nitroalkanes.[184] In a further development, 1-nitroalkanes were replaced by acidic and basic standard drugs with known retention index values. This approach assured very good interlaboratory comparison of results.[185,186] Beside retention index scaling, relative retention times are used for HPLC/DAD libraries. Such a system, based on two reference compounds, is commercially available as REMEDi (Bio-Rad Labs, Hercules, CA). This system comprises automatic extraction of a urine sample, on-line separation, detection, and identification with a library comprising over 800 drugs. The database can be expanded according to need.

The use of LC/MS for toxicological screening was from the beginning an extremely attractive possibility since this technique possesses a much broader detection spectrum than GC-MS and is much more sensitive than HPLC-DAD. However, the establishing of a generally applicable LC/MS library, similar to GC/MS databases, is hindered by large interlaboratory variability in mass spectra. In the study done in three laboratories, mass spectra of identical substances, analyzed on the same instruments in nominally identical conditions, showed large differences in the degree of fragmentation.[187] On the other hand, controlled changes in composition of the mobile phase, i.e., in the percentage of organic modifier or in the molarity of ammonium formate buffer, did not have any relevant influence on mass spectra. Existing LC/MS screening procedures may be divided into two groups: those comprising substances belonging to similar pharmacological classes or used for similar applications (e.g., pesticides) and those applicable for an undirected search, i.e., for a "general unknown" screening. In the first case, the task is much easier, since the number of substances is limited, and the sample preparation is more effective due to similar chemical features of the compounds involved.

2.6.1 Group Screening for Pharmacologically Related Substances

Finnish authors[188] described an automated LC/ESI/MS/MS screening/confirmation method for 16 beta-blocking agents in urine samples. The drugs were tentatively identified on the base of retention time and $(M + H)^+$ value. For confirmation, any qualified compounds were automatically subjected to fragmentation. The product ions were compared with the library data. The limits of identification ranged from 0.02 to 1.2 mg/l.

The same group developed a LC/ESI/MS/MS procedure for screening and quantitation of 18 antihistamine drugs in blood.[189] Since the drugs belong to acidic and basic compounds, two solvent extractions were performed at pH 3 and 11, and the extracts were combined. The chromatography was done in 5 min in an acetonitrile–ammonium acetate gradient. For preliminary identification, protonated quasi-molecular ion, the most intensive fragment, and retention time of each peak were used. The confirmation was done in second run, using the whole product ion spectrum. The limits of identification for each drug were below the lowest therapeutic concentration.

Rittner et al.[190] established a library of mass spectra of 70 various psychoactive drugs and metabolites with LC-ESI-MS. In the preliminary study, the efficiency of various solid-phase extraction methods and various HPLC columns was tested. The best results were obtained with C18 SPE cartridges and C18 columns. Chromatographic separation was performed in ACN–water–methanol–formic acid gradient. Mass spectra of drugs were recorded at two levels of fragmentation energy in full scan mode (*m/z* 100 to 650, positive ions). For many drugs, sodium, potassium, and ACN adducts were observed. The usefulness of the screening procedure was checked on 140 serum samples taken from road traffic offenders. In 9.8% of the cases various drugs, mostly benzodiazepines, were detected.

Thieme et al.[191] established a method for screening and quantitation of 32 diuretic compounds and their metabolites in urine. The method was developed for doping control and was based on LC/ESI/MS/MS. The drugs were extracted from urine using XAD columns and separated on C8 columns in gradient of ACN–ammonium acetate. The library of mass spectra was developed using positive or negative ionization and optimized fragmentation conditions for particular compounds. Since diuretics may belong to acidic and basic drugs, the authors recommended two subsequent chromatographic runs (in positive and negative mode) in the screening procedure. This was superior to the polarity switching in one run.

Lacassie et al.[192] published a procedure for the determination of 61 pesticides of various classes (organophosphates, carbamates, organochlorines, and benzimidazoles) in serum. Forty-seven compounds were determined with GC-MS after SPE on Oasis HLB cartridges, whereas LC/ESI/MS was applied for 14 thermolabile and polar pesticides, such as carbamates and benzimidazoles. These substances were isolated with Oasis MCX cation exchange cartridges. LC/MS was applied for clinical diagnostics in carbofuran and aldicarb self-poisonings.

2.6.2 General Undirected Screening

Studies on the application of LC/MS as a universal screening technique may be divided into two main groups: those performed using a single quadrupole instrument and in-source fragmentation, and those done with an MS/MS technique (triple quadrupole, qTOF, or ion trap). The general strategy of each approach is different; in the case of single quadrupole MS, all substances reaching the analyzer are fragmented and monitored without any preselection. This is in agreement with the basic concept of general screening analysis. On the other hand, in-source fragmentation requires optimized sample pretreatment and thorough chromatographic separation of all compounds present in the sample. The latter is often not feasible, therefore mass spectra of substances may overlap, or ionization suppression may occur. On the contrary, tandem MS gives much better sensitivity and specificity but at the cost of preselection.

2.6.2.1 Undirected Screening Using LC/MS Procedures

The main drawback of LC/MS as an universal identification tool is high variability in the degree of fragmentation of the examined compounds, observed for different instruments or even for similar instruments but used in different labs.[187] As a solution, selected substances were used for tuning the fragmentation energy. Particular applications will be discussed below.

Marquet et al.[193,194] developed a library for over 1500 compounds using positive and negative ionization modes and two fragmentation energies. The substances were separated in a gradient of acetonitrile and ammonium formate buffer. Reconstructed mass spectra (containing quasi-molecular ion and fragments) were used for identification. Glafenin was used as a standard for tuning the degree of fragmentation. About 1100 reconstructed positive and about 500 negative mass spectra were stored. In the next study by the same French group, a complete identification procedure was presented, which comprised sample preparation, optimized chromatographic separation, and ESI/LC/MS detection.[195] The comparison of three extraction methods (Extrelut, Oasis HLB, and Oasis MCX) showed that the last procedure was the best one. Extracted compounds were separated on a C18 column (150 × 1 mm) in a gradient of acetonitrile–ammonium formate buffer with pH 3.0. Mass spectra were stored in positive and negative ionization modes at two fragmentation voltage values. After subtraction of background noise mass spectrum, the reconstructed positive and negative spectra were used for identification. This method was evaluated in real-life conditions in the following way: 51 serum samples from a toxicological clinic were analyzed in parallel with ESI/LC/MS, a standard GC/MS, and HPLC/DAD procedures. The LC/MS procedure identified 75% of the compounds present in the samples, vs. 71% for HPLC/DAD and 66% for GC/MS; 38% of compounds were detected by all three techniques, and 36% by two of them.[196]

Weinmann et al.[197] established databases for ESI-generated mass spectra for 430 compounds. Haloperidol was used as a tuning compound for controlled fragmentation. The breakdown curves of haloperidol, observed on three LC/MS instruments from the same manufacturer and taken under different concentrations and elution conditions, were comparable.

Mass spectra were taken at three fragmentation energy levels. In the next study, Weinmann et al. used tuning compounds for standardization of in-source collision-induced fragmentation. Four drugs — haloperidol, paracetamol, metronidazole, and metamizole — were selected as tune compounds for LC/ESI/MS. Comparative experiments were performed using two LC/MS instruments with different construction of interface (Sciex API 365 and Agilent 1100 MSD)[198] or three instruments of the same manufacturer (Sciex API 365, API 2000 and API 3000). [199] Very similar fragmentation patterns were observed after adjusting of fragmentor voltages of both instruments.

Therefore, the establishment of a general, applicable library of mass spectra obtained with different LC/MS instruments is possible when the fragmentation energy will be adjusted using selected tune compounds. Weinmann et al. combined the LC/ESI/MS library with UV spectra library.[200]

Lips et al.[201] performed a study on the applicability of LC/ESI/MS (positive ions) for the development of mass spectra library based upon in-source collision-induced fragmentation. The influence of mobile-phase composition on the reproducibility of mass spectra of drugs was tested, and data obtained with two instruments of the same brand but different types were compared. The breakdown curves (i.e., fragmentation profiles related to fragmentation energy) of selected drugs were compared with the data of other authors. The authors stated that the concentration of the organic modifier, pH, and molarity of the buffer exerted negligible influence on the mass spectra. This observation is in agreement with the previous finding of Bogusz et al.[187] and Weinmann et al.[197] In order to obtain reproducible mass spectra, the fragmentor voltage was dynamically ramped based on the mass of the substance. The efficiency of identification was tested on over 40 extracts from plasma containing various acidic and basic drugs. All drugs except phenobarbital were correctly recognized. Phenobarbital was not detected at all due to the positive ionization applied. Acidic drugs were properly detected

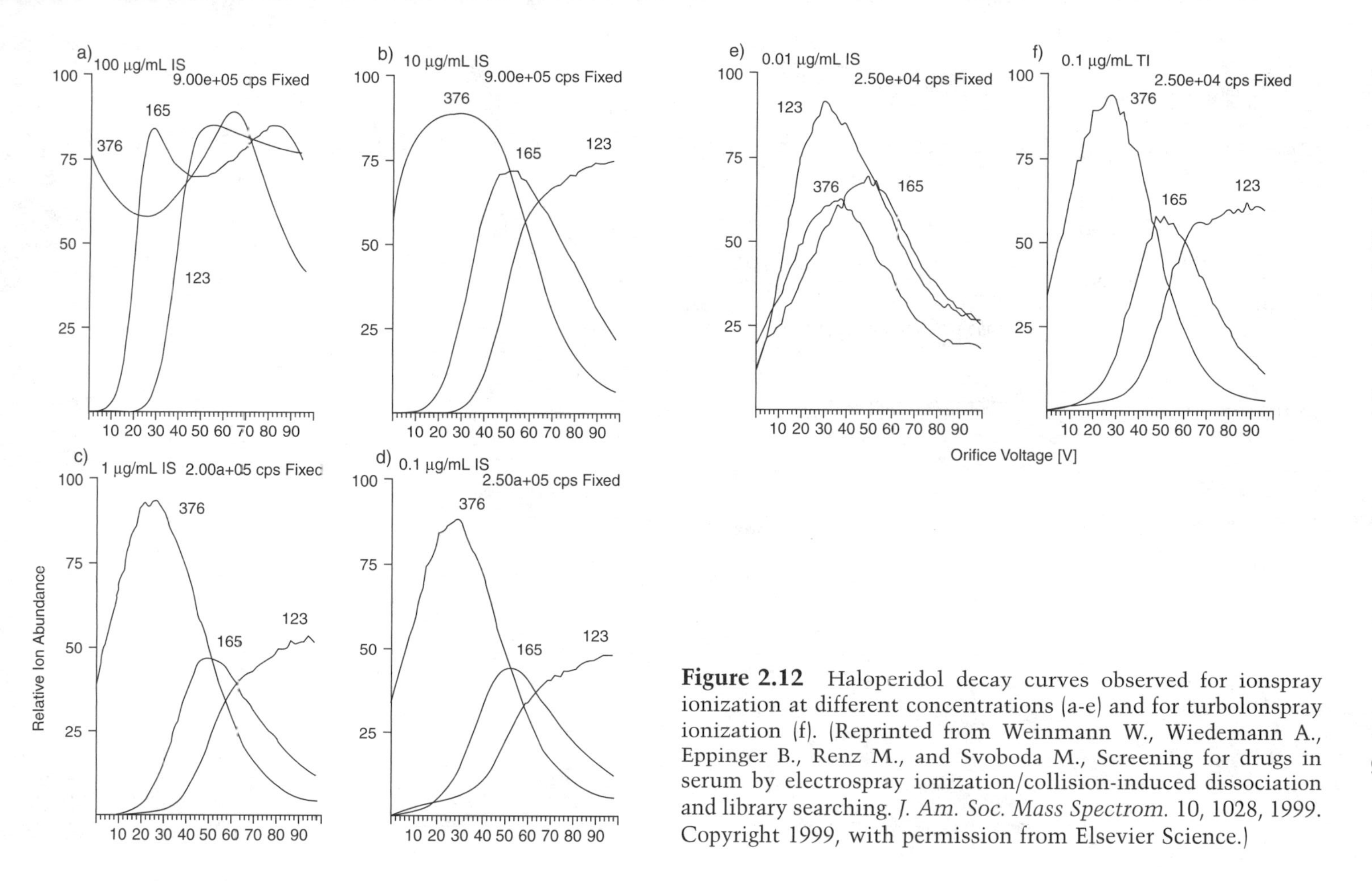

Figure 2.12 Haloperidol decay curves observed for ionspray ionization at different concentrations (a-e) and for turbolonspray ionization (f). (Reprinted from Weinmann W., Wiedemann A., Eppinger B., Renz M., and Svoboda M., Screening for drugs in serum by electrospray ionization/collision-induced dissociation and library searching. *J. Am. Soc. Mass Spectrom.* 10, 1028, 1999. Copyright 1999, with permission from Elsevier Science.)

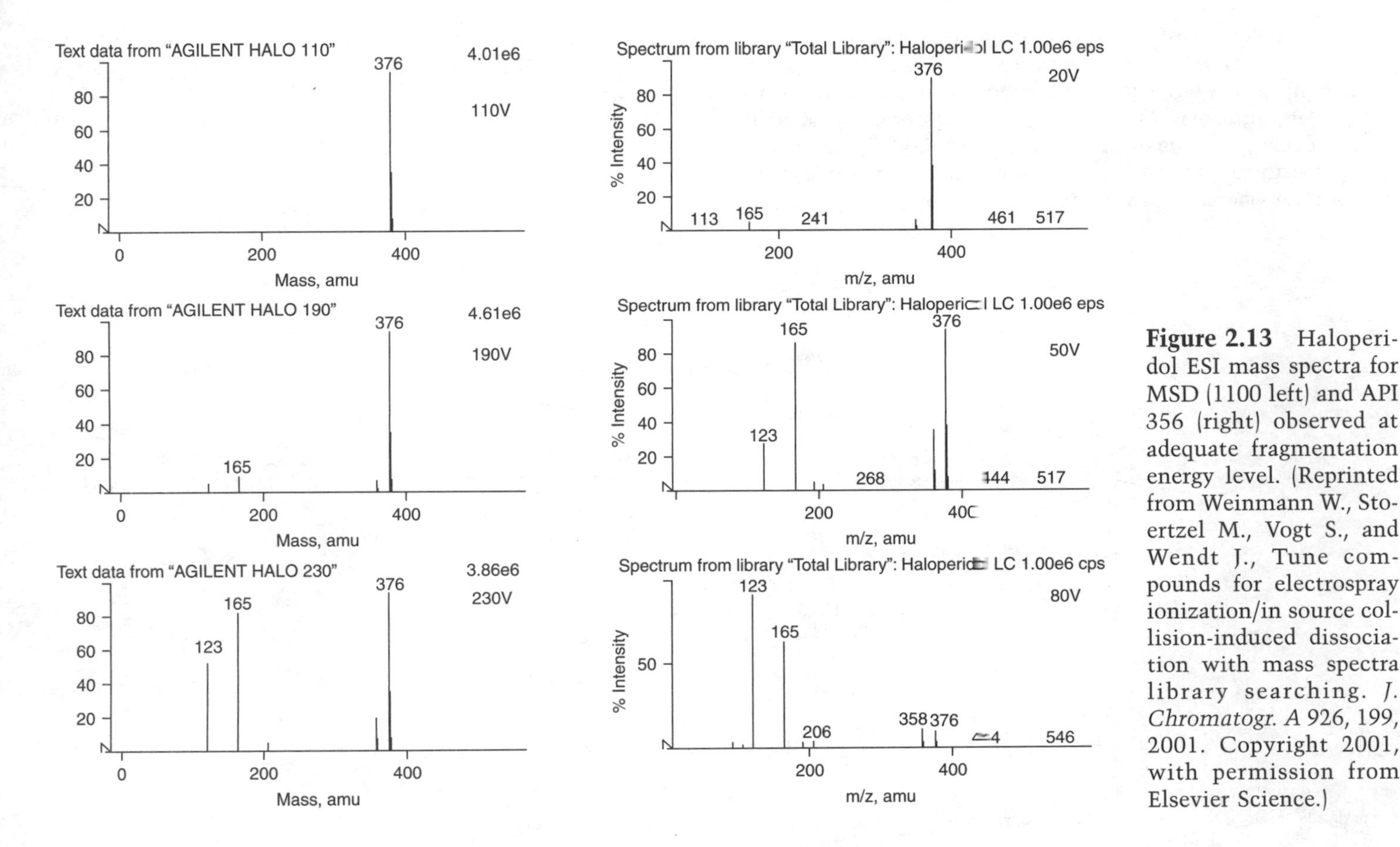

Figure 2.13 Haloperidol ESI mass spectra for MSD (1100 left) and API 356 (right) observed at adequate fragmentation energy level. (Reprinted from Weinmann W., Stoertzel M., Vogt S., and Wendt J., Tune compounds for electrospray ionization/in source collision-induced dissociation with mass spectra library searching. *J. Chromatogr. A* 926, 199, 2001. Copyright 2001, with permission from Elsevier Science.)

in experiments performed in negative ionization mode. Between the two instruments used, better matching results were obtained through the latter mode which was also used for library data collection.

Gergov et al.[202] presented a very straightforward approach for the application of LC/ESI/MS/TOF in toxicological screening. The features of TOF/MS were utilized, i.e., high mass accuracy and high sensitivity over the full spectrum. On this base a library was established containing 433 toxicologically relevant compounds (parent drugs and their metabolites), expanded simply by the calculation of their monoisotopic masses. The mass range for compounds included in the library extended from 105 to 734 Da. This library was used for the identification of drugs in urine extracts, using gradient elution in an ACN–ammonium acetate buffer at pH 3.2. The retention data of drugs were stored but not used for identification in this study. The results of LC/MS/TOF screening were in concordance with the results of GC and TLC screening, run in parallel. According to the authors, the method is very promising and allows screening for compounds with known formula even without reference compounds. This is of practical value, particularly for metabolites, due to the frequent lack of available reference standards. The screening for metabolites using LC/ESI/MS/TOF was done by the same group in consecutive study.[203] Ten previously analyzed autopsy urine samples were subjected to LC/ESI/MS/TOF and the results were compared with the library of theoretical masses of metabolites. Several metabolites of drugs (amitriptyline, methadone, etc.) were identified.

2.6.2.2 Undirected Screening using LC/MS/MS Procedures (QQQ, qTOF, and qIT)

An intralaboratory reproducibility of mass spectra obtained with similar or different instruments is of primary relevance for the establishment and use of mass spectra libraries. Weinmann, Gergov, and Goerner[204] compared mass spectra obtained with two identical (Sciex API 365) and one different (Sciex API 2000) LC/MS/MS in three laboratories. Methadone, benzoylecgonine, and diazepam were used as test substances. Product ion spectra of protonated quasi-molecular ions of more than 400 compounds were similar for all laboratories. Also, in-source CID spectra showed good similarity with the product ion spectra. The last point is very important because in the first step of LC/MS screening, the in-source generated mass spectra are collected and eventually confirmed in MS/MS analysis. Gergov et al.[205] presented a LC/ESI/MS/MS screening procedure for 238 therapeutic and illicit drugs extracted from blood with organic solvents at acidic and basic pH. Only a positive ionization mode was applied. For each compound, the transition from protonated quasi-molecular ion to one most intense or most specific fragment was monitored. Also, retention time and peak area were used for identification. The method was applied in forensic casework. Dams et al.[206] developed a LC/APCI/MS/MS (ion trap) screening method for selected drugs of abuse (opiate agonists, cocaine, and metabolites) in plasma, saliva, and urine. The samples were only subjected to acetonitrile precipitation before analysis. No ion suppression was observed for the compounds involved. In the study of Nordgren and Beck,[207] urine samples were 1:10 diluted with water and directly injected into LC/APCI/MS/MS. A small library of 24 drugs of abuse was established. The procedure was applied in routine casework; among 529 analyzed samples, in 32 cases various drugs were found, mainly psychoactive phenethylamines. Direct analysis of urine was applied also by Fitzgerald et al.,[208] who modified REMEDi HPLC analyzer for use with ion trap MS. A small library of 17 drugs belonging to different classes was established. Urine samples were extracted on-line as for HPLC/DAD analysis.

2.7 Perspectives and Future Trends

Existing needs and observed trends of development allow predicting the main pathways of progress in LC/MS applied for forensic toxicology. This progress may be divided into following categories:

- *Sample preparation for LC/MS:* The problems caused by coeluting matrix compounds (ionization suppression, spectra overlapping) have been well recognized. Therefore, more stress will be put on the development of fast and efficient isolation techniques.
- *HPLC separation:* A trend to shortening of analysis time will be observed, similar to that in clinical toxicology. This may be achieved either through introduction of columns assuring faster and more efficient separation (short, fine-grain or monolithic columns), through application of fast elution (fast gradient, high percentage of organic solvent in mobile phase, high flow rate), or by flow switching of eluent flow during separation process.
- *Application area:* LC/MS will cover practically the whole spectrum of compounds of toxicological relevance. The use of LC/MS as a tool for general screening procedures will be common, and the libraries of mass spectra will be available with any purchased instrument. Most probable will be combined use of in-source CID for preliminary screening with consecutive MS/MS confirmation.
- *Technical aspects:* Detection part of LC/MS, i.e., the mass spectrometer itself, will became smaller without compromising the quality. Bench-top, tandem mass spectrometers at affordable prices already have

appeared on the market and their use will definitely spread in forensic toxicological laboratories.

References

1. Krull I.S. and Cohen S.A. Analytical chemistry and biotechnology. *LC-GC Int.* 11, 139–140, 1998.
2. Willoughby R., Sheehan E., and Mitrovich S. *A Global View on LC/MS*. Global View, Pittsburgh, PA, 1998.
3. Niessen W.M.A. Advances in instrumentation in liquid chromatography-mass spectrometry and related liquid-introduction techniques. *J. Chromatogr. A* 794, 407, 1998.
4. Niessen W.M.A. State-of-the-art in liquid chromatography-mass spectrometry. *J. Chromatogr. A* 856, 179, 1999.
5. Bogusz M.J. Hyphenated liquid chromatographic techniques in forensic toxicology. *J. Chromatogr. B* 733, 65, 1999.
6. Marquet P. and Lachatre G. Liquid chromatography-mass spectrometry: potential in forensic and clinical toxicology. *J. Chromatogr. B* 733, 93, 1999.
7. Bogusz M.J. Liquid chromatography-mass spectrometry as a routine method in forensic science: a proof of maturity. *J. Chromatogr. B* 748, 3, 2000.
8. Marquet P. Progress of liquid chromatography–mass spectrometry in clinical and forensic toxicology. *Ther. Drug Monit.* 24, 255, 2002.
9. Van Bocxlaer J.F., Clauwaert K.M., Lambert W.E., Deforce D.L., Van den Eeckhout E.G., and De Leenheer A.P. Liquid chromatography-mass spectrometry in forensic toxicology. *Mass Spectrom. Rev.* 19, 165, 2002.
10. Weinmann W. and Svoboda M. Fast screening for drugs of abuse by solid phase extraction combined with flow-injection ionspray tandem mass spectrometry. *J. Anal. Toxicol.* 22, 319, 1998.
11. Müller C., Schäfer P., Störtzel M., Vogt S., and Weinmann W. Ion suppression effects in liquid chromatography-electrospray-ionisation transport-region collision induced dissociation mass spectrometry with different serum extraction methods for systematic toxicological analysis with mass spectra libraries. *J. Chromatogr. B* 773, 47, 2002.
12. Zhou S., Larson M.L., Naidong W., and Jiang X. Characterization of matrix suppression effect in quantitative LC/MS-MS analysis of biological samples. Poster on the 50th ASMS Conference on Mass Spectrometry and Allied Topics, Orlando, FL, 2002.
13. Avery M.J. Ion suppression issue in the LC/MS analysis of analytes isolated from human and animal matrices. Poster on the 50th ASMS Conference on Mass Spectrometry and Allied Topics, Orlando, FL, 2002.

14. Tang Y.Q., Coopersmith B.I., Beato B.D., and Naidong W. Strategies for dealing with matrix effects and interferences in bioanalytical LC/MS-MS. Poster on the 50th ASMS Conference on Mass Spectrometry and Allied Topics, Orlando, FL, 2002.
15. Bansal S. and Liang Z. Matrix factor and extraction uniformity: novel quantitative tools for assessing matrix effects in LC/MS-MS during assay validations. Poster on the 50th ASMS Conference on Mass Spectrometry and Allied Topics, Orlando, FL, 2002.
16. Sun W., Coe R., Loin P., and Lee J.W. Matrix suppression resolved using fast dual-column LC/MS-MS in the bioanalysis of nicotine/cotinine in human heparinized plasma. Poster on the 50th ASMS Conference on Mass Spectrometry and Allied Topics, Orlando, FL, 2002.
17. Heinig K. and Bucheli F. Application of column switching liquid chromatography-tandem mass spectrometry for the determination of pharmaceutical compounds in tissue samples. *J. Chromatogr.* B 25, 9, 2002.
18. Temesi D. and Law B. The effect of LC eluent composition on MS responses using electrospray ionization. *LC-GC Int.* 12, 175, 1999.
19. Naidong W., Shou W., Chen Y.L., and Jiang X. Novel liquid chromatographic-tandem mass spectrometric method using silica columns and aqueous-organic mobile phases for quantitative analysis of polar analytes in biological fluids. *J. Chromatogr. B* 754, 387, 2001.
20. Naidong W., Lee J.W., Jiang X., Wehlin M., Hulse J.D., and Lin P.P. Simultaneous assay of morphine, morphine-3-glucuronide, and morphine-6-glucuronide in human plasma using normal-phase liquid chromatography-tandem mass spectrometry with a silica column and an aqueous organic mobile phase. *J. Chromatogr. B* 735, 255, 1999.
21. Shou W.Z., Pelzer M., Addison T., Jiang X., and Naidong W. An automatic 96-well solid phase extraction and liquid chromatography-tandem mass spectrometry method for the analysis of morphine, morphine-3-glucuronide and morphine-6-glucuronide in human plasma. *J. Pharm. Biomed. Anal.* 27, 143, 2002.
22. Shou W.Z., Jiang X., Beato B.D., and Naidong W. A highly automated 96-well solid phase extraction and liquid chromatography/tandem mass spectrometry method for the determination of fentanyl in human plasma. *Rapid Commun. Mass Spectrom.* 15, 466, 2001.
23. Chen Y.L., Hanson G.D., Jiang X., and Naidong W. Simultaneous determination of hydrocodone and hydromorphone in human plasma by liquid chromatography with tandem mass spectrometric detection. *J. Chromatogr. B* 769, 55, 2002.
24. Shou W.Z., Chen Y.L., Eerkes A., Tang Y.Q., Magis L., Jiang X., and Naidong W. Ultrafast liquid chromatography/tandem mass spectrometry bioanalysis of polar analytes using packed silica columns. *Rapid Commun. Mass Spectrom.* 16, 1613, 2002.

25. Naidong W., Shou W.Z., Addison T., Maleki S., and Jiang X. Liquid chromatography/tandem mass spectrometric bioanalysis using normal-phase columns with aqueous/organic mobile phases — a novel approach of eliminating evaporation and reconstitution steps in 96-well SPE. *Rapid Commun. Mass Spectrom.* 16, 1965, 2002.

26. Zweigenbaum J., Heinig K., Steinborner S., Wachs T., and Henion J. High throughput bioanalytical LC/MS/MS determination of benzodiazepines: 1000 samples per 12 hours. *Anal. Chem.* 71, 2294, 1999.

27. Cheng Y.F., Lu Z., and Neue U. Ultrafast liquid chromatography/ultraviolet and liquid chromatography/tandem mass spectrometric analysis. *Rapid Commun. Mass Spectrom.* 15, 141, 2001.

28. Dams R., Benijts T., Gunther W., Lambert W., and De Leenheer A. Influence of the eluent composition on the ionization efficiency for morphine of pneumatically assisted electrospray, atmospheric pressure chemical ionization and sonic spray. *Rapid Commun. Mass Spectrom.* 16, 1072, 2002.

29. Thurman E.M., Ferrer I., and Barcelo D. Choosing between atmospheric pressure chemical ionization and electrospray ionization interfaces for the HPLC/MS analysis of pesticides. *Anal. Chem.* 73, 5441, 2001.

30. Yang C. and Henion J. Atmospheric pressure photoionization liquid chromatographic-mass spectrometric determination of idoxifene and its metabolites in human plasma. *J. Chromatogr. A* 970, 155, 2002.

31. Singh G., Gutierrez A., Xu K., and Blair I.A. Liquid chromatography/electron capture atmospheric pressure chemical ionization/mass spectrometry: analysis of pentafluorobenzyl derivatives of biomolecules and drugs in the attomole range. *Anal. Chem.* 72, 3007, 2000.

32. Clauwaert K.M., Van Bocxlaer J.F., Major H.J., Claereboudt J.A., Lambert W.E., Van den Eeckhout E.M., Van Peteghem C.H., and De Leenheer A.P. Investigation of the quantitative properties of the quadrupole orthogonal acceleration time-of-flight mass spectrometer with electrospray ionization using 3,4-methyylenedioxymethamphetamine. *Rapid Commun. Mass Spectrom.* 13, 1540, 1999.

33. Zhang H. and Henion J. Comparison between liquid chromatography-time-of-flight mass spectrometry and selected reaction monitoring liquid chromatography-mass spectrometry for quantitative determination of idoxifene in human plasma. *J. Chromatogr. B* 757, 151, 2001.

34. Vorce S.P., Sklerov J.H., and Kalasinsky K.S. Assessment of the ion-trap mass spectrometer for routine qualitative and quantitative analysis of drugs of abuse extracted from urine. *J. Anal. Toxicol.* 24, 595, 2000.

35. Dams R., Benijst T., Gunther W., Lambert W., and De Leenheer A. Sonic spray ionization technology: performance study and application to a LC/MS analysis on a monolithic silica column for heroin impurity profiling. *Anal. Chem.* 74, 3206, 2002.

36. Bogusz M.J. Opioid agonists. In Bogusz M.J., Ed., *Forensic Science*, Vol. 2, *Handbook of Analytical Separations*. Elsevier, Amsterdam, 2000, 3.

37. Zuccaro P., Ricciarello R., Pichini S., Altieri R.I., Pellegrini M., and D'Aszenzo G. Simultaneous determination of heroin, 6-monoacetylmorphine, morphine, and its glucuronides by liquid chromatography-atmospheric pressure ionspray-mass spectrometry. *J. Anal. Toxicol.* 21, 268, 1997.

38. Bogusz M.J., Maier R.D., and Driessen S. Morphine, morphine-3-glucuronide, morphine-6-glucuronide and 6-monoacetylmorphine determined by means of atmospheric pressure chemical ionization-mass spectrometry-liquid chromatography in body fluids of heroin victims. *J. Anal. Toxicol.* 21, 346, 1997.

39. Bogusz M.J., Maier R.D., Erkens M., and Driessen S. Determination of morphine and its 3-and 6-glucuronide, codeine, codeine-6-glucuronide and 6-monoacetylmorphine in body fluids by liquid chromatography-atmospheric pressure chemical ionization mass spectrometry. *J. Chromatogr. B* 703, 115, 1997.

40. Katagi M., Nishikawa M., Tatsuno M., Miki A., and Tsushihashi H. Column switching high-performance liquid chromatography-electrospray ionization mass spectrometry for identification of heroin metabolites in human urine. *J. Chromatogr. B* 751, 177, 2001.

41. Schanzle G., Li S., Mikus G., and Hofmann U. Rapid, highly sensitive method for the determination of morphine and its metabolites in body fluids by liquid chromatography-mass spectrometry. *J. Chromatogr. B* 721, 65, 1999.

42. Dienes-Nagy A., Rivier L., Giroud G., Augsburger M., and Mangin P. Method for quantification of morphine and its 3- and 6-glucuronides, codeine, codeine glucuronide, and 6-monoacetylmorphine in human blood by liquid chromatography-electrospray mass spectrometry for routine analysis in forensic toxicology. *J. Chromatogr. A* 854, 109, 1999.

43. Blanchet M., Bru G., Guerret M., Bromet-Petit M., and Bromet N. Routine determination of morphine, morphine 3-β-D-glucuronide and morphine 6 β-D-glucuronide in human serum by liquid chromatography coupled to electrospray mass spectrometry. *J.Chromatogr. A* 854, 03, 1999.

44. Slawson M.H., Crouch D.J., Andrenyak D.M., Rollins D.E., Lu J.K., and Bailey P.L. Determination of morphine, morphine-3-glucuronide, and morphine-6-glucuronide in plasma after intravenous and intratecal morphine administration using HPLC with electrospray ionization and tandem mass spectrometry. *J. Anal. Toxicol.* 23, 116, 1999.

45. Bogusz M.J. LC/MS as a tool for case interpretation in forensic toxicology. 49th ASMS Conference on Mass Spectrometry and Allied Topics, Chicago, May 27–31, 2001.

46. Huizer, H. Analytical studies on illicit heroin. Ph.D. thesis, Technical University Delft, The Netherlands, 1988.

47. O'Neal C.L. and Poklis A. Simultaneous determination of acetylcodeine, monoacetylmorphine and other opiates in urine by GC-MS. *J. Anal. Toxicol.* 21, 427, 1997.
48. O'Neal C.L. and Poklis A. The detection of acetylcodeine and 6-acetylmorphine in opiate-positive urines. *Forensic Sci. Int.* 95, 1, 1998.
49. Staub C., Marset M., Mino A., and Mangin P. Detection of acetylcodeine in urine as an indicator of illicit heroin use: method validation and results of a pilot study. *Clin. Chem.* 47, 301, 2001.
50. Bogusz M.J., Maier R.D., Erkens M., and Kohls U. Detection of non-prescription heroin markers in urine with liquid chromatography-atmospheric pressure chemical ionization mass spectrometry. *J. Anal. Toxicol.* 25, 431, 2001.
51. Gaulier J.M., Marquet P., Lacassie E., Dupuy J.L., and Lachatre G. Fatal intoxication following self-administration of a massive dose of buprenorphine. *J. Forensic Sci.* 45, 226, 2000.
52. Polettini A. and Huestis M.A. Simultaneous determination of buprenorphine, norbuprenorphine, and buprenorphine-glucuronide in plasma by liquid chromatography-tandem mass spectrometry. *J. Chromatogr. B* 754, 447, 2001.
53. Moody D.E., Laycock J.D., Spanbauer A.C., Crouch D.J., Foltz R.L., Josephs J.L., and Bickel W.K. Determination of buprenorphine in human plasma by gas chromatography-positive ion chemical ionization mass spectrometry and liquid chromatography-tandem mass spectrometry. *J. Anal. Toxicol.* 21, 406, 1997.
54. Moody D.E., Slawson M.H., Strain E.C., Laycock J.D., Spanbauer A.C., and Foltz R.L. A liquid chromatographic-electrospray ionization tandem mass spectrometric method for determination of buprenorphine, its metabolite norbuprenorphine, and a coformulant, naloxone, that is suitable for *in vivo* and *in vitro* metabolism studies. *Anal. Biochem.* 306, 31, 2002.
55. Hoja H., Marquet P., Verneuil B., Lofti H., Dupuy J.L, and Lachatre G. Determination of buprenorphine and norbuprenorphine in whole blood by liquid chromatography-mass spectrometry. *J. Anal. Toxicol.* 21, 160, 1997.
56. Bogusz M.J., Maier R.D., Kruger K.D., and Kohls U. Determination of common drugs of abuse in body fluid using one isolation procedure and liquid chromatography-atmospheric pressure chemical ionization mass spectrometry. *J. Anal. Toxicol.* 22, 549, 1998.
57. Breindahl T. and Andreasen K. Validation of urine drug-of-abuse testing for ketobemidone using thin-layer chromatography and liquid chromatography-electrospray mass spectrometry. *J. Chromatogr. B* 736, 103, 1999.
58. Sundstrom I., Bondesson U., and Hedeland M. Identification of phase I and phase II metabolites of ketobemidone in patient urine using liquid chromatography-electrospray tandem mass spectrometry. *J. Chromatogr. B* 763, 121, 2001.
59. Sundstrom I., Hedeland M., Bondesson U., and Andren P.E. Identification of glucuronide conjugates of ketobemidone and its phase I metabolites in human urine utilizing accurate mass and tandem time-of-flight mass spectrometry. *J. Mass Spectrom.* 37, 14, 2002.

60. Ceccato A., Vanderbist F., Pabst J.Y., and Streel B. Automated determination of tramadol enantiomers in human plasma using solid-phase extraction in combination with chiral liquid chromatography. *J. Chromatogr. B* 748, 65, 2000.
61. Juzwin S.J, Wang D.C., Anderson N.J., and Wong F.A. The determination of RWJ-38705 (tramadol-*N*-oxide) and its metabolites in preclinical pharmacokinetic studies using LC-MS/MS. *J. Pharm. Biomed. Anal.* 22, 469, 2000.
62. Zheng M., McErlane K.M., and Ong M.C. LC-MS-MS analysis of hydromorphone and hydromorphone metabolites with application to a pharmacokinetic study in the male Sprague-Dawley rat. *Xenobiotica* 32, 141, 2002.
63. Kintz P., Eser H.P., Tracqui A., Moeller M. Cirimele V., and Mangin P. Enantioselective separation of methadone and its main metabolite in human hair by liquid chromatography/ion spray-mass spectrometry. *J. Forensic Sci.* 42, 291, 1997.
64. Ortelli D., Rudaz S., Chevalley A.F., Deglon J.J., Balant L., and Veuthey J.L. Enantioselective analysis of methadone in saliva by liquid chromatography-mass spectrometry. *J. Chromatogr. A* 871, 163, 2000.
65. Dale O., Hoffer C., Sheffels P., and Kharasch E.D. Disposition of nasal, intravenous, and oral methadone in healthy volunteers. *Clin.Pharmacol.Ther.* 72, 536, 2002.
66. Shulgin A. and Shulgin A. *Pihkal, A Chemical Love Story*, 1st ed. Transform Press, Berkeley, CA, 1995.
67. Henry A.S., Jeffrys K.J., and Dawling S. Toxicity and deaths from 3,4-methylenedioxymethamphetamine ("ecstasy"). *Lancet* 340, 384, 1992.
68. Lora-Tamayo A., Tena T., and Rodriguez A. Amphetamine derivative related deaths. *Forensic Sci. Int.* 85:149, 1997.
69. Schifano F., Oyefeso A., Webb L., Corkery J., and Ghodse A.H. Review of deaths related to taking ecstasy, England and Wales, 1997–2000. *Brit. Med. J.* 326, 80, 2003.
70. Zhao H., Brenneisen R., Scholer A., McNally A.J., ElSohly M.A., Murphy T.P., and Salamone S.J. Profiles of urine samples taken from ecstasy users at rave parties: analysis by immunoassays. *J. Anal. Toxicol.* 25, 258, 2001.
71. Bogusz M.J., Kala M., and Maier R.D. Determination of phenylisothiocyanate derivatives of amphetamine and its analogues in biological fluids by HPLC-APCI-MD or DAD. *J. Anal. Toxicol.* 21, 59, 1997.
72. Bogusz M.J., Kruger K.D., and Maier R.D. Analysis of underivatized amphetamines and related phenethylamines with high-performance liquid chromatography-atmospheric pressure chemical ionization mass spectrometry. *J. Anal. Toxicol.* 24, 77, 2000.
73. Kataoka H., Lord H.L., and Pawliszyn J., Simple and rapid determination of amphetamine, methamphetamine, and their methylenedioxy derivatives in urine by automated in-tube solid-phase microextraction coupled with liquid chromatography-electrospray ionization mass spectrometry. *J. Anal. Toxicol.* 24, 257, 2000.

74. Clauwaert K.M., VanBocxlaer J.F., De Letter E.A., Van Calenbergh S., Lambert W.E., and De Leenheer A.P. Determination of the designer drugs 3,4-methylenedioxymethamphetamine, 3,4-methylenedioxyethylamphetamine, and 3,4-methylenedioxyamphetamine in whole blood and fluorescence detection in whole blood, serum, vitreous humor, and urine. *Clin. Chem.* 46, 1968, 2000.

75. Mortier K.A., Dams R., Lambert W.E., De Letter E., Van Calenbergh S., and De Leenheer A.P. Determination of paramethoxyamphetamine and other amphetamine-related designer drugs by liquid chromatography/sonic spray mass spectrometry. *Rapid Comm. Mass Spectrom.* 16, 865, 2002.

76. Sato M., Mitsui T., and Nagase H. Analysis of benzphetamine and its metabolites in rat urine by liquid chromatography-electrospray ionization mass spectrometry. *J. Chromatogr. B* 751, 277, 2001.

77. Katagi M., Tatsuno M., Miki A., Nishikawa M., and Tsushihashi H. Discrimination of dimethylamphetamine and methamphetamine use: simultaneous determination of dimethylamphetamine-*N*-oxide and other metabolites in urine by high-performance liquid chromatography-electrospray ionization mass spectrometry. *J. Anal. Toxicol.* 24, 354, 2000.

78. Katagi M., Tatsuno M., Miki A., Nishikawa M., Nakajima K., and Tsushihashi H. Simultaneous determination of selegiline-*N*-oxide, a new indicator for selegiline administration, and other metabolites in urine by high performance liquid chromatography-electrospray ionization mass spectrometry. *J. Chromatogr. B* 759, 125, 2001.

79. Katagi M., Tatsuno M., Tsutsumi H., Miki A., Kamata T., Nishioka H., Nakajima K., Nishiwaka M., and Tsushihashi H. Urinary excretion of selegiline-N-oxide, a new indicator for selegiline administration in man. *Xenobiotica* 32, 823, 2002.

80. Wood M., De Boeck D., Samyn N., Cooper D., and Morris M. Development of a rapid and sensitive method for the quantitation of amphetamines in human saliva. Poster presented on 49th ASMS Conference, Chicago, 2001.

81. Samyn N., De Boeck G., Wood M., Lamers C.T., De Waard D., Brookhuis K.A., Verstraete A.G., and Riedel W.J. Plasma, oral fluid and sweat wipe ecstasy concentrations in controlled and real life conditions. *Forensic Sci. Int.* 128, 90, 2002.

82. Stanaszek R., Piekoszewski W., Karakiewicz B., and Kozielec T. Correlation between self-reported drug use and the results of hair analysis in detoxification and methadone treatment patients. *Przegl. Lek.* 59, 347, 2002.

83. Jufer R.A., Darwin W.D., and Cone E.J. Current methods for the separation and analysis of cocaine analytes. In Bogusz M.J., Ed., *Forensic Science*, Vol. 2, *Handbook of Analytical Separations*. Elsevier, Amsterdam, 2000, 67.

84. Sosnoff C.S., Ann Q., Bernert J.T., Jr., Powell M.K., Miller B.B., Henderson L.O., Hannon W.H., and Sampson E.J. Analysis of benzoylecgonine in dried blood spots by liquid chromatography-atmospheric pressure chemical ionization mass spectrometry. *J. Anal. Toxicol.* 20, 179, 1996.

85. Singh G., Arora V., Fenn P.T., Mets B., and Blair I.A. A validated stable isotope dilution liquid chromatography tandem mass spectrometry assay for the trace analysis of cocaine and its major metabolites in plasma. *Anal. Chem.* 71, 2021, 1999.

86. Jeanville P.M., Estape E.S., Needham S.R., and Cole M.J. Rapid confirmation/quantitation of cocaine and benzoylecgonine in urine utilizing high performance liquid chromatography and tandem mass spectrometry. *J. Am. Soc. Mass Spectrom.* 11, 257, 2000.

87. Jeanville P.M., Estape E.S., Torres-Negron I., and Marti A. Rapid confirmation/quantitation of ecgonine methyl ester, benzoylecgonine, and cocaine using on-line extraction coupled with fast HPLC and tandem mass spectrometry. *J. Anal. Toxicol.* 25, 69, 2001.

88. Jeanville P.M., Woods J.H., Baird T.J., and Estape E.S. Direct determination of ecgonine methyl ester and cocaine in rat plasma, utilizing on-line sample extraction coupled with rapid chromatography/quadrupole orthogonal acceleration time-of-flight detection. *J. Pharm. Biomed. Anal.* 23, 987, 2000.

89. Skopp G., Klingmann A., Potsch L., and Mattern R. In vitro stability of cocaine in whole blood and plasma including ecgonine as a target analyte. *Ther. Drug Monit.* 23, 174, 2001.

90. Klingmann A., Skopp G., and Aderjan R. Analysis of cocaine, benzoylecgonine, ecgonine methyl ester, and ecgonine by high pressure liquid chromatography-API mass spectrometry and application to a short-term degradation study of cocaine in plasma. *J. Anal. Toxicol.* 25, 425, 2001.

91. Lin S.N., Moody D.E., Bigelow G.E., and Foltz R.E. A validated liquid chromatographic-atmospheric pressure chemical ionization tandem-mass spectrometry method for quantitation of cocaine and benzoylecgonine in human plasma. *J. Anal. Toxicol.* 25, 497, 2001.

92. Clauwaert K.M., Van Bocxlaer J.F., Labmert W.E., Van den Eeckhout E., Lemiere F., Esmans E.L., and De Leenheer A.P. Narrow-bore HPLC in combination with fluorescence and electrospray mass spectrometric detection for the analysis of cocaine and metabolites in human hair. *Anal. Chem.* 70, 2336, 1998.

93. Xia Y., Wang P., Bartlett M.G., Solomon H.M., and Busch K.L. An LC-MS-MS method for the comprehensive analysis of cocaine and cocaine metabolites in meconium. *Anal. Chem.* 15, 764, 2000.

94. Srinivasan K., Wang P.P., Eley A.T., White C.A., and Bartlett M.G. Liquid chromatography-tandem mass spectrometry analysis of cocaine and its metabolites from blood, amniotic fluid, placenta; and fetal tissues: study of the metabolism and distribution of cocaine in pregnant rats. *J. Chromatogr. B* 745, 287, 2000.

95. Rustichelli C., Ferioli V., Vezzalini F., Rossi M.C., and Gamberini G. Simultaneous separation and identification of hashish constituents by coupled liquid chromatography-mass spectrometry (HPLC-MS). *Chromatographia* 43, 129, 1996.

96. Bäckström B., Cole M.D., Carrott M.J., Jones D.C., Davidson G., and Coleman K. A preliminary study on the analysis of *Cannabis* by supercritical fluid chromatography with atmospheric pressure chemical ionization mass spectroscopic detection. *Sci. Just.* 37, 91, 1997.

97. Mireault P. Analysis of Δ^9-tetrahydrocannabinol and its two major metabolites by APCI-LC/MS. Poster at the 46th ASMS conference, Orlando, FL, 1998.

98. Picotte P., Mireault P., and Nolin G. A rapid and sensitive LC/APCI/MS/MS method for the determination of Δ^9-tetrahydrocannabinol and its metabolites in human matrices. Poster at the 48th ASMS conference, Long Beach, CA, 2000.

99. Breindahl T. and Andreasen K. Determination of 11-nor-delta9-tetrahydrocannabinol-9-carboxylic acid in urine using high-performance liquid chromatography and electrospray ionization mass spectrometry. *J. Chromatogr. B* 732, 155, 1999.

100. Tai S.S. and Welch M.J. Determination of 11-nor-delta9-tetrahydrocannabinol-9-carboxylic acid in a urine-based standard reference material by isotope-dilution liquid chromatography-mass spectrometry with electrospray ionization. *J. Anal. Toxicol.* 24, 385, 2000.

101. Weinmann W., Vogt S., Goerke R., Muller C., and Bromberger A. Simultaneous determination of THC-COOH and THC-COOH-glucuronide in urine samples by LC/MS/MS. *Forensic Sci. Int.* 113, 381, 2000.

102. Weinmann W., Goerner M., Vogt S., Goerke R., and Pollak S. Fast confirmation of 11-nor-9-carboxy-Δ^9-tetrahydrocannabinol (THC-COOH) in urine by LC/MS/MS using negative atmospheric pressure chemical ionization (APCI). *Forensic Sci. Int.* 121, 103, 2001.

103. Siddik Z.H., Barnes R.D., Dring L.G., Smith R.L., and Williams R.T. The fate of lysergic acid di[14C] ethylamide in rat, guinea pig and rhesus monkey and of [14C] iso-LSD in rat. *Biochem. Pharmacol.* 28, 3093, 1979.

104. Webb K.S., Baker P.B., Cassells N.P., Francis J.M., Johnston D.E., Lancaster S.L., Minty P.S., Reed G.D., and White S.A. The analysis of lysergide (LSD): the development of novel enzyme immunoassay and immunoaffinity extraction procedures with an HPLC-MS confirmation procedure. *J. Forensic Sci.* 41, 938, 1996.

105. White S.A., Kidd A.S., and Webb K.S. The determination of lysergide (LSD) in urine by high-performance liquid chromatography-isotope dilution mass spectrometry (IDMS), *J. Forensic Sci.* 44, 375, 1999.

106. Hoja H., Marquet P., Verneuil B., Lofti H., Dupuy J.L., and Lachatre G. Determination of LSD and *N*-demethyl-LSD in urine by liquid chromatography coupled to electrospray ionization mass spectrometry. *J. Chromatogr. B* 692, 329, 1997.

107. de Kanel J., Vickery W.E., Waldner B., Monahan R.M., and Diamond F.X. Automated extraction of lysergic acid diethylamide (LSD) and *N*-demethyl-LSD from blood, serum, plasma, and urine samples using the Zymark RapidTrace with LC/MS/MS confirmation. *J. Forensic Sci.* 43, 622, 1998.

108. Bodin K. and Svensson J.O. Determination of LSD in urine with liquid chromatography-mass spectrometry. *Ther. Drug Monit.* 23, 389, 2001.

109. Nelson C.C. and Foltz R.L. Chromatographic and mass spectrometric methods for determination of lysergic acid diethylamide (LSD) and metabolites in body fluids. *J. Chromatogr.* 580, 97, 1992.

110. Poch G.K., Klette K.L., Hallare D.A., Manglicmot M.G., Czarny R.J., McWhorter L.K., and Anderson C.J. The detection of metabolites of LSD in human urine specimens: 2-oxo-3-hydroxy-LSD, a prevalent metabolite of LSD. *J. Chromatogr.* 724, 23, 1999.

111. Poch G.K., Klette K.L., and Anderson C.J. The quantitation of 2-oxo-3-hydroxy LSD in human urine specimens, a metabolite of LSD: comparative analysis using liquid chromatography-selected ion monitoring mass spectrometry and liquid chromatography-ion trap mass spectrometry. *J. Anal. Toxicol.* 24, 170, 2000.

112. Sklerov J.H., Magluilo J., Shannon K.K., and Smith M.L. Liquid chromatography-electrospray ionization mass spectrometry for the detection of lysergide and a major metabolite, 2-oxo-3-hydroxy LSD, in urine and blood. *J. Anal. Toxicol.* 24, 543, 2000.

113. Canezin J., Caillcux A., Turcant A., Le Bouil A., Harry P., and Allain P. Determination of LSD and its metabolites in human biological fluids by high-performance liquid chromatography with electrospray tandem mass spectrometry. *J. Chromatogr. B* 765, 12, 2001.

114. Klette K.L., Horn C.K., Stout P.R., and Anderson C.J. LC-MS analysis of human urine specimens for 2-oxo-3-hydroxy LSD: method validation for potential interferants and stability study for 2-oxo-3-hydroxy LSD under various storage conditions. *J. Anal. Toxicol.* 26, 193, 2002.

115. Mash D.C., Kovera C.A., Pablo J., Tynale R.F., Evin F.D., Williams E.C., Singleton E.G., and Mayor M. Ibogaine complex pharmacokinetics, concerns for safety, and preliminary efficacy measures. *Ann. N.Y. Acad. Sci.* 914, 394, 2000.

116. Glick S.D., Maisonneuve I.M., and Szumlinski K.K. 18-methoxycoronaridine (18-MC) and ibogaine: comparison of antiaddictive efficacy, toxicity, and mechanism of action. *Ann. N.Y. Acad. Sci.* 914, 369, 2000.

117. Zhang W., Ramamoorthy Y., Tyndale R.F., Glick S.D., Maisonneuve I.M., Kuehne M.E., and Sellers E.M. Metabolism of 18-methoxycoroaridine, an ibogaine analog, to 18-hydroxycoronaridine by genetically variable CYP2C219. *Drug Metab. Disp.* 30, 663, 2002.

118. Forstrom T., Tuominen J., and Karkkainen J. Determination of potentially hallucinogenic *N*-dimetylated indoleamines in human urine by HPLC/ESI-MS-MS. *Scand. J. Clin. Lab. Invest.* 61, 547, 2001.

119. Barker S.A., Littlefield-Chabaud M.A., and David C. Distribution of the hallucinogens *N,N*-diethyltryptamine and 5-methoxy-*N,N*-dimetryltryptamine in rat brain following intraperitoneal, injection: application of a new solid-phase extraction LC-APCI-MS-MS-isotope dilution method. *J. Chromatogr. B* 751, 376, 2001.

120. Brenneisen R. and Stalder A.B. Psilocybe. In *Hagers Handbuch der Pharmazeutischen Praxis*, Vol. 6. Hansel R., Keller K., Romples H., and Schneider G., Eds., Springer-Verlag, Berlin, 1994, 287.

121. Bogusz M.J., Maier R.D., Schäfer A.T., and Erkens M. Honey with Psilocybe mushrooms: a revival of a very old preparation on the drug market? *Int. J. Legal Med.* 111, 147, 1998.

122. Peel H.W. and Jeffrey W.K. A report on the incidence of drugs and driving in Canada. *Can. Soc. Forensic Sc. Ji.* 23, 75, 1990.

123. Schiwy-Bochat K.H., Bogusz M., Vega J.A., and Althoff H. Trends in occurence of drugs of abuse in blood and urine of arrested drivers and drug traffickers in the border region of Aachen. *Forensic Sci. Int.* 71, 33, 1995.

124. Cailleux A., Le Bouil A., Auger B., Bonsergent G., Turcant A., and Allain P. Determination of opiates and cocaine and its metabolites in biological fluids by high performance liquid chromatography with elecrospray tandem mass spectrometry. *J. Anal. Toxicol.* 23, 620, 1999.

125. Mortier K.A., Clauwaert K.M., Lambert W.E., Van Bocxlaer J.F., Van den Eeckhout E.G., Van Peteghem C.H., and De Leenheer A.P. Pitfalls associated with liquid chromatography/electrospray tandem mass spectrometry in quantitative bioanalysis of drugs of abuse in saliva. *Rapid Commun. Mass Spectrom.* 15, 1773, 2001.

126. Mortier K., Maudens K., Lambert W., Clauwaert K., Van Bocxlaer J., Deforce D., Van Peteghem C., and De Leenheer A. Simultaneous, quantitative determination of opiates, amphetamines, cocaine and benzoylecgonine in oral fluid by liquid chromatography quadrupole time-of-flight mass spectrometry. *J. Chromatogr. B* 779, 321, 2002.

127. LeBeau M.A. and Mozayani A., Eds., *Drug-Facilitated Sexual Assault.* Academic Press, San Diego, CA, 2001.

128. Bogusz M.J., Maier R.D., Krüger K.D., and Früchtnicht W. Determination of flunitrazepam in blood by high-performance liquid chromatography-atmospheric pressure chemical ionization mass spectrometry. *J. Chromatogr. B* 713, 361, 1998.

129. LeBeau M.A., Montgomery M.A., Wagner J.R., and Miller M.L. Analysis of biofluids for flunitrazepam and metabolites by electrospray liquid chromatography/mass spectrometry. *J. Forensic Sci.* 45, 1133, 2000.

130. Darius J. and Banditt P. Validated method for the therapeutic drug monitoring of flunitrazepam in human serum using liquid chromatography-atmospheric pressure chemical ionization tandem mass spectrometry with a ion trap detector. *J. Chromatogr. B* 738, 437, 2000.
131. Yuan H., Mester Z., Lord H., and Pawliszyn J. Automated in-tube solid-phase microextraction coupled with liquid chromatography-electrospray ionization mass spectrometry for the determination of selected benzodiazepines. *J. Anal. Toxicol.* 24, 718, 2000.
132. Kollroser M. and Schober C. Simultaneous analysis of flunitrazepam and its major metabolites in human plasma by high performance liquid chromatography tandem mass spectrometry. *J. Pharm. Biomed. Anal.* 28, 1173, 2002.
133. Marquet P., Baudin O., Gaulier J.M., Lacassie E., Dupuy J.L, Francois B., and Lachatre G. Sentitive and specific determination of midazolam and 1-hydroxymidazolam in human serum by liquid chromatography-electrospray mass spectrometry. *J. Chromatogr. B* 734, 137, 1999.
134. Ware M.L. and Hidy B.J. LC/MS/MS determination of midazolam and 1-hydroxymidazolam in human plasma. Poster at the 49th ASMS Conference, Chicago, 2001.
135. Smith K.M. Drugs used in acquaintance rape. *J. Am. Pharm. Assoc.* 39, 519, 1999.
136. Gaulier J.M., Canal M., Pradeille J.L., Marquet P., and Lachatre G. New drugs at "rave parties": ketamine and prolintane. *Acta Clin. Belg. Suppl.* 1, 41, 2002.
137. Moore K.A., Sklerov J., Levine B., and Jacobs A.J. Urine concentrations of ketamine and norketamine following illegal consumption. *J. Anal. Toxicol.* 25, 583, 2001.
138. Ramos L., Bakhtiar R., and Tse F. Application of liquid chromatography atmospheric pressure chemical ionization tandem mass spectrometry in the quantitative analysis of glyburide (glibenclamide) in human plasma. *Rapid Commun. Mass Spectrom.* 13, 2439, 1999.
139. Magni F., Marazzini L., Pereira S., Monti L., and Galli Kienle M. Identification of sulfonylureas in serum by electrospray mass spectrometry. *Anal. Biochem.* 282, 136, 2000.
140. Maurer H.H., Kratzsch, Kraemer T., Peters F.T., and Weber A.W. Screening, library-assisted identification and validated quantification of oral antidiabetics of the sulfonylurea-type in plasma by atmospheric pressure chemical ionization liquid chromatography-mass spectrometry. *J. Chromatogr. B* 773, 63, 2002.
141. Stocklin R., Vu L., Vadas L., Cerini F., Kippen A.D., Offord R.E., and Rose K. A stable isotope dilution assay for the *in vivo* determination of insulin levels in humans by mass spectrometry. *Diabetes* 46, 44, 1997.

142. Darby S.M., Miller M.L., Allen R.O., and LeBeau M. A mass spectrometric method for quantitation of intact insulin in blood samples. *J. Anal. Toxicol.* 25, 8, 2001.

143. Zhu Y., Meyer D., Lem Z., and Chien B. Simultaneous quantitation of human insulin, its analogue, and their catabolites in plasma. Poster at the 49th ASMS Conference, Chicago, 2001.

144. Farenc C., Enjalbal C., Sanchez P., Bressolle F., Audran M., Martinez J., and Aubagnac J.L. Quantitative determination of rocuronium in human plasma by liquid chromatography-electrospray ionization mass spectrometry. *J. Chromatogr. A* 910, 61, 2001.

145. Ballard K.D., Vickery W.E., Citrino R.D., Diamond F.X., and Rieders F. Analysis of quarternary ammonium neuromuscular blocking agents in a forensic setting using LC-MS/MS with internal standardization of a Q-TOF. Poster presented at 49th ASMS Conference, Chicago, 2001.

146. Kerskes C.H., Lusthof K.J., Zweipfenning P.G., and Franke J.P. The detection and identification of quaternary nitrogen muscle relaxants in biological fluids and tissues by ion-trap LC-ESI-MS. *J. Anal. Toxicol.* 26, 29, 2002.

147. Kumazawa T., Seno H., Suzuki-Watanabe K., Hattori H., Ishii A., Sato K., and Suzuki O. Determination of phenothiazines in human body fluids by solid phase microextraction and liquid chromatography-tandem mass spectrometry. *J. Mass Spectrom.* 35, 1091, 2000.

148. Zhang H., Heinig K., and Henion J. Atmospheric pressure ionization time-of-flight mass spectrometry coupled with fast liquid chromatography for quantitation and accurate mass measurement of five pharmaceutical drugs in human plasma. *J. Mass Spectrom.* 35, 423, 2000.

149. Kollroser M. and Schober C. Simultaneous determination of seven tricyclic antidepressant drugs in human plasma by direct-injection HPLC-APCI-MS-MS with an ion trap detector. *Ther. Drug Monit.* 24, 537, 2002.

150. Müller C., Vogt S., Goerke R., Kordon A., and Weinmann W. Identification of selected psychopharmaceuticals and their metabolites in hair by LC/ESI-CID/MS and LC/MS/MS. *Forensic Sci. Int.* 113, 415, 2000.

151. McClean S., O'Kane E.J., and Smyth W.F. Electrospray ionization-mass spectrometric characterization of selected anti-psychotic drugs and their detection and determination in human hair samples by liquid chromatography-tandem mass spectrometry. *J. Chromatogr. B* 740, 141, 2000.

152. Hoskins J.M., Gross A.S., Shenfield G.M., and Rivory L.P. High-performance liquid chromatography-electrospray ionization mass spectrometry method for the measurement of moclobemide and two metabolites in plasma. *J. Chromatogr. B* 754, 319, 2001.

153. Sutherland F.C., Badenhorst D., de Jager A.D., Scanes T., Hundt H.K. Swart K.J., and Hundt A.F. Sensitive liquid chromatographic-tandem mass spectrometric method for the determination of fluoxetine and its primary active metabolite norfluoxetine in human plasma. *J. Chromatogr. A* 914, 45, 2001.
154. Shen Z., Wang S., and Bakhtiar R. Enantiomeric separation and quantification of fluoxetine (Prozac) in human plasma by liquid chromatography/tandem mass spectrometry using liquid-liquid extraction in 96-well plate format. *Rapid Commun. Mass Spectrom.* 16, 332, 2002.
155. Li C., Nan F., Shao Q., Liu P., Dai J., Zhen J., Yuan H., Xu F., Cui J., Huang B., Zhang M, and Yu C. Liquid chromatography/tandem mass spectrometry for the determination of fluoxetine and its main active metabolite norfluoxetine in human plasma with deuterated fluoxetine as internal standard. *Rapid Commun. Mass Spectrom.* 16, 1844, 2002.
156. Berna M., Shugert R, and Mullen J. Determination of olanzapine in human plasma and serum by liquid chromatography/tandem mass spectrometry. *J. Mass Spectrom.* 33, 1003, 1998.
157. Berna M., Ackermann B., Ruterbories K., and Glass S. Determination of olanzapine in human blood by liquid chromatography-tandem mass spectrometry. *J. Chromatogr. B* 767, 163, 2002.
158. Bogusz M.J., Krüger K.D., Maier R.D., Erkwoh R., and Tuchtenhagen F. Monitoring of olanzapine in serum by liquid chromatography-atmospheric pressure chemical ionization mass spectrometry. *J. Chromatogr. B* 732, 257, 1999.
159. Aravagiri M. and Marder S.R. Simultaneous determination of clozapine and its *N*-desmethyl and *N*-oxide metabolites in plasma by liquid chromatography/electrospray tandem mass spectrometry and its application to plasma level monitoring in schizophrenic patients. *J. Pharm. Biomed. Anal.* 26, 301, 2001.
160. Aravagiri M. and Marder S.R. Simultaneous determination of risperidone and 9-hydroxyrisperidone in plasma by liquid chromatography/electrospray tandem mass spectrometry. *J. Mass Spectrom.* 35, 718, 2000.
161. Goeringer K.E., McIntyre I.M., and Drummer O.H. Postmortem tissue concentrations of venlafaxine. *Forensic Sci. Int.* 121, 70, 2001.
162. Ramos L., Bakhtiar R., Majumdar T., Hayes M., and Tse F. Liquid chromatography/atmospheric pressure chemical ionization tandem mass spectrometry enantiomeric separation of *dl-threo*-methylphenidate, (Ritalin®) using a macrocyclic antibiotic as the chiral selector. *Rapid Commun. Mass Spectrom.* 13, 2054, 1999.
163. Ramos L., Bakhtiar R., and Tse F. Liquid-liquid extraction using 96-well plate format in conjunction with liquid chromatography/tandem mass spectrometry for quantitative determination of methylphenidate (Ritalin®) in human plasma. *Rapid Commun. Mass Spectrom.* 14, 740, 2000.

164. Oertel R., Richter K., Ebert U., and Kirch W. Determination of scopolamine in human serum and microdialysis samples by liquid chromatography-tandem mass spectrometry. *J. Chromatogr. B* 750, 12, 2001.

165. Xia Y., McGuffey J.E., Wang L., Sosnoff C.S., and Bernert J.T. Analysis of urinary nicotine metabolite profiles by LC atmospheric pressure ionization tandem mass spectrometry. Poster at the 49th ASMS Conference, Chicago, 2001.

166. Goldmeier D. and Lamba H. Prolonged erections produced by dihydrocodeine and sildenafil. *Brit. Med. J.* 324, 1555, 2002.

167. Weinmann W., Lehmann N., Miller C., Wiedmann A., and Svoboda M. Postmortem detection and identification of sildenafil (Viagra) and its metabolites by LC/MS and LC/MS/MS. *Int. J. Legal Med.* 114, 252, 2001.

168. Dumestre-Toulet V., Cirimele V., Ludes B., Beloousoff T., Gromb S., and Kintz P. Last performance with sildenafil (Viagra). Poster at the 39th International TIAFT Meeting, Prague, 2001.

169. Rudy J.A., Uboh C.E., Soma L., Birks E., Kahler M., Tsang D., Watson A., and Teleis D. Identification of sildenafil and metabolites in equine plasma and urine by LC-MS-MS. Poster presented at 49th ASMS Conference, Chicago, 2001.

170. Guan F., Ishii A., Seno H., Watanabe-Suzuki K., Kumazawa T., and Suzuki O. Identification and quantification of cardiac glycosides in blood and urine sample by HPLC/MS/MS. *Anal. Chem.* 71, 4034, 1999.

171. Lacassie E., Marquet P., Martin-Dupont S., Gaulier J.M., and Lachatre G. A non-fatal case of intoxication with foxglove, documented by means of liquid chromatography-electrospray mass spectrometry. *J. Forens. Sci.* 45, 1154, 2000.

172. Gaillard Y. and Pepin G. LC-EI-MS determination of veratridine and cevadine in two fatal cases of *Veratrum album* poisoning. *J. Anal. Toxicol.* 25, 485, 2001.

173. Wang X., Plomley J.B., Newman R.A., and Cisneros A. LC/MS/MS analysis of an oleander extract for cancer treatment. *Anal. Chem.* 72, 3547, 2000.

174. Tracqui A., Kintz P., Branche F., and Ludes B. Confirmation of oleander poisoning by HPLC/MS. *Int. J. Legal Med.* 111, 32, 1998.

175. Arao T., Fuke C., Takaesu H., Nakamoto M., Morinaga Y., and Miyazaki T. Simultaneous determination of cardenolides by sonic spray ionization liquid chromatography-ion trap mass spectrometry — a fatal case of oleander poisoning. *J. Anal. Toxicol.* 26, 222, 2002.

176. Reilly C.A., Crouch D.J., Yost G.S., and Fatah A.A. Determination of capsaicin, dihydrocapsaicin, and nonivamide in self-defense weapons by liquid chromatography-mass spectrometry and liquid chromatography-tandem mass spectrometry. *J. Chromatogr. A* 912, 259, 2001.

177. Reilly C.A., Crouch D.J., Yost G.S., and Fatah A.A. Determination of capsaicin, nonivamide, and dihydrocapsaicin, in blood and tissue by liquid chromatography-tandem mass spectrometry. *J. Anal. Toxicol.* 26, 313, 2002.

178. Darby S.M., Miller M.L., and Allen R.O. Forensic determination of ricin and the alkaloid marker ricinine from castor bean extracts. *J. Forensic Sci.* 46, 1033, 2001.

179. Baselt R.C. *Disposition of Toxic Drugs and Chemicals in Man*. Chemical Toxicology Institute, Foster City, CA, 2000, p. 215.

180. Tracqui A., Kintz P., Ludes B., Rouge C., Douibi H., and Mangin P. High-performance liquid chromatography coupled to ion spray mass spectrometry for the determination of colchicine at ppb levels in human biofluids. *J. Chromatogr. B* 675, 235, 1996.

181. Jones G.R., Singer P.P., and Bannach B. Application of LC-MS analysis to a colchicine fatality. *J. Anal. Toxicol.* 26, 365, 2002.

182. Maurer H.H., Schmitt C.J., Weber A.A., and Kraemer T. Validated electrospray liquid chromatographic-mass spectrometric assay for the determination of the mushroom toxins alpha- and beta-amanitin in urine after immunoaffinity extraction. *J. Chromatogr. B* 748, 125, 2000.

183. Smith R.M. Retention index scales used in high-performance liquid chromatography. In R.M. Smith, Ed., *Retention and Selectivity in Liquid Chromatography, Journal of Chromatography Library*, Vol. 57, Elsevier, Amsterdam, 1995, p. 93.

184. Bogusz M.J. Application of nitroalkanes and secondary retention index standards for the identification of drugs, in R.M. Smith, Ed., *Retention and Selectivity in Liquid Chromatography, Journal of Chromatography Library*, Vol. 57, Elsevier, Amsterdam, 1995, p. 171.

185. Bogusz M.J., Hill D.W., and Rehorek A. Comparability of RP-HPLC retention indices of drugs in three databases. *J. Liq. Chromatogr.* 19, 1291, 1996.

186. Elliott S.P. and Hale K.A. Application of an HPLC-DAD drug screening system based on retention indices and UV spectra. *J. Anal. Toxicol.* 22, 279, 1998.

187. Bogusz M.J., Maier R.D., Krüger K.D., Webb K.S., Romeril J., and Miller M.L. Poor reproducibility of in-source collisional atmospheric pressure ionization mass spectra of toxicologically relevant drugs. *J. Chromatogr. B* 844, 409, 1999.

188. Gergov M., Robson J.N., Duchoslav E., and Ojanpera I. Automated liquid chromatographic/tandem mass spectrometric method for screening beta-blocking drugs in urine. *J. Mass Spectrom.* 35, 912, 2000.

189. Gergov M., Robson J.N., Ojanpera I., Heinonen O.P., and Vuori E. Simultaneous screening and quantitation of 18 antihistamine drugs in blood by liquid chromatography ionspray tandem mass spectrometry. *Forensic Sci. Int.* 121, 108, 2001.

190. Rittner M., Pragst F., Bork W.R., and Neumann J. Screening method for seventy psychoactive drugs or drug metabolites in serum based on high-performance liquid chromatography-electrospray ionization mass spectrometry. *J. Anal. Toxicol.* 25, 112, 2001.

191. Thieme D., Grosse J., Lang R., Müller R.K., and Wahl A. Screening, confirmation and quantitation of diuretics in urine for doping control analysis by high-performance liquid chromatography-atmospheric pressure ionization tandem mass spectrometry. *J. Chromatogr. B* 757, 49, 2001.

192. Lacassie E., Marquet P., Gaulier J.M., Dreyfuss M.F., and Lachatre G. Sensitive and specific multiresidue methods for the determination of pesticides of various classes in clinical and forensic toxicology. *Forensic Sci. Int.* 121, 116, 2001.

193. Marquet P., Dupuy J.L., and Lachaitre G. Development of a general unknown screening procedure using liquid chromatography-ionspray-mass spectrometry. Poster at the 46th ASMS Conference, Orlando, FL, 1998.

194. Marquet P., Venisse N., Lacassie E., and Lachatre G., In-source CID mass spectral libraries for the "general unknown" screening of drugs and toxicants. *Analysis* 28, 41, 2000.

195. Venisse N., Marquet P., Duchoslav E., Dupuy J.L., and Lachatre G. A general unknown procedure for drugs and toxic compounds in serum using liquid chromatography-electrospray-single quadrupole mass spectrometry. *J. Anal. Toxicol.* 7, 7, 2003.

196. Saint-Marcoux F., Lachatre G., and Marquet P. Evaluation of an improved general unknown screening procedure using liquid chromatography-electrospray mass spectrometry by comparison with gas chromatography and high performance liquid chromatography-diode array detection. *J. Am. Soc. Mass Spectrom.* 14, 14, 2003.

197. Weinmann W., Wiedemann A., Eppinger B., Renz M., and Svoboda M. Screening for drugs in serum by electrospray ionization/collision-induced dissociation and library searching. *J. Am. Soc. Mass Spectrom.* 10, 1028, 1999.

198. Weinmann W., Stoertzel M., Vogt S., and Wendt J. Tune compounds for electrospray ionization/in source collision-induced dissociation with mass spectra library searching. *J. Chromatogr. A* 926, 199, 2001.

199. Weinmann W., Stoertzel M., Vogt S., Svoboda M., and Wendt J. Tuning compounds for electrospray ionization/in source collision-induced dissociation and mass spectra library searching. *J. Mass Spectrom.* 36, 1013, 2001.

200. Mueller C., Schaefer P., Vogt S., and Weinmann W. Combining an ESI-CID mass spectra and a UV-spectra library of drugs with an Access database for clinical and forensic-toxicological analysis. Poster at 50th ASMS Conference, Orlando, FL, 2002.

201. Lips A.G., Lameijer W., Fokkens R.H., and Nibbering N.M. Methodology for the development of a drug library based upon collision-induced fragmentation for the identification of toxicologically relevant drugs in plasma samples. *J. Chromatogr. B* 759, 191, 2001.

202. Gergov M., Boucher B., Ojanpera I., and Vuori E. Toxicological screening of urine for drugs by liquid chromatography/time-of-flight mass spectrometry with automated target library search based on elemental formulas. *Rapid Commun. Mass Spectrom.* 15, 521, 2001.

203. Pelander A., Ojanpera I., Gergov M., and Vuori E. Qualitative screening analysis of autopsy urine samples by improved LC/TOFMS method. Poster at the 40th TIAFT Meeting, Paris, 2002.
204. Weinmann W., Gergov M., and Goerner M. MS/MS-libraries with triple quadrupole-tandem mass spectrometers for drug identification and drug screening. *Analysis* 28, 934, 2000.
205. Gergov M., Ojanpera I., and Vuori E. Simultaneous screening for 238 drugs in blood by liquid chromatography-ionspray/tandem mass spectrometry with multiple action monitoring. *J. Chromatogr. B*, 795, 41, 2003.
206. Dams R., Murphy C.M., Choo R., Lambert W., and Huestis M. Development of a rugged and sensitive LC-APCI-MS/MS method combined with as simplified sample preparation for the comprehensive analysis of drugs of abuse of biological fluids. Poster at the 50th ASMS Conference, Orlando, FL, 2002.
207. Nordgren H. and Beck O. Use of LC-MS-MS for direct detection of drugs of abuse in diluted urine. Poster at the 40th TIAFT Meeting, Paris, 2002.
208. Fitzgerald R.L., Rivera J.D., and Herold D.A. Broad spectrum drug identification directly from urine, using liquid chromatography-tandem mass spectrometry. *Clin. Chem.* 45, 1224, 1999.

Substance Abuse in Sports: Detection of Doping Agents in Hair by Mass Spectrometry

3

MARION VILLAIN
VINCENT CIRIMELE
PASCAL KINTZ

Contents

3.1 Introduction

It is generally accepted that chemical testing of biological fluids is the most objective means of diagnosis of drug use. The presence of a drug analyte in a biological specimen can be used as evidence of recent exposure. The standard in drug testing is immunoassay screening, followed by gas chromatographic–mass spectrometric (GC/MS) confirmation conducted on a urine sample. More recently, a variety of body specimens other than urine, such as saliva, sweat, meconium, or hair have been proposed to document drug exposure.

0-8493-1522-0/04/$0.00+$1.50

In the 1960s and 1970s, hair analysis was used to evaluate exposure to toxic heavy metals, such as arsenic, lead, or mercury; this was due to atomic absorption that allowed detection in the nanogram range. At that time, examination of hair for organic substances, especially drugs, was not possible because analytical methods were not sensitive enough.

Examination by means of drugs marked with radioactive isotopes, however, established that these substances can move from blood to hair and are deposited there. Ten years after these first investigations, it became possible both in the U.S. and West Germany to denominate various organic drugs by means of the radioimmunologic assay (RIA) technique.

In 1979, Baumgartner and colleagues[1] published the first report on the detection of morphine in the hair of heroin abusers using RIA. They found that differences in the concentration of morphine along the hair shaft correlated with the time of drug use.

Today, gas chromatography coupled with mass spectrometry (GC/MS) is the method of choice for hair analysis, and this technology is routinely used to document repetitive drug exposure in forensic sciences, traffic medicine, occupational medicine, clinical toxicology, and more recently in sports.

The major practical advantage of hair testing compared to urine or blood testing for drugs is that it has a larger surveillance window (weeks to months, depending on the length of the hair shaft, against 2 to 4 d for most xenobiotics). For practical purposes, the two tests complement each other. Urinalysis and blood analysis provide short-term information of an individual's drug use, whereas long-term histories are accessible through hair analysis. While analysis of urine and blood specimens cannot distinguish between chronic use or single exposure, hair analysis can offer this distinction. Table 3.1 summarizes major characteristics of each specimen in regard to its place in doping control.

Table 3.1 Comparison between Urine and Hair for Testing Doping Agents in Sports

Parameters	Urine	Hair
Drugs	All, except some peptidic hormones	All, except hormones
Major compound	Metabolites	Parent drug
Detection period	2–5 d, except anabolic steroids	Weeks, months
Type of measure	Incremental	Cumulative
Screening	Yes	No
Invasiveness	High	Low
Storage	–20°C	Ambient temperature
Risk of false negative	High	Low
Risk of false positive	Low	Undetermined
Risk of adulteration	High	Low
Control material	Yes	Needed

It is generally proposed that drugs can enter into hair by two processes: adsorption from the external environment and incorporation into the growing hair shaft from blood supplying the hair follicle. Drugs can enter the hair from exposure to chemicals in aerosols, smoke, or secretions from sweat and sebaceous glands. Sweat is known to contain drugs present in blood. Because hair is very porous and can increase its weight up to 18% by absorbing liquids, drugs may be transfered easily to hair via sweat.

Drugs appear to be incorporated into the hair during at least three stages: from the blood during hair formation, from sweat and sebum, and from external environment.

From various studies, it has been demonstrated that after the same dosage, black hair incorporates much more of a drug than blond hair.[2,3] This has resulted in discussions about a possible racial bias of hair analysis and is still under evaluation.

The possibility of racial bias due to differences in melanin concentrations or in hair porosity is still under discussion. Melanins are responsible for the color of hair. Two types of melanin are present: eumelanin (with low sulfur content) and pheomelanin (with high sulfur content). Black and brown hair contain more eumelanin than red and blond hair. It appears that it is not simply the concentration of drugs in blood that determines the concentration in hair. Numerous factors may influence the incorporation of drugs into hair, such as the nature of the compounds (pKa, lipid solubility, and metabolism pattern) and variation in hair growth cycles. Until these mechanisms are elucidated, the quantitative results and extrapolation to the amount of drug intake from such a hair analysis should be considered with extreme caution.[4]

3.2 Procedures

3.2.1 Specimen Collection

Collection procedures for hair analysis for drugs have not been standardized. In most published studies, the samples are obtained from random locations on the scalp. Hair is best collected from the area at the back of the head, called the *vertex posterior.* Compared with other areas of the head, this area has less variability in the hair growth rate, the number of hairs in the growing phase is more constant, and the hair is less subject to age- and sex related influences. Hair strands are cut as close as possible from the scalp, and the location root-tip must be mentioned. Storage is achieved at ambiant temperature in aluminium foil, an envelope, or a plastic tube. The sample size varies considerably among laboratories and depends on the drug to be analyzed and the test methodology. For example, when nandrolone or betamethasone is investigated, a 100 mg sample is recommended. Sample sizes reported in the literature range from a single hair to 200 mg, cut as close to the scalp as possible. When sectional analysis is performed, the hair is cut into segments of about 1, 2, or 3 cm, which corresponds to about 1, 2, or 3 months' growth.

3.2.2 Decontamination Procedures

Contaminants of hair would be a problem if they were drugs of abuse or their metabolites, or if they interfered with the analysis and interpretation of the test results. It is unlikely that people would intentionally or accidentally apply anything to their hair that would contain a drug of abuse. The most crucial issue facing hair analysis is the avoidance of technical and evidentiary false-positives. Technical false-positives are caused by errors in the collection, processing, and analysis of specimens, while evidentiary false-positives are caused by passive exposure to the drug. Various approaches for preventing evidentiary false-positives due to external contamination of the hair specimens have been described.

Most but not all laboratories use a wash step; however, there is no consensus or uniformity in the washing procedures. Among the agents used in washing are detergents such as Prell shampoo, surgical scrubbing solutions, surfactants such as 0.1% sodium dodecylsulfate, phosphate buffer, or organic solvents such as acetone, diethyl ether, methanol, ethanol, dichloromethane, hexane, or pentane of various volumes for various contact times. Generally, a single washing step is realized; sometimes a second identical wash is performed. As opposed to crack, cannabis, or smoked heroin, decontamination when testing for doping agents does not appear to be a critical need.

3.3 Analysis of Doping Agents

3.3.1 Detection of Anabolic Steroids

Athletes use both endogenous (testosterone, DHEA) or exogenous (nandrolone, stanozolol, mesterolone, etc.) anabolic steroids because it has been claimed that they increase lean body mass, increase strength, increase aggressiveness, and lead to a shorter recovery time between workouts.

The first data available for endogenous steroids in hair were given late in 1995 by the German group of Scherer and Reinhardt,[5] who determined by GC/MS androstenediol (9 to 19 pg/mg), testosterone (13 to 24 pg/mg), androstenedione (5 to 15 pg/mg), DHEA (21 to 56 pg/mg), dihydrotestosterone (2 to 8 pg/mg) and 17alpha-hydroxy-progesterone (1 to 7 pg/mg). Some years later, Kintz et al.[6,7] established the physiological concentrations of both testosterone and DHEA with a distinction between the hair of male and female subjects. After decontamination with dichloromethane, 100 mg of hair was incubated in 1 *M* NaOH in the presence of testosterone-d_3. After neutralization, the extract was purified using solid-phase extraction with Isolute C18 columns followed by liquid–liquid extraction with pentane. After silylation, the drugs were analyzed by GC/MS. Concentrations for DHEA were in the range 1 to 7 pg/mg (mean 4 pg/mg) and 0.5 to 11 pg/mg (mean 5 pg/mg) for the males (n = 15) and females (n = 12), respectively. Concentrations for testosterone were in the range 0.5 to 12 pg/mg (mean 4 pg/mg) and not detected to 2 pg/mg for the males (n = 41) and females (n = 12), respectively.

Unlike testosterone in urine, the interpretation of concentration findings in hair can be difficult and critical. The range between physiological concentrations of testosterone and those found in abusers seems to be rather small. Therefore, in complement of testosterone determination, the identification of unique testosterone esters in hair enables an unambiguous charge for doping because the esters are certainly exogenous substances. This was largely developed by Thieme et al.[8] and Gaillard et al.[9] Recently Rivier[10] claimed that although hair analysis alone cannot be useful for screening purposes, it could become in the future a possibly useful technique for obtaining additional information on long-term testosterone abuse.

Thieme et al.[8] published in 2000 a complete analytical strategy for detecting anabolics in hair. The preparation of the sample was carried out by a methanol extraction step with sonication for all the anabolics except for stanozolol which was incubated in NaOH. Extensive cleanup procedures were employed such as HPLC and solid-phase extraction, followed by derivatization to form the enol-TMS derivatives. Drugs were identified either by GC/MS/MS or GC/HRMS. Metandienone and its metabolite, 6β-hydroxy-metandienone, stanozolol and its metabolite 3′-hydroxy-stanozolol, mesterolone, metenolone enantate, and nandrolone decanoate, and several testosterone esters such as propionate, isocaproate, decanoate, and phenylpropionate were identified in the hair of several bodybuilders.

Gaillard et al.[9] developed a method for testing both the anabolic steroids and their esters. A 100-mg amount of powdered hair was first treated with methanol for extraction of esters, then alkaline digested with 1 *M* NaOH for the recovery of the other drugs. These preparations were extracted with ethyl acetate, pooled, then finally highly purified using a twin solid-phase extraction on amino and silica cartridges. After silylation, drugs were detected by GC/MS/MS. Figure 3.1 shows the MS/MS chromatograms of an extract of a human hair spiked with 50 pg/mg of each one of a series of drugs. Nandrolone and testosterone undecanoate were identified in the hair of 2 athletes at 5.1 and 15.2 pg/mg, respectively.

A sensitive, specific, and reproducible method for the quantitative determination of stanozolol in human hair was developed by Cirimele et al.[11] The

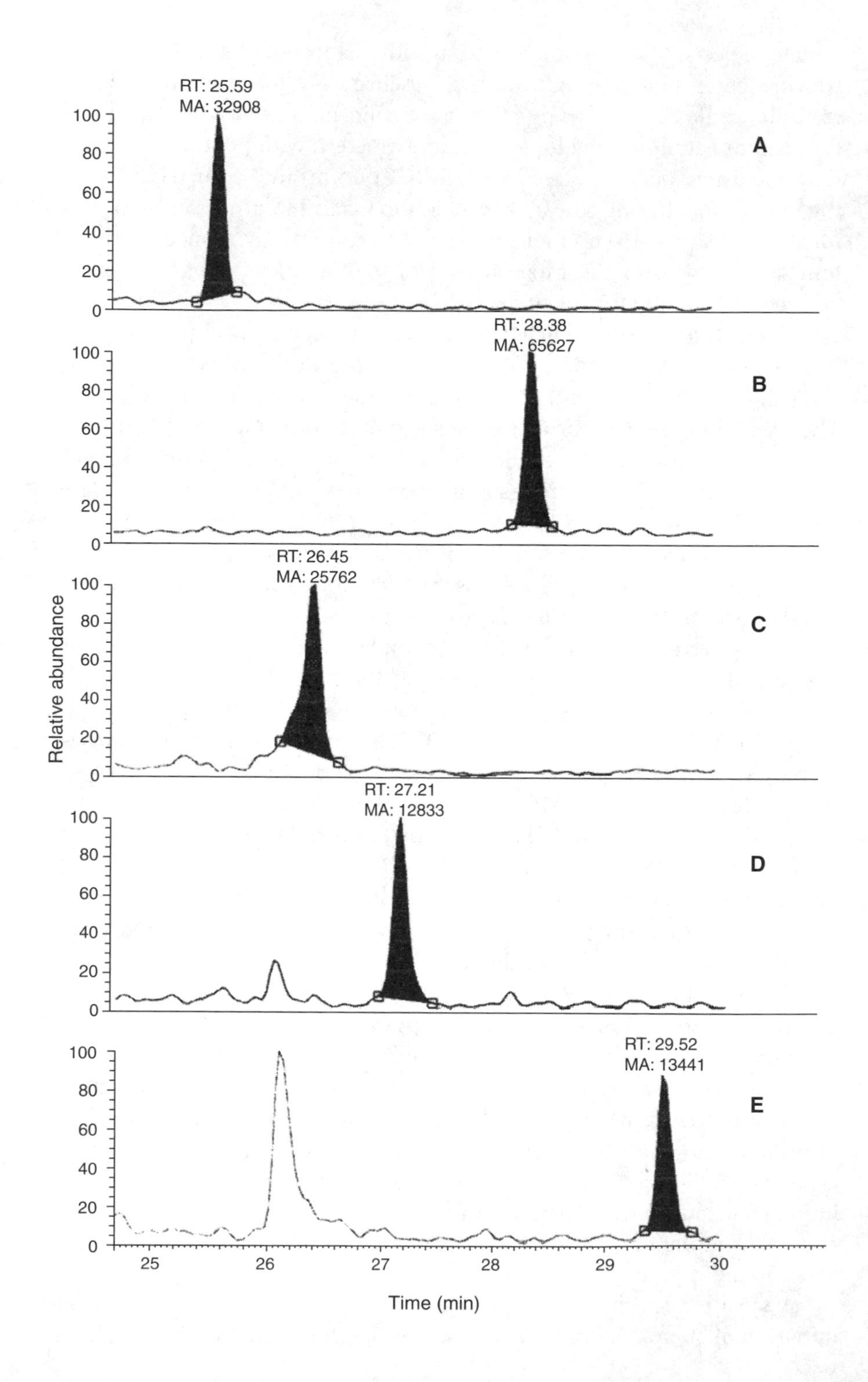
RT: 25.59
MA: 32908
A
RT: 28.38
MA: 65627
B
RT: 26.45
MA: 25762
C
RT: 27.21
MA: 12833
D
RT: 29.52
MA: 13441
E
Relative abundance
Time (min)

sample preparation involved a decontamination step of the hair with methylene chloride and the sonication in methanol of 100 mg of powdered hair for 2 h. After elimination of the solvent, the hair sample was solubilized in 1 ml 1 *N* NaOH, 15 min at 95°C, in the presence of 10 ng stanozolol-d_3 used as internal standard. The homogenate was neutralized and extracted using consecutively a solid phase (Isolute C18) and a liquid–liquid (pentane) extraction. After evaporation of the final organic phase, the dry extract was derivatized using 40 µl MBHFA/TMSI (1000:20, v/v), incubated for 5 min at 80°C, followed by 10 µl of MBHFBA, incubated for 30 min at 80°C. The derivatized extract was analyzed by a Hewlett–Packard GC/MS system with a 5989 B Engine operating in negative chemical ionization (NCI) mode of detection. Figure 3.2 shows the mass spectrum of an HFB-derivative of stanozolol in the NCI mode. The assay was able to detect 2 pg of stanozolol per mg of hair when approximately 100 mg hair material were processed. The analysis of a 3-cm-long hair strand, obtained from a young bodybuilder (27 years old) declaring to be a regular user of Winstrol® (stanozolol, 2 mg), revealed the presence of stanozolol at the concentration of 15 pg/mg (Figure 3.3).

More recently, a method for the quantitative determination of methenolone in human hair was developed.[12] The sample preparation involved a decontamination step of the hair with methylene chloride. The hair sample (about 100 mg) was solubilized in 1 ml NaOH 1 *N*, 15 min at 95°C, in the presence of 1 ng testosterone-d_3 used as internal standard. The homogenate was neutralized and extracted using consecutively a solid phase (Isolute C18 eluted with methanol) and a liquid–liquid (pentane) extraction. The residue was derivatized by adding 50 µl MSTFA/NH_4I/2-mercaptoethanol (1000:2:5, v/v/v), then incubated for 20 min at 60°C. A 1.5-µl aliquot of the derivatized extract was injected into the column (HP5-MS capillary column, 5% phenyl-95% methylsiloxane, 30 m × 0.25 mm i.d. × 0.25 mm film thickness) of a Hewlett–Packard (Palo Alto, CA) gas chromatograph (6890 Series). Methenolone was detected by its parent ion at *m/z* 446 and daughter ions at *m/z* 208 and 195 by means of a Finnigan TSQ 700 MS/MS system. The assay was capable of detecting 1 pg/mg of methenolone when approximately 100 mg

---

Figure 3.1 **(See figure on facing page.)** MS/MS chromatograms of an extract of a human hair spiked with 50 pg/mg each of a series of drugs. Peaks: A = testosterone cypionate, sum of ions *m/z* 209 + 469, daughter ions of *m/z* 484; B = testosterone phenyl propionate, sum of ions *m/z* 209 + 477, daughter ions of *m/z* 492; C = nandrolone decanoate, sum of ions *m/z* 182 + 194, daughter ions of *m/z* 500; D = testosterone decanoate, sum of ions *m/z* 209 + 499, daughter ions of *m/z* 514; E = testosterone undecanoate, sum of ions *m/z* 209 + 513, daughter ions of *m/z* 528. (Reprinted from Gaillard, Y. et al., *J. Chromatogr. B*, 735, 189, 1999. Copyright 1999, with permission from Elsevier Science.)

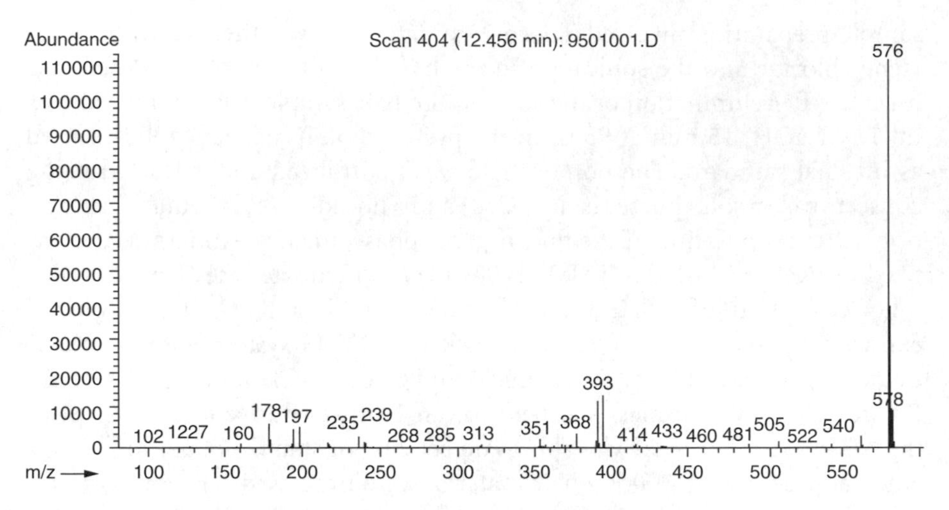

Figure 3.2 Mass spectrum of an HFB-derivative of stanozolol in the NCI mode, with methane. (Reprinted from Cirimele, V., Kintz, P., and Ludes, B., *J. Chromatogr. B*, 740, 265, 2000. Copyright 2000, with permission from Elsevier Science.)

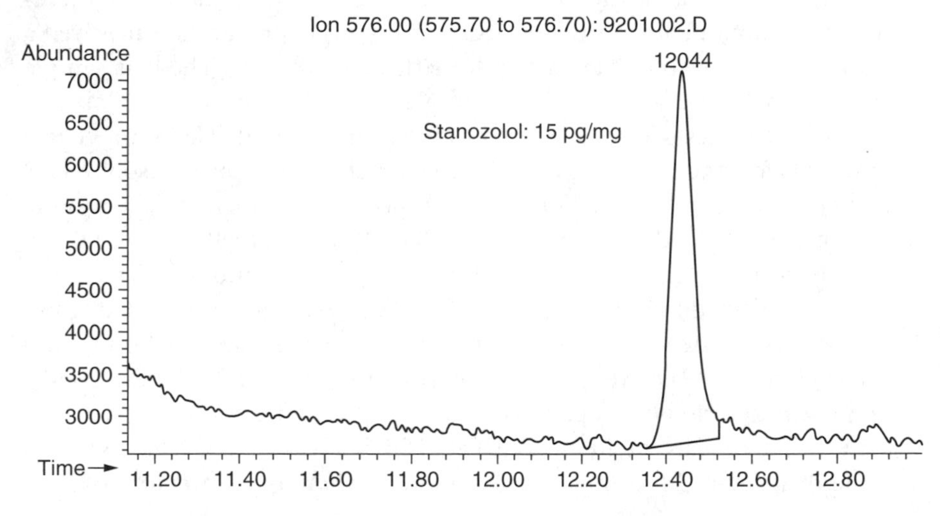

Figure 3.3 SIM mass chromatogram of a hair specimen of a bodybuilder, revealing the presence of stanozolol. (Reprinted from Cirimele, V., Kintz, P., and Ludes, B., *J. Chromatogr. B*, 740, 265, 2000. Copyright 2000, with permission from Elsevier Science.)

of hair material were used. The analysis of a strand of hair obtained from two bodybuilders revealed the presence of methenolone at the concentration of 7.3 and 8.8 pg/mg. Figure 3.4 shows the MS/MS-single reaction monitoring (SRM) mass chromatograms of a 100-mg hair specimen of one of the bodybuilders, revealing the presence of methenolone.

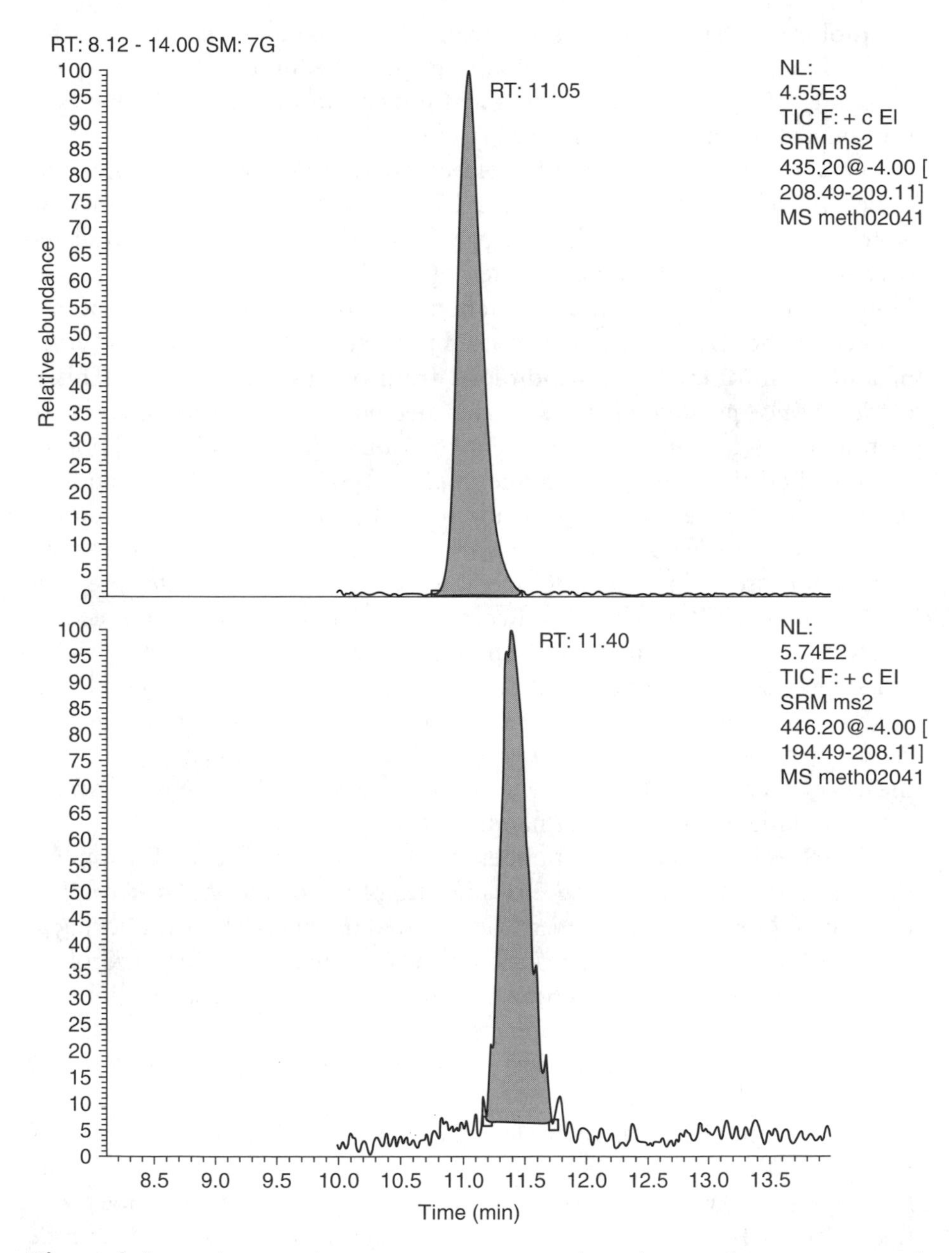

Figure 3.4 MS/MS-single reaction monitoring (SRM) mass chromatograms of a 100-mg hair specimen of a bodybuilder, revealing the presence of methenolone. (Reprinted from Kintz, P. et al., *J. Chromatogr. B*, 766, 161, 2001. Copyright 2001, with permission from Elsevier Science.)

Using quite the same method (except for the internal standard where nandrolone-d_3 was used), the same group developed a procedure to test for nandrolone, the most abused anabolic agent.[13] The limit of detection of the assay was 0.5 pg/mg. Nandrolone tested positive in the hair of 3 athletes at the concentration of 1, 3.5 and 7.5 pg/mg.

Nandrolone is metabolized to norandrosterone and noretiocholanolone. Other 19-norsteroids, such as norandrostenedione or norandrostenediol, which are classified as anabolic androgenic steroids by the IOC, are available over the counter or through the Internet and have the same metabolites as nandrolone. Although norandrostenediol and norandrostenedione are banned by the IOC, there is a great need in forensic science and for a survey of athletes to discriminate nandrolone from other 19-norsteroids. This is obviously not possible in urine, as the metabolites are common. Hair can identify the exact nature of the parent compound (e.g., nandrolone, norandrostenediol, or norandrostenedione, in case of positive urine for norandrosterone), as it has been accepted by the scientific community that the parent compound is the major analyte that is incorporated in hair. Thus, hair analysis would discriminate nandrolone abuse from over-the-counter preparations containing 19-norsteroids. Recently, our laboratory was requested by an attorney to evaluate potential doping practices by an athlete. The 30-year-old subject tested positive for norandrosterone in urine at 230 ng/ml. The analysis was done in an accredited laboratory, but the athlete denied the result. The analysis of a strand of hair obtained from the athlete revealed the presence of 19-norandrostenedione at the concentration of 7 pg/mg,[14] making an unique distinction with nandrolone doping.

In 1999,[15] two male bodybuilders were arrested by French Customs in possession of 2050 tablets and 251 ampules of various anabolic steroids. It was claimed that the steroids were for personal use and not for trafficking as suggested by the police. Hair from both males were positive for nandrolone (196 and 260 pg/mg), testosterone (46 and 71 pg/mg), and stanozolol (135 and 156 pg/mg), clearly indicating steroids abuse.

In a series of 7 steroid abusers, Deng et al.[16] identified nandrolone (20 pg/mg) and methyltestosterone (170 pg/mg).

More recently, Dumestre-Toulet et al.,[17] in a case of trafficking of doping agents, identified by GC/MS nandrolone (1 to 7.5 pg/mg, n = 3), stanozolol (2 to 84 ng/mg, n = 4), methenolone (17 and 34 pg/mg), testosterone enanthate (0.6 to 18.8 ng/mg, n = 5), and testosterone cypionate (3.3 and 4.8 ng/mg) in the hair of bodybuilders.

As a full example, the procedure to test for endogenous anabolics in this laboratory is described in detail.

The hair was decontaminated twice using 5 ml of methylene chloride for 2 min at room temperature, and then was pulverized in a ball mill. A total of 100 mg of decontaminated hair were incubated in 1 ml 1 *N* NaOH, 15 min at 95°C, in presence of 1 ng of testosterone-d_3 used as internal standard (IS). After cooling, the homogenate was neutralized with 1 ml 1 *M* HCl, and 2 ml of 0.2 *M* phosphate buffer (pH 7.0) were added.

The Isolute C18 columns were conditioned with 3 ml of methanol, followed by 2 ml of deionized water. After sample addition, the columns were washed twice with 1 ml of deionized water. After column drying, analyte elution occured with the addition of 3 aliquots of 0.5 ml of methanol. The eluant was evaporated to dryness under nitrogen flow, and the residue reconstitued in 1 ml of 0.2 *M* phosphate buffer (pH 7.0). A further purification step was achieved by the addition of 100 mg of $Na_2CO_3/NaHCO_3$ (1:10, w/w) and 2 ml of pentane. After agitation and centrifugation, the organic phase was removed and evaporated to dryness. The residue was derivatized by adding 50 μl MSTFA/NH_4I/2-mercaptoethanol (1000/2/5; v/v/v) and then incubated for 20 min at 60°C.

A 2-μl aliquot of the derivatized extract was injected into the column of a Hewlett–Packard gas chromatograph (6890 Series). The flow of carrier gas through the column (HP5-MS capillary column, 5% phenyl-95% methylsiloxane, 30 m × 0.25 mm i.d. × 0.25 mm film thickness) was 1 ml/min.

The injector temperature was 270°C and splitless injection was employed with a split valve off-time of 1 min. The column oven temperature was programmed to rise from an initial temperature of 150°C, maintained for 1 min, to 295°C at 5°C/min.

The detector was a Finnigan TSQ 700 operated in the electron impact mode and in selected reaction monitoring. The parent ions, *m/z* 417, 432, 432, 434, and 435 for DHEA, testosterone, epitestosterone, DHT, and the IS, respectively, were selected in the first quadrupole. The corresponding daughter ions, *m/z* 237 and 327, 196 and 209, 196 and 209, 143 and 195, and 209 for DHEA, testosterone, epitestosterone, DHT, and the IS, respectively, were selected in the third quadrupole after collision with argon at a cell pressure at 0.6 mTorr. The collision offset voltage was –8 V. The electron multiplier was operated at 1900 V.

Results shown in Table 3.2 were obtained from about one hundred hair samples.

Table 3.2 Compendium of Results for Endogenous Anabolics in Hair

Compounds	Mean (pg/mg)	Mini (pg/mg)	Maxi (pg/mg)
Testosterone	8.4	1.5	64.2
Epitestosterone	2.4	0.5	17.6
DHEA	16.9	0.8	94.2
DHT	1.8	0.5	4.2

Figure 3.5 shows a chromatogram representative of an authentic hair specimen with the corresponding concentrations: 2.5, 3.2, 10, and 0.6 pg/mg for testosterone, epitestosterone, DHEA, and DHT, respectively.

3.3.2 Detection of Corticosteroids

Cortisone and hydrocortisone, naturally occuring hormones, influence metabolism, inflammation, and electrolyte and water balance. Their synthetic derivatives are used in therapeutic medicine for their antiinflammatory and immunosuppressive actions. They are used in certain sports to improve the performances of the athletes (euphoria, motor activity).

Cirimele et al.[18] published in 1999 the first identification of such a drug, in this case, prednisone, in the hair of a subject treated for years. A 50-mg hair specimen was incubated overnight in Sorensen buffer, then extracted by solid-phase extraction using an Isolute C18 column. Prednisone was detected by LC/MS at 1280 pg/mg. Using the same preparation technique, and cortisol-d_3 as an internal standard, the same group published[19] several months later a screening procedure for 10 corticosteroids, with detection limits in the range of 30 to 170 pg/mg. Two applications were documented for prednisone and beclomethasone, identified in hair at 140 and 230 pg/mg, respectively.

Using a 2 mm i.d. column, Cirimele et al.[20] demonstrated in 10 patients treated with prednisone a low but not insignifiant correlation ($R^2 = 0.578$, $p < 0.03$) between the total amount of ingested drug and the measured concentrations in hair. The procedure was sensitive enough to detect prednisone in the hair of patients treated with a low 5 mg/d dose.

Repetitive abuse of corticosteroids by athletes can be demonstrated by segmental analysis along the hair shaft, in contrast to ponctual urinalysis. A single treatment of about 1 week will make positive a single 1-cm segment, while long-term abuse will lead to the identification of the corticoid(s) in several segments. For such an application, particularly in the case of longitudinal surveys of athletes, hair analysis appears as the solution of choice to document doping practices. It has been demonstrated[21] that a single oral therapeutic treatment with 4-mg/d betamethasone for 9 consecutive days is detectable through hair analysis. The drug tested at a concentration of 4.7 pg/mg in the corresponding hair segment, whereas no betamethasone could be identified in the distal hair strand. Extraction of the drug was classic from this group, however, to enhance sensitivity a MIC 15 CP Nucleosil C18, 150 × 1 mm i.d. column was used.

A confirmatory method was developed[22] for the quantitative determination in hair of the most common corticosteroids used as doping agents by athletes. Drugs were extracted from 50 mg of powdered hair by methanolic extraction followed by a solid-phase extraction on a C18 column. Detection was performed with an electrospray ionization mass spectrometer in the

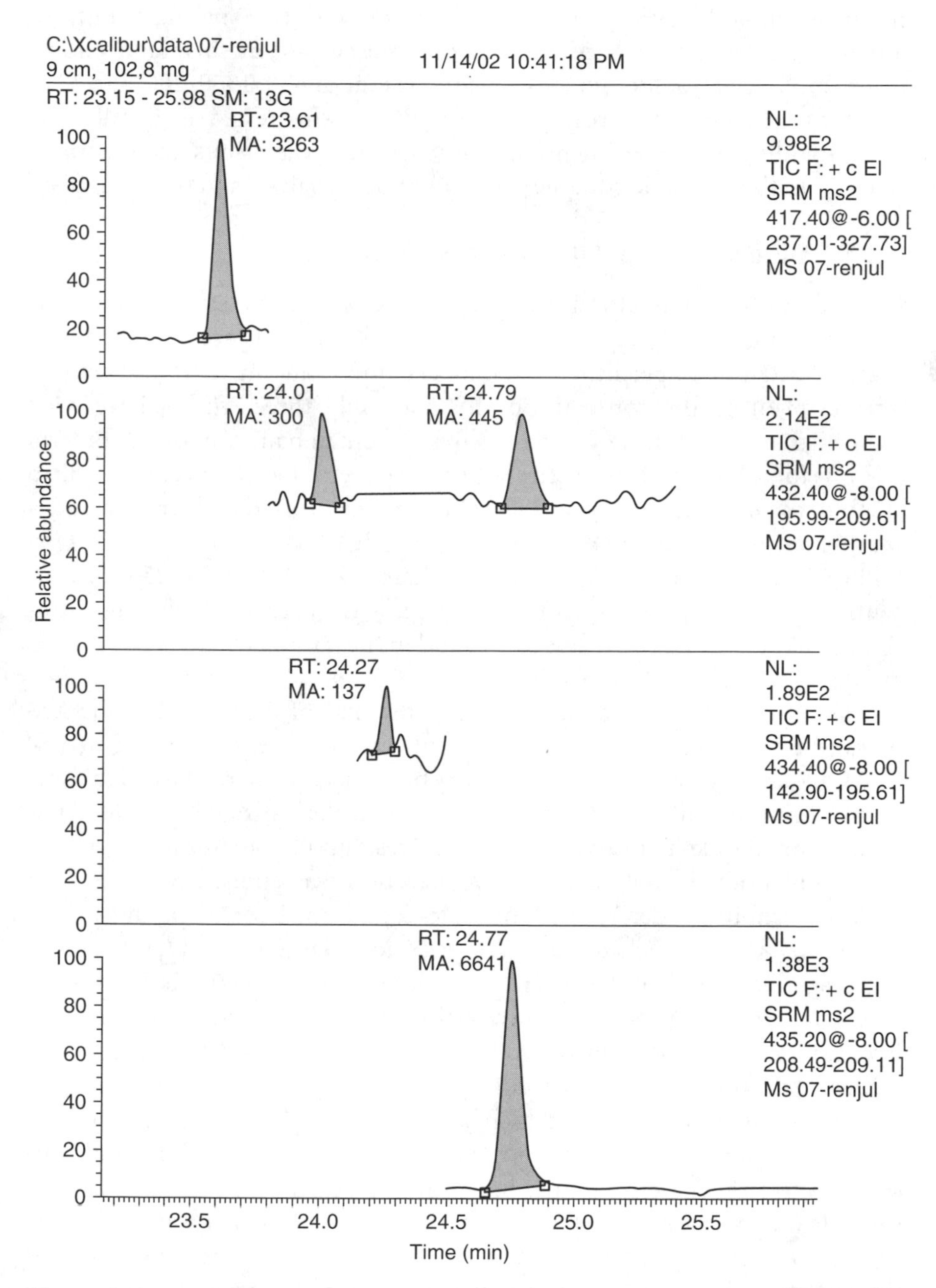

Figure 3.5 GC/MS/MS mass chromatograms of a hair specimen with the corresponding concentrations: 2.5, 3.2, 10.0, and 0.6 pg/mg for testosterone (RT: 24.79 min), epitestosterone (RT: 24.01 min), DHEA (23.61 min) and DHT (24.27 min), respectively.

negative ion mode. The limits of sensitivity were about 100 pg/mg. Hair from athletes revealed the presence of hydrocortisone acetate, methylprednisolone, triamcinolone acetonide, and dexa/betametasone at 430, 1350, 280, and 1310 pg/mg, respectively. According to the authors, who tested in parallel the corresponding urine specimens, the comparison of the results demonstrated once again the dramatic complementary use of urinalysis and hair analysis.[23]

3.3.3 Detection of β-Adrenergic Stimulants

Because of their sympatomimetic properties (stimulant effects) and their activity as anabolic agents at higher dosages, β2-agonists are banned. However, salbutamol is permitted by inhalers only and must be declared in writing prior to the competition. To date, only three studies have been published in the literature for these drugs in human hair, two for clenbuterol and one for salbutamol, respectively. In their paper,[24] Gleixner et al. identified clenbuterol after incubation in 1,4-dithiothreitol, NaOH, and tertiary butylmethyl ether by enzyme immunoassay (EIA), with confirmation by HPLC/EIA. Clenbuterol accumulated in hair after 10 μg/d for 25 d at concentrations ranging from 23 to 161 pg/mg, with relatively high concentrations in dark hair. The drug was also found in the hair from two bodybuilders at 50 and 92 pg/mg, respectively.

Machnik et al.[25] tested clenbuterol in the hair of four females who had therapeutically taken the drug as a tocolyticum. Hair was incubated in 1 *M* KOH, and the drug extracted with tertiary butylmethyl ether, followed in some cases by immunoaffinity chromatography, then derivatizated with MSTFA-ammonium iodide-TMS ethanethiol. High-resolution mass spectrometry was used to identify clenbuterol. Limit of detection was about 0.8 pg/mg. The levels of clenbuterol determined in hair ranged from 2 to 236 pg/mg.

A screening procedure was developed[26] for simultaneous identifications of β2-agonists and β-blockers. The procedure involved overnight incubation in 0.1 *M* HCl, followed by neutralization, solid-phase extraction with an Isolute C18 column, derivatization with trimethylboroxine/ethyl acetate, and GC/MS detection. Limits of detection were 2 pg/mg for both salbutamol and clenbuterol. In nine asthmatic patients, the salbutamol concentrations in hair were in the range 27 to 192 pg/mg. In two asthma deaths, salbutamol concentrations in hair were 210 and 87 pg/mg, respectively. Finally, the laboratory identified salbutamol in the hair of a swimmer, which was positive in urine, at a level of 71 pg/mg. Figure 3.6 shows the mass chromatogram and the electron ionization mass spectrum of the extracted hair specimen.

Most of the positive specimens are reported for salbutamol (46% of the total urines in 1999 for the IOC laboratory in France). However, as this drug is permitted for specific therapeutic purposes, together with a medical prescription, it appears very easy to evade the test, and almost all cases are

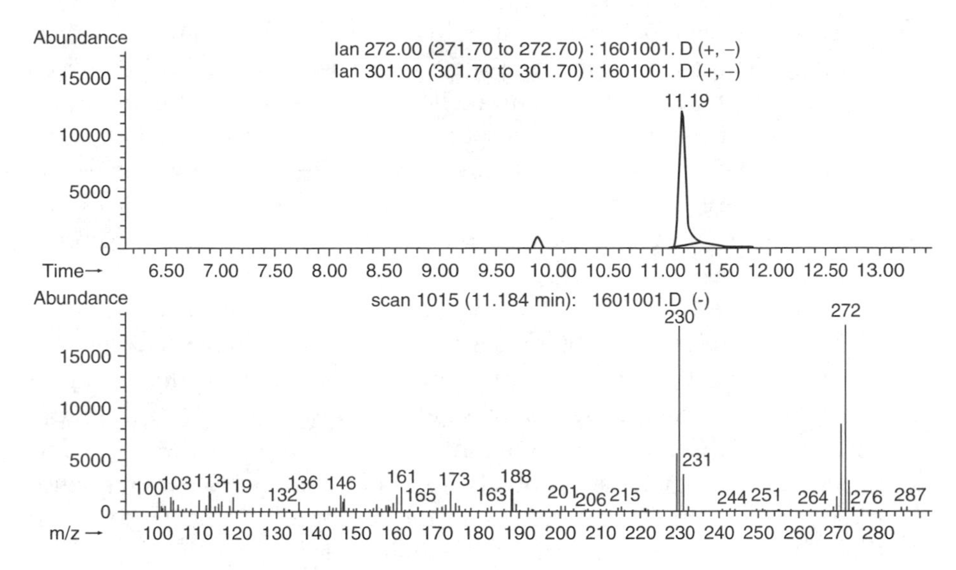

Figure 3.6 Mass chromatogram and electron ionization mass spectrum of an extracted hair specimen, revealing the presence of salbutamol. (From Kintz, P. et al., *J. Forensic Sci.*, 45, 170, 2000. © ASTM International. Reprinted with permission.)

considered as justified, even if it is known that the drug can be used to enhance performance. By comparison with concentrations measured in asthmatic patients, and segmental analysis (repetitive positive segments), hair would document unambiguously a doping attitude from the athlete.

3.4 Discussion

There are essentially three types of problems with urinalysis drug testing: false-positives when not confirmed with GC/MS, degradation of observed urine collection, and evasive maneuvers, including adulteration. These problems can be greatly mitigated or eliminated through hair analysis. It is always possible to obtain a fresh, identical hair sample if there is any claim of a specimen mix-up or breach in the chain of custody. This makes hair analysis essentially fail-safe, in contrast to urinalysis, since an identical urine specimen cannot be obtained at a later date.

Another potential use of hair analysis is to verify accidental or unintentional ingestion of drinks or foods that have been laced with drugs. In case of a single use, the hair will not test positive, particularly for anabolics or corticosteroids, that are badly incorporated in hair. Its greatest use, however, may be in identifying false-negatives, since neither abstaining from a drug for a few days or nor trying to "beat the test" by diluting urine will alter the concentration in hair. Urine does not indicate the frequency of drug intake in subjects who might deliberately abstain for several days before biomedical screenings. While analysis of urine specimens cannot distinguish between chronic use or single exposure, hair analysis can make this distinction.

Although hair is not yet a valid specimen for the International Olympic Committee (IOC), it is accepted in most courts of justice. During the period 1998–2002, some conflicting results were observed, all involving athletes that tested positive in urine in accredited IOC laboratories and negative in hair in forensic certified laboratories.

A lot of experience has been aquired in the detection of opiates and cocaine in hair. In contrast, there is a serious lack of suitable references to interpret the analytical findings for doping agents. In hair, doping agent concentrations for such drugs as anabolic steroids, corticosteroids, or β2-agonists are in the range of picograms per milligram, whereas cocaine, amphetamines, or opiates are generally found in the range of several nanograms per milligram. Therefore, it was the feeling of the Society of Hair Testing to obtain a consensus on hair testing for doping agents.[27]

This consensus is as follows:

1. Hair analysis can essentially contribute to doping analysis in special cases, in addition to urine.
2. Hair specimens are not suitable for general routine control.
3. In case of positive urine results, the negative hair result cannot exclude the administration of the detected drug and cannot overrule the positive urine result.
4. In case of negative urine result, the positive hair result demonstrates drug exposure during the period prior to the sample collection.
5. Before using hair analysis for doping control, sample collection and analytical methods have to be harmonized with respect to the sophisticated requirements already valid for urine.
6. The Society feels responsible to support efforts that lead to this harmonization.
7. This statment was adopted on June 16, 1999 by the Society of Hair Testing.

It is clear that there is a great deal of research to be performed before the scientific questions and curiosity surrounding hair drug testing is satisfied. Some of this is due to a lack of consensus among the active investigators on how to interpret the results on an analysis of hair. Among the unanswered questions, five are of critical importance: (1) What is the minimal amount of drug detectable in hair after administration? (2) What is the relationship between the amount of the drug used and the concentration of the drug or its metabolites in hair? (3) What is the influence of hair color? (4) Is there any racial bias in hair testing? (5) What is the influence of cosmetic treatments? Several answers were recently provided[28] on these specific topics.

It is always possible to obtain a fresh identical hair sample if there is any claim of a specimen mix-up or breach in the chain of custody. This makes hair analysis essentially fail-safe, in contrast to urine analysis, since an identical urine specimen cannot be obtained at a later date. Clearly, hair analysis can thus function as a "safety net" for urine analysis.

Unfortunately, according to the International Olympic Committee (IOC), basic scientific knowledge in hair biology is still lacking to make scalp hair analysis now a valid specimen in the field of doping control and the following points should be resolved before applications: (1) analytical methods are missing for several doping compounds, such as diuretics, (2) peptide hormones are not extractable, (3) hair washing, discoloring, tinting, and hair color (resulting in potential ethnic discrimination) appear to influence the concentration of drug measured in hair, and (4) drug incorporation within the hair longitudinal axis and upon time is not proved to be regular on all occasions.

3.5 Conclusions

It appears that the value of hair analysis for the identification of drug users is steadily gaining recognition. This can be seen from its growing use in preemployment screening, in forensic sciences, and in clinical applications. Hair analysis may be a useful adjunct to conventional drug testing in doping control. Specimens can be more easily obtained with less embarrassment, and hair can provide a more accurate history of drug use.

Although there are still controversies on how to interpret the results, particularly concerning external contamination, cosmetic treatments, ethnic bias, or drug incorporation, pure analytical work in hair analysis has reached a sort of plateau, having solved almost all the analytical problems.

Though GC/MS is the method of choice in practice, GC/MS/MS or LC/MS are today used in several laboratories, even for routine cases, particularly to target low dosage compounds like corticoids.

In the case of doping control, drugs are screened in urine specimens according to validated standard operating procedures in accredited laboratories. As forensic laboratories can be involved in testimony dealing with doping agents, the idea of using hair for doping control has emerged as hair analysis has been accepted in court in other cases. Courts can request additional information on the pattern of use of doping substances, such as during the 1998 cycling Tour de France where blood, urine, and hair were simultaneously collected. Hair can both confirm repetitive abuse and identify the exact nature of the parent compound (e.g., nandrolone, norandrostendiol, or norandrostendione, in the case of positive urine for norandrosterone). Moreover, long-term use (over several months) of restricted compounds (only authorized under specific conditions and for a short period), such as salbutamol or corticoids, can be documented through hair analysis. The determination of testosterone esters in hair should allow a definitive unambiguous confirmation of the administration of exogenous testosterone.

However, some issues have to be discussed before considering hair as a valid specimen by the IOC and the International Sport Federations. The relationship between urine and hair results is not yet established and negative hair result does not mean "no doping." The potential ethnic discrimination must be evaluated to avoid inequality during doping control.

In contrast with the problems associated with cosmetic treatments or the absence of specimen (bald or fully shaved subject), external contamination does not constitute a major trouble when testing for doping agents.

References

1. Baumgartner, A.M. et al., Radioimmunoassay of hair for determining opiate-abuse histories, *J. Nuclear Med.*, 20, 748, 1979.
2. Borges, C.R., Wilkins, D.G., and Rollins, D.E., Amphetamine and *N*-acetyl-amphetamine incorporation into hair: an investigation of the potential role of drug basicity in hair color bias, *J. Anal. Toxicol.*, 25, 221, 2001.
3. Henderson, G.L., Harkey, M.R., and Zhou, C., Incorporation of isotopically labeled cocaine into human hair: race as a factor, *J. Anal. Toxicol.*, 22, 156, 1998.
4. Nakahara, Y. et al., Hair analysis for drug abuse III. Movement and stability of methoxyphenamine (as a model compound of methamphetamine) along hair shaft with hair growth, *J. Anal. Toxicol.*, 16, 253, 1992.
5. Scherer, C.R. and Reinhardt, G., Nachweis sechs endogener Steroide in menschlichen Haaren mit GC/MS und Isotopenverdünnungsanalyse. In Althoff H., Ed., Abstract from 74. *Jahrestagung der Deutschen Gesellschaft für Rechtsmedizin. Aachen,* 1995, 55.
6. Kintz, P. et al., Identification of testosterone and testosterone esters in human hair, *J. Anal. Toxicol.*, 23, 352, 1999.
7. Kintz, P., Cirimele, V., and Ludes, B., Physiological concentrations of DHEA in human hair, *J. Anal. Toxicol.*, 23, 424, 1999.
8. Thieme, D. et al., Analytical strategy for detecting doping agents in hair, *Forensic Sci. Int.*, 107, 335, 2000.
9. Gaillard, Y. et al., Gas chromatographic-tandem mass spectrometric determination of anabolic steroids and their esters in hair. Application in doping control and meat quality control, *J. Chromatogr. B*, 735, 189, 1999.

10. Rivier, L., New trends in doping analysis, *Chimia*, 56, 84, 2002.

11. Cirimele, V., Kintz, P., and Ludes, B., Testing of the anabolic stanozolol in human hair by gas chromatography–negative ion chemical ionization mass spectrometry, *J. Chromatogr. B*, 740, 265, 2000.

12. Kintz, P. et al., Doping control for methenolone using hair analysis by gas chromatography–tandem mass spectrometry, *J. Chromatogr. B*, 766, 161, 2002.

13. Kintz, P. et al., Doping control for nandrolone using hair analysis, *J. Pharm. Biomed. Anal.*, 24, 1125, 2001.

14. Kintz, P., Cirimele, V., and Ludes, B., Discrimination of the nature of doping with 19-norsteroids through hair analysis, *Clin. Chem.*, 46, 2020, 2000.

15. Kintz, P. et al., Testing for anabolic steroids in hair from two bodybuilders, *Forensic Sci. Int.*, 101, 209, 1999.

16. Deng, X.S., Kurosu, A., and Pounder, D.J., Detection of anabolic steroids in head hair, *J. Forensic Sci.*, 44, 343, 1999.

17. Dumestre-Toulet, V. et al., Hair analysis of seven bodybuilders for anabolic steroids, ephedrine and clenbuterol, *J. Forensic Sci.*, 47, 211, 2002.

18. Cirimele, V. et al., First identification of prednisone in human hair by liquid chromatography–ionspray mass spectrometry, *J. Anal. Toxicol.*, 23, 225, 1999.

19. Cirimele, V. et al., Identification of ten corticosteroids in human hair by liquid chromatography-ionspray mass spectrometry, *Forensic Sci. Int.*, 107, 381, 2000.

20. Cirimele, V. et al., Prednisone concentrations in human hair, *J. Anal. Toxicol.*, 26, 110, 2002.

21. Raul, J.S. et al., A single therapeutic treatment with betamethasone is detectable in hair, *J. Anal. Toxicol.*, 26, 582, 2002.

22. Bévalot, F. et al., Analysis of corticosteroids in hair by liquid chromatography-electrospray ionization mass spectrometry, *J. Chromatogr. B*, 740, 227, 2000.

23. Gaillard, Y., Vaysette, F., and Pépin, G., Compared interest between hair analysis and urinalysis in doping controls. Results for amphetamines, corticosteroids, and anabolic steroids in racing cyclists, *Forensic Sci. Int.*, 107, 361, 2000.

24. Gleixner, A., Sauerwein, H., and Meyer, H.H.D., Detection of the anabolic β2-adrenoceptor agonist clenbuterol in human scalp hair by HPLC/EIA, *Clin. Chem.*, 42, 1869, 1996.

25. Machnik, M. et al., Long-term detection of clenbuterol in human scalp hair by gas chromatography-high resolution mass spectrometry, *J. Chromatogr. B*, 723, 147, 1999.

26. Kintz, P. et al., Doping control for β-adrenergic compounds through hair analysis, *J. Forensic Sci.*, 45, 170, 2000.

27. Sachs, H. and Kintz, P., Consensus of the Society of Hair Testing on hair testing for doping agents, *Forensic Sci. Int.*, 107, 3, 2000.
28. Kintz, P., Cirimele, V., and Ludes, B., Pharmacological criteria that can affect the detection of doping agents in hair, *Forensic Sci. Int.*, 107, 325, 2000.

Forensic Applications of Isotope Ratio Mass Spectrometry

4

WOLFRAM MEIER-AUGENSTEIN
RAY H. LIU

Contents

0-8493-1522-0/04/$0.00+$1.50

4.1 Introduction

Stable isotope ratio mass spectrometry (IRMS) has undergone about 50 years of development and applications. It is now often adopted as the basis for (1) monitoring the fate of selected compounds in the biosphere[1]; (2) labeling experiments used in biological[2–4] and organic reaction mechanistic studies[5]; (3) quantitative analysis by the "isotope dilution technique" (IDT)[6]; (4) sample differentiation based on isotope composition distributions[7]; and (5) archaeological studies.[8]

Isotope dilution techniques are now routinely used in forensic toxicology laboratories for the quantitation of drugs and their metabolites in biological specimens. These applications are based on variations in *intermolecular isotope abundances* at an enriched level. A general purpose mass spectrometer is adequately used in these applications. Sample differentiations in forensic applications are typically based on *intra-molecular isotope composition* variations at natural abundance level in the samples of interest. An isotope ratio mass spectrometer (IRMS) is required in these and archaeological studies.

In this chapter, we shall first address some essential conventions adopted in IRMS and the theoretical basis and instrumentation of IRMS, followed by a survey on the fundamental aspects of IDT and a review of the applications of IRMS in forensic science.

4.1.1 Convention

Differences in stable isotope contents are typically expressed in delta notations in reference to a standard. For example, ^{13}C-enrichment levels are universally expressed as[9]:

$$\delta^{13}C(‰) = \left[\frac{R_{sample}}{R_{standard}} - 1\right] \times 1000$$

where R is the ratio of the number of atoms of the minor isotope to that of the major isotope.

The ^{13}C-enrichment level found in PDB (calcium carbonate — a fossil of *Belemnitella americana* from the Peedee formation in South Carolina) is universally used as the reference.[10] Thus, the $\delta^{13}C$ (‰) found in the samples of interest are compared to this standard. PDB originates from a marine carbonate shell and as such contains one of the heavier varieties of carbon in the terrestrial environment. Compared with carbon from PDB ($^{13}C/^{12}C$ = 0.0112372), most natural carbon gives a negative delta value. The range of

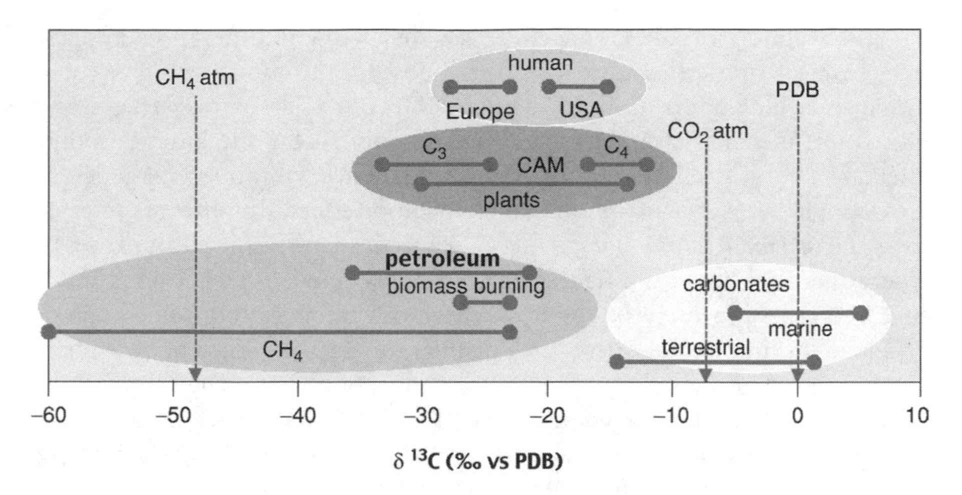

Figure 4.1 Variation in the natural abundance of ^{13}C in CH_4, CO_2, organic materials, and carbonates. ($\delta^{13}C$ value of 0.00‰ for PDB corresponds to a ^{13}C abundance of 1.1112 atom%, whereas a $\delta^{13}C$ value of −50.00‰ is equivalent to a ^{13}C abundance of 1.0563 atom%.) (From Meier-Augenstein, W., *LC/GC Int.*, 10(1), 17–25, 1997. With permission from Advanstar Communications.)

carbon isotope variations in selected carbon cycle reservoirs is summarized in Figure 4.1.[11]

Similarly, SMOW (Standard Mean Ocean Water),[12] and air[13] are commonly used as the primary standards for D/H and $^{15}N/^{14}N$ ratios. PDB and SMOW are also used as the primary standards for $^{18}O/^{16}O$ and $^{17}O/^{16}O$ ratios.

4.1.2 Variations in Stable Isotope Contents

In spite of the fact that the whole earth isotope ratios are fixed since they were determined at the time of our planet's formation, compartmental isotope ratios are not but are in a continuous state of flux due to mass discriminatory effects of biological, biochemical, chemical, and physical processes. In principle, two different types of isotope effects can cause isotopic fractionation — kinetic isotope effects and thermodynamic isotope effects.[14] In general, *kinetic isotope effects* are caused by differences in vibration energy levels of bonds involving heavier isotopes as compared to bonds involving lighter isotopes. This difference in bond strength can lead to different reaction rates for a bond when different isotopes of the same element are involved. The most significant kinetic isotope effect is the primary isotope effect, whereby a bond containing the atom or its isotope in consideration is broken or formed in the rate-determining step of the reaction. Rieley presented an excellent in-depth discussion of kinetic isotope effects, and associated theoretical considerations in 1994.[15]

The second kind of isotope effect is associated with differences in physico–chemical properties such as infrared absorption, molar volume, vapor pressure, boiling point, and melting point. Of course, these properties are all linked to the same parameters as those mentioned for the kinetic isotope effect, i.e., bond strength, reduced mass, and, hence, vibration energy levels. However, to set it apart from the kinetic isotope effect, this effect is referred to as the *thermodynamic isotope effect* because it manifests itself in processes where chemical bonds are neither broken nor formed. Typical examples for such processes in which the results of thermodynamic isotope effects can be observed are infrared spectroscopy, distillation, and any kind of two-phase partitioning. This thermodynamic isotope effect or physico–chemical isotope effect is the reason for the higher infrared absorption of $^{13}CO_2$ as compared to $^{12}CO_2$, the enrichment of ocean surface water with $H_2{}^{18}O$, and the isotopic fractionation observed during chromatographic separations.[16]

Due to the high precision of isotope abundance measurements by modern IRMS instruments, even minute changes in the isotopic composition of a given compound can be reliably detected, irrespective of whether these minute changes have been caused by kinetic isotope effects associated with enzyme-mediated biochemical reactions or batch-to-batch variations of reaction conditions during its chemical synthesis. The resulting variation in the natural abundance of, e.g., ^{13}C can be as high as 0.1 atom%. This wide range reflects the varying degree of mass discrimination associated with the different pathways of carbon assimilation and fixation, but can also reflect different growing conditions due to climatic and geographic differences.

4.1.3 Discrimination Power

Analytical methods currently applied in support of law enforcement agencies establish a degree of identity between one substance and another by means of identifying their constituent elements, functional groups, and by elucidating their chemical structures. Should the spectroscopic data of two compounds correspond, it may be concluded that they are chemically identical. However, an argument can be brought forward as a defense that although two substances in question are chemically identical, they are not the same, i.e., not sharing the same origin, hence coming from a different source. The analytical technique at the heart of this chapter, IRMS, permits this contention to be tested and provides the resolution to the question of whether two compounds or substances are truly identical. With the help of stable isotope finger-printing (or stable isotope "DNA"), forensic scientists will be able to link a person to an event, a crime scene, or a criminal organization (such as a terrorist group) based on a unique characteristic of physical evidence.

For example, research carried out on cocaine base has confirmed the potential probative power of the data provided by this analytical technique by analyzing the isotopic composition of trimethoxycocaine and truxilline extracted from coca leaves sampled at different geographic locations in South America.[17] This work confirmed that the stable isotope make-up of a substance is a function of its origin. In other words, two substances that are chemically the same may have different stable isotope compositions if either their origin and/or (bio)chemical history differ. This finding will remove the defense of substances that are chemically the same not necessarily originating from identical sources and will thus be a significant advance in forensic science, intelligence gathering, and crime detection and reduction.

Stable isotope finger-printing or "chemical DNA," in conjunction with other spectroscopic data, will significantly increase the probative power of analytical results for drugs, flammable liquids, explosives, fibers, textiles, paints, varnishes, papers, inks, plastics, adhesives, and organic materials in general — indeed any nonbiological physical evidence. Multidimensional isotope analysis (or multiisotope analysis) of a given material can provide a combined specificity on a scale hitherto known only for DNA analysis. If one assumes that two given organic compounds (e.g., truxilline and trimethoxycocaine) made up of C, N, H, and O would occur naturally in only 14 significantly different isotopic variations (in fact, this number would be much higher), the combined specificity of a 4-dimensional isotope analysis (^{2}H, ^{13}C, ^{15}N, and ^{18}O) would thus be $1/\{(14^2)^4\}$ or $1/14^{(2*4)}$, i.e., 1 in 1.47 billion.

4.2 Instrumentation

4.2.1 Introduction

Mass spectrometry is best known among chemists as one of the most important analytical techniques for the characterization of molecular structure and amount, elemental composition, and spatial arrangement.[18] In contrast, natural or life scientists such as biomedical, biological, food, and geochemical scientists regard MS as a tool that reveals origin and genesis, or the state of complex systems through high precision measurements of isotopic abundances. The latter type of MS is more aptly named IRMS to denote differences in instrumental design between IRMS and scanning MS systems, as well as the fact that IRMS determines isotope abundances by measuring isotope ratios to reveal this other dimension of information locked into matter, which is usually not available from structural or quantitative studies.

In contrast to single collector MS that yield structural information by scanning a mass range over several hundred Dalton for characteristic fragment ions, IRMS instruments achieve highly accurate and precise measurement of isotopic abundance at the expense of the flexibility of scanning MS. However, scanning mass spectrometers universally employ single detector electron multipliers (EM) and therefore cannot simultaneously detect particular isotope pairs for continuous isotope ratio measurement. For isotope ratio measurement, a MS has to be operated in a selected ion monitoring mode (SIM) to optimize sensitivity to selected masses. Even in SIM mode, limited accuracy and precision of such isotope ratio measurements impose a minimum working enrichment for ^{13}C and ^{15}N of at least 0.5 atom% excess (APE).[19,20] In other words, single collector scanning MS cannot provide reliable quantitative information on isotopic composition at natural abundance level.

Isotope ratio mass spectrometry systems have been designed to measure isotopic composition at low enrichment and natural abundance levels. In an IRMS, molecular ions emerge from the ion source and are separated by a magnetic sector set to a single field strength throughout the experiment. No energy filtering is used so as to optimize transmission near unity, giving a typical absolute sensitivity of 10^{-3} (about 1000 molecules enter the ion source per ion detected) and better. Mass-filtered ions are focused onto dedicated Faraday cup (FC) detectors positioned specifically for the masses of interest. For example, IRMS instruments equipped to determine CO_2 have three FCs for measurement of *m/z* 44, 45, and 46, positioned so that the ion beam of each mass falls simultaneously on the appropriate cup. Each FC has a dedicated amplifier mounted on the vacuum housing to minimize noise pickup, and dedicated counters/recorders for continuous and simultaneous recording of all relevant ion beams. FCs are the detectors of choice for IRMS due to two major considerations. First, the absolute precision required for IRMS determinations is at least 10^{-4}, which is attainable based on counting statistics with at least 10^8 particles detected. Ion currents that achieve these levels are well within the range detectable by FCs. For instance, the major ion beam current for atmospheric CO_2 analysis, *m/z* 44, will typically be around 10 nA, with the minor *m/z* 46 beam around 40 pA. Second, FCs are highly stable and rugged, and rarely need replacement, compared to EMs, whose sensitivity degrades with use and age. This means that minute variations in very small amounts of the heavier isotope are detected in the presence of large amounts of the lighter isotope. The abundance A_s of the heavier isotope n_2 in a sample, given in atom%, is defined as:

$$A_s = R_s/(1 + R_s) \times 100 \text{ [atom\%]} \quad (1)$$

where R_s is the ratio n_2/n_1 of the two isotopes for the sample. The enrichment of an isotope in a sample as compared to a standard value (A_{std}) is given in atom% excess (APE):

$$APE = A_s - A_{std} \quad (2)$$

Since the small variations of the heavier isotope habitually measured by IRMS are of the order of –0.07 to +0.02 APE, the δ-notation in units of per mil (‰) has been adopted to report changes in isotopic abundance as a per mil deviation compared to a designated isotopic standard:

$$\delta_s = ([R_s - R_{std}]/R_{std}) \times 1000\ [‰] \quad (3)$$

where R_s is the measured isotope ratio for the sample and R_{std} is the measured isotope ratio for the standard. To give a convenient rule-of-thumb approximation, in the δ-notation, a ^{13}C abundance in the range of –0.033 to +0.0549 APE corresponds to a $\delta^{13}C$ value range of –30‰ to +50‰. A change of +1‰ is approximately equivalent to a change of 0.001 APE and 0.0003 APE for ^{13}C and ^{15}N, respectively.

Natural abundance applications for the most part require the high-precision techniques introduced in the late 1940s; these techniques dominated IRMS for four decades. They focused initially on the analysis of bulk materials of high chemical complexity such as plant extracts or petroleum reduced to simple gases by combustion to CO_2 or conversion to N_2 or H_2. Analysis of chemically pure materials was much more common for compounds that occur in relatively pure form in nature, such as water, because off-line sample preparation is a time-consuming affair fraught with the risk of contamination and subtle isotopic fractionation.

4.2.2 Dual-Inlet Isotope Ratio/Mass Spectrometry (IRMS)

However, until the commercial availability of continuous-flow isotope ratio mass spectrometer (CF-IRMS) systems enabling on-line isotope analysis of organic compounds in the late 1980s, initial attempts to exploit the information locked into stable isotope ratios at natural abundance level for forensic purposes were confined to employing dual-inlet IRMS systems. Typically, off-line preparation involves multiple steps on custom-designed vacuum lines equipped with high-vacuum and sample-compression pumps, concentrators using cryogenic or chemical traps, reactions in furnaces using catalysts or true reagents, and microdistillation steps. Contamination and isotopic fractionation are a constant threat at any step, and, in general, manual off-line methods are slow and tedious, usually requiring large sample amounts.[7] The quality of the results depends considerably on operator skill and dedication.

The dual-adjustable volume inlet system facilitates sample/standard comparison under the most nearly identical of circumstances. It is necessary for highest precision to compensate for (a) normal fluctuations in instrumental response and (b) ion source nonlinearity which produces isotope ratios that depend on the source gas pressure. The latter phenomenon can be understood from consideration of the physical arrangement of an EI ion source. The source includes magnets that collimate the electron flow from the filament through the ion box. Once formed, ions are accelerated across magnetic field lines which induce minor but measurable mass selectivity. As the mean-free path of ions depends on the relatively high pressure in the closed ion source, the mean position from which ions escape the ion box depends on pressure, as well as on mass. This translates into a subtle dependence of isotope ratio on pressure, which becomes apparent in the precision of these measurements. The severity of this effect in any particular source defines the relative importance of sample/standard pressure matching. Continuous-flow IRMS instruments have only become available because modern ion sources have been designed for maximum linearity and because the He carrier gas flow maintains the ion source pressure nearly constant and independent of the level of sample gas flowing into the source.

4.2.3 Continuous-Flow IRMS

4.2.3.1 Bulk Stable Isotope Analysis (BSIA)

In the IRMS inlet system referred to as "continuous flow," a He carrier gas passes continuously into the ion source and sweeps bands or peaks of analyte gas into the source for analysis. This approach overcomes the sample size requirements for viscous flow because the He carrier gas stream maintains viscous conditions independent of sample size. Although originally developed for on-line coupling of a gas chromatograph (GC) to an IRMS, this approach has also been successfully used to connect IRMS to a variety of automated sample processing devices such as an elemental analyzer.

With the increasing use of bulk stable isotopes analysis (BSIA) in fields such as geochemistry, ecology, bio-medicine, and nutritional sciences, a great demand existed for the rapid analysis of large numbers of solid samples. In 1983, Preston and Owens[21] demonstrated the first CF interface to a multi- (dual) collector IRMS for the bulk analysis of nitrogen from solid samples. Precisions of about SD (δ^{15}N) = 0.7‰ were obtained for 3.5 μmol aliquots of N_2 derived from urea in CF mode after flash combustion and separation in a GC equipped with a gas separation column. These authors observed degradation in the precision due to ion source nonlinearities when analyzing samples of varying size, as is now observed in demanding CF analyses. This system was the first high-precision multicollector CF-IRMS. Two years later, these authors extended the instrument to the analysis of C with similar reproducibility.[22]

Modern commercially available elemental analyzers provide an automated means for on-line high-precision isotope ratios for bulk analysis of solid and nonvolatile liquid samples. Samples are placed in a capsule, typically silver or tin, and loaded into a carousel for automated analysis. The sample is dropped into a heated reactor that contains an oxidant, such as copper and chromium oxide for C or S analysis, where combustion takes place in an He atmosphere with an excess of oxygen. Combustion products are transported by flowing He through a reduction furnace for removal of excess oxygen and conversion of nitrous oxides into N_2. A drying tube is used to remove any excess water in the system. The gas-phase products are separated on a PLOT column under isothermal conditions, and detected nondestructively by thermal conductivity before introduction to the IRMS.

Oxygen and hydrogen are the two most recent elements for which elemental analyzer data for bulk compounds have been presented. Oxygen-containing samples are converted on-line to CO by pyrolytic reaction with carbon (the "Unterzaucher reaction") as first shown by Brand et al.,[23] using a GC-based system; several other reports using this principle subsequently appeared, showing an elemental analyzer[24] or direct injection analysis.[25] The report of Farquhar et al.[26] demonstrates the automated on-line conversion of the oxygen in water or nitrogen-containing plant dry matter to CO, using a pyrolysis-based reaction on carbonized nickel at about 1100°C. CO is separated from N_2, using a GC with a molecular sieve column; precisions of SD ($\delta^{18}O$) = 0.2‰ are obtained. Begley and Scrimgeour have shown the analysis of oxygen ($\delta^{18}O$) and hydrogen ($\delta^{2}H$) on a single sample by measuring the isotope ratio of H_2 gas produced in the pyrolytic reactor with a high mass dispersion IRMS, capable of fully resolving analyte HD (*m/z* 3) from excess He carrier (*m/z* 4). Precisions for water, urine, and volatile organic compounds are about SD (δD) = 2‰ and SD ($\delta^{18}O$) = 0.3‰.[25]

Nowadays, thanks to advances in electronics and instrument design, dual $^{13}C/^{15}N$ analysis of the same sample in one analytical run is possible on a combustion/reduction EA-IRMS. Allowing for a good separation of the N_2 peak from the CO_2 peak to permit a high precision magnetic field jump, total analysis time can be as fast as 7 min per sample. Similarly, dual $^{2}H/^{18}O$ analysis from the same sample can be carried out using a high temperature thermal conversion EA (TC/EA). In addition to the aforementioned reaction on carbonized nickel at 1100°C, high temperature conversion on glassy carbon at 1400°C is used as an alternative. Both solid and liquid samples, the latter by means of a special liquid injector, can be analyzed for ^{2}H and ^{18}O simultaneously, with total analysis time being as fast as 6 min per sample. In conclusion, EA-IRMS or TC/EA-IRMS would appear to be the method of choice for many forensic applications (drugs, explosives, hair, fingernails, etc.), at least as a quickly performed initial measurement to concentrate the efforts of more elaborate techniques and analyses on samples seemingly identical, based on their bulk isotopic composition.

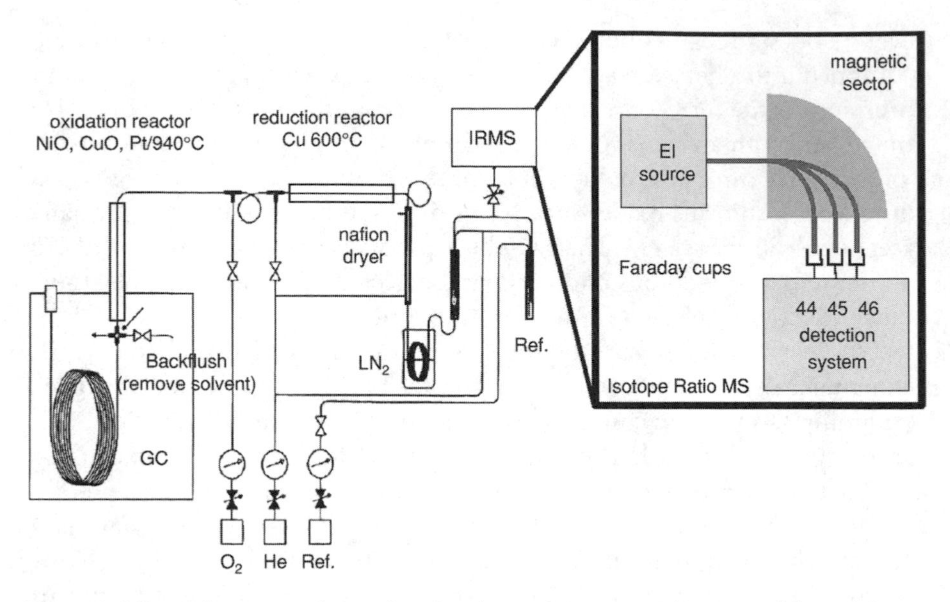

Figure 4.2 Schematic of a modern GC/C-IRMS system. (Reprinted from Meier-Augenstein, W., *Anal. Chim. Acta*, 465, 63, 2002. Copyright 2002, with permission from Elsevier Science.)

4.2.3.2 Compound Specific Isotope Analysis (CSIA)

Accurate and precise stable isotope analysis depends on careful sample preparation and, in the case of on-line compound specific isotope analysis (CSIA), on high-resolution capillary gas chromatography (HRcGC).[16,27] Demands on sample size, sample derivatization, quality of GC separation, interface design, and isotopic calibration have been discussed in a number of reviews.[11,28–31] However, we will briefly discuss the instrumental set-up of gas chromatography/conversion interface-IRMS (GC/C-IRMS) since this set-up results in a phenomenon unique to GC/C-IRMS (Figure 4.2).

The need for sample conversion into simple analyte gases has prompted the design of an interface adapted to the particular sample size and carrier gas flow requirements of gas chromatography (GC). In a set-up for either ^{13}C or ^{15}N CSIA, the GC effluent is fed into a combustion reactor, either a quartz glass or ceramic tube, filled with CuO/Pt or CuO/NiO/Pt wires and maintained at a temperature of approximately 820 or 940°C, respectively.[32,33] The influence of combustion tube packing on the analytical performance of GC/C-IRMS has been reported by Eakin et al.[34] To convert NO_x generated during the combustion process into N_2, a reduction tube, filled with Cu and held at 600 °C, is positioned behind the combustion reactor. Next in line is a water trap to remove water vapor generated during combustion, and most instrument manufacturers employ a Nafion™ tube for this purpose. Nafion™ is a fluorinated polymer that acts as a semipermeable membrane through which H_2O passes freely while all the other combustion products are retained in the carrier gas stream. Quantitative water removal prior to admitting the combustion gases into the ion source is essential because any water residue would lead to protonation of CO_2 to produce HCO_2^+, which interferes with analysis of $^{13}CO_2$ (isobaric interference). A detailed study of this effect has been reported by Leckrone and Hayes.[35]

In dual-inlet systems, the analyte gas comes from an adjustable reservoir (bellow) and only travels a short distance prior to entering the ion source. For this reason, the gas pulses result in rectangular signals. In contrast, in CF-IRMS systems used for gas isotope analysis, on-line gas purification steps and overall interface length lead to bell-shaped signals. This is, evidently, even more the case in GC/C-IRMS systems, where analyte peaks eluting from the GC column are fed into an on-line microchemical reactor to produce, e.g., CO_2 peaks. However, due to the chromatographic isotope effect[36,37] the *m/z* 45 signal ($^{13}CO_2$) precedes the *m/z* 44 signal ($^{12}CO_2$) by 150 ms on average,[32] an effect not observed in ordinary CF-IRMS systems. This time displacement depends on the nature of the compound and on chromatographic parameters such as polarity of the stationary phase, column temperature, and carrier gas flow.[38] Therefore, loss of peak data due to unsuitably-set time windows for peak detection and, hence, partial peak integration will severely compromise the quality of the isotope ratio measurement by GC/C-IRMS, as will traces of peak data from another sample compound due to close proximity, resulting in peak overlap with the sample peak to be analyzed. Due to the fact that isotope ratios cannot be determined accurately from the partial examination of a GC peak, HRcGC resulting in true baseline separation for adjacent peaks is of paramount importance for accurate and precise CSIA.

4.2.3.3 Hyphenated CSIA Systems

In recent years, the research efforts of different groups working in the field of GC/C-IRMS have focused on extending the scope of on-line CSIA towards the measurement of $^{18}O/^{16}O$ and $^{2}H/^{1}H$ isotope ratios of organic compounds. In a first step towards on-line measurement of ^{2}H isotope signatures of organic compounds, Prosser and Srimgeour coupled a high mass dispersion IRMS to a GC via a pyrolysis interface including a 5 Å molecular sieve PLOT column to achieve CSIA for ^{2}H of fatty acids.[39] Employing this instrumental set-up, $\delta^{2}H$-values for 16:0, 18:1, and 22:6 fatty acids (as methyl esters) from tuna oil were reported as −148.5 ± 4.1‰, −155.3 ± 1.0‰, and −147.7 ± 1.2‰ (vs. VSMOV), respectively.[40] IRMS manufacturers now offer GC coupled IRMS systems where the GC effluent is fed into a pyrolysis (Py)/thermal conversion (TC) reactor, or even systems with a dual reactor set-up, i.e., combustion and thermal conversion, giving the user the freedom to connect the capillary coming from the GC column to either reactor, depending on the analytical task.

Another hyphenated technique for position specific isotope analysis (PSIA) of fatty acids using an on-line pyrolysis system was described in detail for the first time by Corso and Brenna.[41] They coupled one GC (GC-1) for sample separation prior to pyrolysis to a second GC (GC-2), separating pyrolytic products of the selected sample compound. Furthermore, they installed a valve into GC-2 to permit separated pyrolysis fragments to be admitted to an organic MS for structure analysis of these fragments.

Even more important, from a forensic point of view, is another type of hyphenated CSIA system. Since organic compounds have to be converted into simple analyte gases isotopically representative of the parent material, naturally all structural information is lost that would otherwise be used to confirm the identity of the organic compound whose isotopic signature has been measured. One "solution" to overcoming this dilemma is to analyze another aliquot of the same sample on a scanning GC/MS system employing the same chromatographic conditions that have been used for GC/C-IRMS analysis. This approach relies ultimately on a mere comparison of retention index or retention time of a given peak, a *modus operandi* that could cause severe problems if such results would have to be put forward as evidence in a court of law. This problem was recognized by one of the authors who, in collaboration with an instrument manufacturer, developed a hyphenated mass spectrometric hybrid system that enables CSIA, while at the same time recording a conventional mass spectrum of the target compound to aid its unambiguous identification.[42] To this end, a GC/C-IRMS system was interfaced with an Ion Trap mass spectrometer to facilitate splitting of the GC effluent to the conversion interface with simultaneous admission to the ion source of the organic MS yet without incurring isotopic fractionation.[43,44]

4.3 Isotope Dilution

The use of mass spectrometry (MS)–based methodologies for quantitative analysis is now a routine practice in forensic science laboratories. In most applications, isotope-labeled analogs (ILA) of the analytes are used as the internal standards (IS), and the MS is operated in selected ion monitoring (SIM) mode. Although ^{2}H-analogs are most commonly used, recent comparative studies suggested ^{13}C-analogs (see Figure 4.3) could be more effective.[45] With practically identical chemical properties and MS fragmentation characteristics, an ILA is a preferred IS because it offers the following advantages.

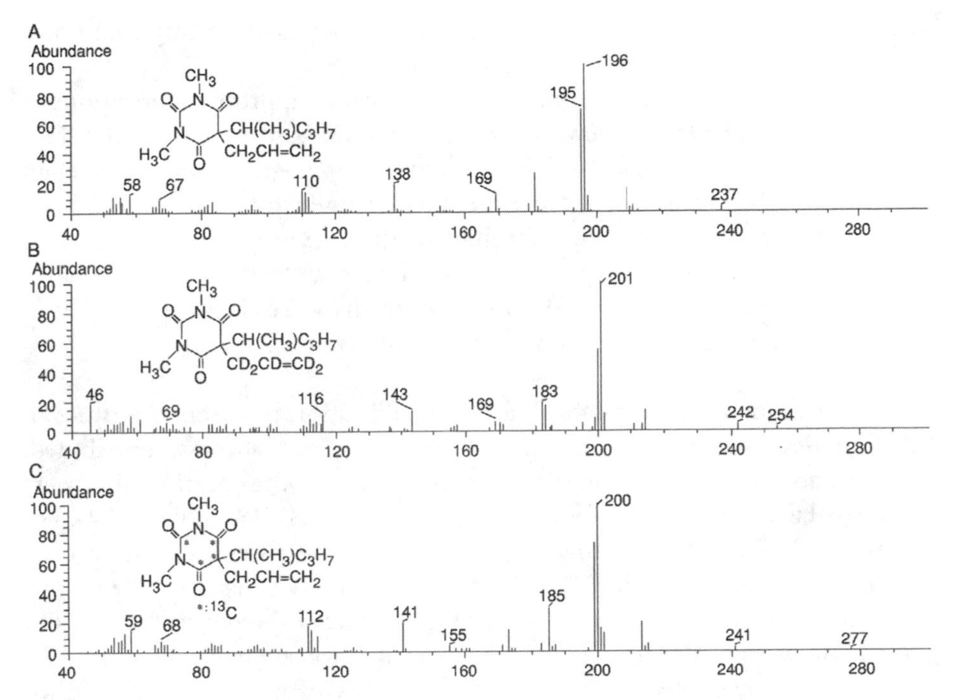

Figure 4.3 Mass spectra and chemical structures of secobarbital (A), $^{2}H_{5}$-secobarbital (B), and $^{13}C_{4}$-secobarbital (C) (all as methyl-derivatives). (Reprinted, with permission, from the *Journal of Forensic Sciences,* Vol. 47, © ASTM International, 100 Bar Harbor Dr., West Conshohocken, PA 19428.)

1. Errors derived from incomplete recovery of the analyte in the sample preparation process or varying GC and MS conditions are compensated for.
2. The presence of interfering materials (or mechanisms) which affects the detection or quantitation of the analyte will result in the absence of the IS in the final chromatogram or altered response and ion intensity ratios thus alerting the analyst to conduct further investigation.

However, when an ILA of the analyte is adopted as the IS, several important parameters should be carefully considered[6,45]:

1. The ILA should be labeled with a sufficient number of atoms of a selected isotope so that the corresponding ions selected from the IS and the analyte will have a significant mass difference. This is especially critical when trimethylsilyl additive or sulfur atoms are present.
2. The ILA should be manufactured with sufficient isotopic purity, otherwise, the addition of the IS may result in the observation of a

significant amount of the analyte in a true negative sample and may also introduce errors in quantitation.

3. The labeling isotopes must be positioned at appropriate locations in the molecular framework of the compounds so that, after the fragmentation process, sufficient number of high-mass ions (that retain the labeling isotopes) are present with significant intensities and will not contribute to the intensities of the corresponding ions derived from the analyte. These ions and their counterparts in the analyte may then be monitored for ion ratio evaluation to facilitate qualitative compound identifications and quantitative determinations.

The critical matter is to establish a linear calibration line; the intensities of the ions designated for the analyte and the IS must not be cross-contributed. For example, in a secobarbital/$^{13}C_4$-secobarbital study, calibration lines were evaluated using two pairs of ions, *m/z* 196/200 and 181/185 (Figure 4.4). For *m/z* 196 (designed for the analyte), 0.23% of the measured intensity is contributed by the IS, while 0.017% of the measured intensity of *m/z* 200 (designed for the IS) is contributed by the analyte. On the other hand, 1.6% of the measured intensity of *m/z* 181 (designed for analyte) is contributed by the IS, while 0.29% of the measured intensity of *m/z* 185 (designed for the IS) is contributed by the analyte. Figure 4.4 clearly reflects the difference in linearity of the calibration lines established by these two ion pairs. When cross-contribution occurs, the nonlinear (such as hyperbolic) model can best describe the observed data and should be seriously considered.

4.4 Sample Differentiation

Analytical work performed in forensic science laboratories primarily aim for the *identification* and *comparison* of various samples with the intention of linking the samples of concern to a specific person or event. *Identifications* are often achieved through the characterization of specific compounds, while *comparisons* often involve the identification and quantitation of multiple components in the samples of interest. With automated instrumentation widely available, highly specific MS–based technologies are now, in most modern forensic science laboratories, the most valuable tool used to achieve the analytical goals: compound identification, or sample comparison.

Since sample decomposition, successive dilution, and contamination during the storage and distribution process may hinder sample differentiation based solely on component identification and quantitation, isotope ratio measurement provides a complementary approach. The inherent strength of this approach is that as long as the source of the chemicals used to prepare

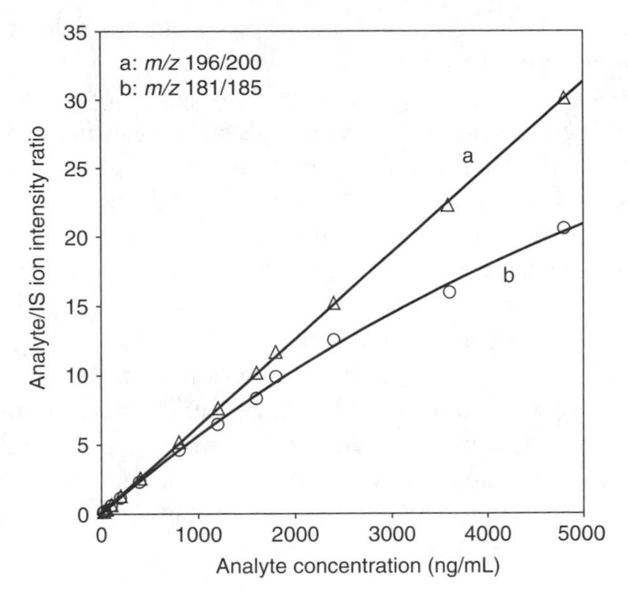

Figure 4.4 Linear and hyperbolic calibration results using ion-pairs with different degrees of cross-contribution — Secobarbital (SB)/$^{13}C_4$-secobarbital ($^{13}C_4$-SB): *m/z* 196/200, 181/185. A total of 0.23% of the measured intensity of *m/z* 196 (designed for SB) is contributed by $^{13}C_4$-SB; 0.017% of the measured intensity of *m/z* 200 (designed for ^{13}C4-SB) is contributed by SB; 1.6% of the measured intensity of *m/z* 181 (designed for SB) is contributed by $^{13}C_4$-SB; and 0.29% of the measured intensity of *m/z* 185 (designed for $^{13}C_4$-SB) is contributed by SB. (Reprinted with permission from Liu, R.H. et al., *Anal. Chem.*, 74, 618A, 2002. © 2002 American Chemical Society.)

the sample of interest is the same, partial degradation or addition of foreign materials will not alter the analytical results. Many applications of this concept are highly relevant to forensic science and will be discussed in this chapter. Although stable isotope compositions of oxygen, nitrogen, and hydrogen have also been utilized, carbon isotope measurement will be emphasized for its simplicity in sample preparation and the potential for online measurement and practical application.

4.4.1 Detection of Adulteration in Food Products

Atmospheric carbon dioxide is fixated by plants for their biosynthesis of organic matters. Although the resulting organic matter is generally depleted in ^{13}C as compared to atmospheric carbon dioxide, the extent of depletion depends on the primary carboxylating enzymatic reaction adopted by the plant.[46,47] As shown in Figure 4.1, the δ^{13}C values found in C3 plants range from about –22 to about –35‰, while that of C4 plants range from –8 to about –20‰. Values for crassulacean acid metabolism (CAM) plants, which use features of both the C3 and C4 pathways, spread throughout most of the range of values found for C3 and C4 plants.

The possibility of characterizing a plant by its $^{13}C/^{12}C$ ratio has lead to some interesting applications in food regulation relevant to forensic science.[48,49] Most of the plants used in agriculture are C3 plants with two major exceptions; corn and cane are C4 plants. Cane sugar and high-fructose corn syrup derived from these two plants are lower-priced sweeteners and $^{13}C/^{12}C$ analysis allows the detection of the illicit addition of these sweeteners to food items such as orange, grape, apple juices, maple syrup, and honey. The carbohydrates in honey are normally derived from flowering plants[50] which are almost exclusively C3 plants. The distributions of $\delta^{13}C$ values in honey,[50] maple syrup,[51] apple,[52] grape,[51] and orange[53] juices have been thoroughly studied. The delta ^{13}C values of related food items are listed in Table 4.1. This isotope ratio information is a basis for the detection of cane sugar and high-fructose corn syrup in honey;[54] apple,[55] orange,[56] and lemon[57] juices; candied pineapple and papaya;[58] and maple syrup.[51]

Table 4.1 Carbon Isotope Composition of Common Sugars, Syrups, and Juices

Item	$\delta^{13}C$ (‰)
Sugar and Syrup	
High-fructose corn syrup	–9.5 to –10.7
Beet sugar	–24.2
Maple syrup	–23.5 to –24.3
Maple sugar	–23.4
Sorghum syrup	–12.0
Cane sugar	–11.5
Fruit juice	
Apricot	–24.1
Concord grape	–26.2
Cranberry	–25.0
Grapefruit	–25.3
Pineapple	–12.5
Plum	–25.0
Prune	–26.0
Strawberry	–24.3
Lemon	–24.2

Source: Parker, P.L., The chemical basis of the use of $^{13}C/^{12}C$ ratios to detect the addition of sweeteners to fruit juice concentrates, in *Proc. Symp. Technol. Probl. Fruit Juice Concentrates*, Oregon State University, Portland, OR, 1981, p. 50.

There are, however, complicating factors. For example, the $\delta^{13}C$ values of honey range from –22.5 to –27.4‰ (standard deviation = 0.98), while those derived from CAM plants with $\delta^{13}C$ values range from –12 to –18‰ and require special limits in determining the authenticity.[54] The Association of Official Analytical Chemists has established isotopic analysis procedures for providing legal proof of adulteration of honey and apple juice. Considering possible variations of the natural isotopic composition of a particular product, a conservative approach is adopted for the interpretation of analytical results; usually a limit of four standard deviations (1 in 25,000 error probability) of the natural variability of the product is set for concluding that the product is adulterated. Thus, –21.5‰ was used as the limit which could allow as much as 20% of added C4 sugars. An internal standard carbon isotope ratio analysis (ISCIRA) approach was used to improve the sensitivity and objectivity in testing honey with $\delta^{13}C$ values less negative than –23.5‰ (2 standard deviation less negative than the mean). This method compared the $\delta^{13}C$ values of the protein isolated from the honey and that of the honey itself. A difference greater than –1‰ indicated the presence of significant amount of added C4 sugars. This index (1‰) is equivalent to 7% of added material.[59]

Using a similar approach, it is possible to prove whether a food sample is of synthetic or of biogenic origin. Examples of these studies include the differentiation of:

1. Vanillin extracted from vanilla beans of different geographic origins, and that synthesized from lignin, eugenol (clove oil), or guaiacol (a coal tar)[60]
2. Vinegars obtained by biogenic fermentation and by synthetic process[61]
3. Beers brewed with barley alone, and those with the addition of maize or petrochemical carbon dioxide derived from natural gas[62]

To detect fraudulent adulteration of high premium olive oils, Angerosa et al. compared $\delta^{13}C$ values of the aliphatic alcoholic oil fractions and found those of the adulterant pomace oil to be significantly more negative than those of virgin and refined olive oils.[63] In a subsequent study, Angerosa et al. employed both ^{13}C and ^{18}O isotope analysis to determine geographical origin of olive oils according to climatic regions in different Mediterranean countries such as Greece, Morocco, and Spain.[64]

4.4.2 Dietary Composition as a Basis for Sample Differentiation

A more specific study has been reported for coffee and tea. Carbon, hydrogen and oxygen isotope compositions of chemically pure caffeine derived from

Table 4.2 Carbon, Hydrogen, and Oxygen Isotope Compositions of Caffeine Derived from Various Sources

Origin	Source	$\delta^{13}C$[a]		δD[a]		$\delta^{18}O$[a]	
		Ave.	Std. dev.	Ave.	Std. dev.	Ave.	Std. dev.
Jamaica	Coffee	–28.8	0.6	–132.5	3.8	9.6	1.8
Kenya		–29.8	0.6	–136.5	3.5	3.6	0.6
Brazil		–28.2	0.2	–157.3	3.9	4.9	0.7
Sri Lanka	Tea	–31.7	0.8	–223.6	2.8	1.8	0.2
Darjeeling		–29.6	0.2	–195.9	2.5	–4.3	0.8
China		–32.4	0.6	–226.8	4.1	1.2	0.3
BDH lab grade	Unknown	–35.8	0.2	–237.1	1.7	13.0	0.3

[a] These values are reported in ‰ with respect to PDB (carbon) and Standard Mean Ocean Water [SMOW] (hydrogen and oxygen).

Source: Reprinted with permission from Dunbar, J. and Wilson, A.T., Determination of geographic origin of caffeine by stable isotope analysis, *Anal. Chem.*, 54, 590, 1982. © 1982 American Chemical Society.

coffee and tea from different geographic locations are shown in Table 4.2.[65] Although total differentiations of all samples based on isotope compositions are not likely, the Brazilian D/H ratio can be distinguished from the other two D/H values, as can be the Jamaican $^{18}O/^{16}O$ ratio from the other two $^{18}O/^{16}O$ values.

The lower ^{18}O content observed from the Darjeeling sample is expected due to the higher altitude of this mountainous country. The isotope composition of the water in this region is expected to be depleted in both deuterium and ^{18}O. The D/H ratio of this sample is more enriched than the other two caffeine extracts from tea. The D/H ratio of caffeine derived from tea, –196 to –227‰, is distinctly lower than that derived from coffee, –132 to –157‰. The distinct results obtained from the commercial preparation are considered an indication of synthetic preparation rather than extract from biological sources.

Deuterium and ^{18}O composition of (alleged) natural compounds is strongly linked to the ^{2}H and ^{18}O signature of the groundwater in the region where the parent plant has grown.[66] If adulteration (or fractionation during isolation and purification) has not taken place, $\delta^{2}H$ and $\delta^{18}O$ values are linked via the equation that links $\delta^{2}H$ and $\delta^{18}O$ for the meteoric water line, i.e.,

$$\delta^{2}H = 8 \times \delta^{18}O + 10 \tag{4}$$

For natural compounds isolated from plant material, the gradient of this equation stays almost the same (it can be 7.35 or 7.5 instead of 8), only the off-set changes. For example, the equation describing the relation of ^{2}H and ^{18}O for sugar from genuine American honeys is:

$$\delta^{2}H = 7.35 \times \delta^{18}O - 254 \tag{5}$$

Table 4.3 Carbon Isotope Composition of Dietary Protein and Human Hair for the United States, Japan, and Germany

	United States		Japan		Germany	
Protein source	Weighing factor	$\delta^{13}C$ (‰)	Weighing factor	$\delta^{13}C$ (‰)	Weighing factor	$\delta^{13}C$ (‰)
Meats	0.430	–14.5	0.128	–15.2	0.334	–21.8
Fish, seafood	0.030	–19.9	0.235	–16.6	0.032	–21.5
Milk, cheese	0.252	–18.5	0.089	–21.6	0.232	–24.6
Eggs	0.044	–14.7	0.053	–13.3	0.057	–16.6
Legumes, nuts	0.030	–24.8	0.086	–26.0	0.016	–25.1
Cereals	0.135	–21.2	0.280	–25.7	0.216	–25.3
Potatoes, staples	0.027	–27.2	0.011	–26.8	0.061	–28.1
Vegetables	0.054	–26.2	0.114	–24.9	0.028	–28.6
Fruits	0.007	–26.9	0.006	–27.2	0.023	–26.8
Total diet		–18.1[a]		–21.2[a]		–23.6[a]
Hair (mean; SD)[b]	–16.4; 0.9		–18.0; 0.8		–20.4; 0.5	
Range	–14.4 to –17.5		–16.5 to –20.0		–19.7 to –21.7	

[a] Mean values, weighted to relative consumption of various products within each group of food.
[b] Mean for 15 subjects.

Source: Nakamura, K. et al., Geographical variations in the carbon isotope composition of the diet and hair in contemporary man, *Biomed. Mass Spectrom.*, 9, 390, 1982. © 1982 John Wiley & Sons Limited. Reproduced with permission.

In other words, this graph runs almost parallel to the graph describing the meteoric water line. Any data whose linear regression analysis would yield gradients <7 or >9 should be treated as suspicious.

An interesting study[67] shows that changes in the carbon isotopic composition of beard hair in two individuals traveling from Munich to Chicago and Tokyo, reflect the changes in dietary composition consumed. The change in the isotopic composition became apparent about 6 to 12 d after arriving at the new locale. The carbon isotopic composition of common foodstuffs purchased in Chicago, Tokyo, and Munich and the weighted averages are listed in Table 4.3, together with the carbon isotopic composition of human hair for the three populations. The $\delta^{13}C$ values of hair of these three populations correlate well with the calculated values of the dietary protein, but had a 2 to 3‰ higher ^{13}C content. Although the carbon isotope composition of dietary protein and hair are not in absolute agreement, the isotope differences between the populations could be identified. The large isotopic differences between the different dietary components within each population, as well as the systematic isotopic differences between like components in the different diets, demonstrate the possibilities of using isotopic composition analysis for sample differentiation of related samples.

One would not expect bulk $\delta^{13}C$ values of diet A of bulk $\delta^{13}C$ values from the organism B feeding on diet A to be the same. Moving up in the food chain, i.e., going from one trophic level to the next one up, is associated with a mass discrimination (= isotopic fractionation) favoring the heavier isotopes. For ^{13}C, moving up in the food chain by 1 level is associated with bulk $\delta^{13}C$ values in the feeding organism to be 1 to 3‰ higher (i.e., more enriched in ^{13}C) as compared to those found in its diet. The same is true for $\delta^{15}N$ values; here the trophic mass discrimination leads to bulk $\delta^{15}N$ values being generally higher by 2 to 4‰. As a matter of fact, this relation of isotopic fraction and feeding level can be exploited to prove that salmon sold as wild salmon is indeed wild and not farmed. Bulk $\delta^{13}C$ values of wild salmon should be of the order of –16 to –17‰, i.e., 2 to 3‰ higher than its favorite food, crustaceans (e.g., prawns) that shows bulk $\delta^{13}C$ values of about –19 to –20‰. In contrast, farmed salmon usually exhibits bulk $\delta^{13}C$ values of –22 to –23‰ thus betraying its origin. So, finding a difference in the isotopic composition of diet and feeding organism is (1) expected and (2) a very useful analytical tool.[68] This is further confirmed by a recent study on two strains of bacteria (*Bacillus globigii* and *Erwinia agglomerans*).[69]

4.4.3 Drugs

Drug sample differentiation based on natural stable isotope variation can rely on isotope composition determination in the bulk sample or in specific compounds. These approaches are more commonly used to differentiate products from different geographical regions. It has also been advocated to differentiate products from different manufacturing batches.

4.4.3.1 Marijuana

Data shown in Table 4.4 show variations in $^{13}C/^{12}C$ ratios derived from the combustion of bulk marijuana samples.[70] These results indicate that ambient air seems to control the isotopic composition of the plant. The ^{13}C content in the samples analyzed decreases in the order of A, B (from rural area), C (from metropolitan area), D, F (indoor plant). This order coincides with the order of the ^{13}C content in the ambient air under which these plants are grown. Since a heater using natural gas containing less-than-usual amounts of ^{13}C is used in the greenhouse, the carbon dioxide inside the greenhouse would be the lightest. The reabsorption of carbon dioxide derived from plants confined in the greenhouse might have led to further depletion in ^{13}C content.[71] Due to the heavy use of petroleum products, it seems reasonable to assume that the ^{13}C content in metropolitan areas are less than that of rural areas. Whether the difference in samples D (a male plant) and F (a female plant) is due to sex difference is not certain as the origin of the seeds is not well established. This preliminary study certainly provides interesting information, but a more extended survey and more specific analysis of pure chemical components are needed for better understanding.

Table 4.4 Carbon Isotope Composition of *Cannabis Sativa* L. Male and Female Samples

Sample	Sample description	$\delta^{13}C$ (‰)	Average
A	Leaf, male, rural	–28.11	–27.99
		–27.82	
		–28.05	
		–28.85	
B	Leaf, male, rural	–28.76	–28.80
		–28.79	
C	Leaf, metropolitan	–29.85	–30.25
		–30.64	
D	Leaf, male, indoor	–31.64	–31.72
		–31.79	
		–34.15	
E	Flower, male, indoor	–34.26	–34.13
		–33.97	
		–32.82	
F	Leaf, female, indoor	–32.79	–32.80
		–32.78	

Source: Reprinted, with permission, from the *Journal of Forensic Sciences,* Vol. 47, © ASTM International, 100 Bar Harbor Dr., West Conshohocken, PA 19428.)

4.4.3.2 Coca Leaves

Coca leaves from South America were found to vary in their $\delta^{13}C$ (–32.4 ‰ to –25.3‰) and $\delta^{15}N$ (0.1–13.0‰) values.[17] Humidity levels and the length of the rainy season and differences in soils were thought to affect the fixation processes and cause the observed subtle variations in ^{13}C and ^{15}N contents, respectively. In conjunction with the variations of trace alkaloids (truxilline and trimethoxycocaine) contents found in cocaine, researchers were able to correctly identify 96% of 200 cocaine samples originated from the regions studied (Figure 4.5).

4.4.3.3 Heroin, Morphine, and Cutting Agents

Differentiations of ^{13}C-enrichment levels have been applied to samples of forensic science interest on the molecular level.[72–76] Specifically, the ^{13}C-enrichment level in heroin samples derived from various geographic regions are measured. Reported data are converted into graphical presentation by this author and shown in Figure 4.6. Since the $\delta^{13}C$ values in a heroin sample are linked to the geographical origin of the papaver and the $\delta^{13}C$ value of the acetic anhydride used to convert morphine to heroin, this method may be effective for differentiating heroin batches, but may not be used directly for assigning geographic origins of samples with specific $\delta^{13}C$ values.[77] For geographical region differentiation, heroin samples were hydrolyzed to morphine,

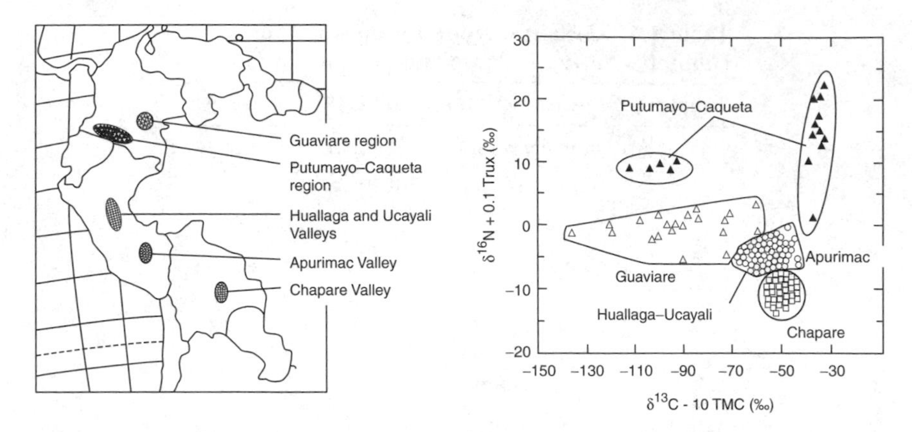

Figure 4.5 Regional grouping of cocaine samples based on $^{15}N/^{14}N$ ratio and truxilline (Trux) content and $^{13}C/^{12}C$ ratio and trimethoxycocaine (TMC) content. (Reprinted with permission from *Nature,* Vol. 408, pp. 311. © 2000 Macmillan Publishers Ltd.)

which was then analyzed for carbon isotopic ratios.[73] This study was not successful as the $\delta^{13}C$ values scattered in a narrow range (–30.5 to –28.6‰).

As the conversion of morphine into heroin does not involve nitrogen, studies were done to explore the possibility in making a $\delta^{15}N$ value determination for direct geographic origin assignments. Unfortunately, three studies indicated a narrow range of $\delta^{15}N$ value scattering: –3.6 to 1.7‰, –1.6 to 1.3‰, and –4.30 to 0.40‰.[77]

Studies have also been done in the determinations of $\delta^{13}C$ and $\delta^{15}N$ values of cutting agents found in heroin samples. Thus, acetaminophen was analyzed for its $\delta^{13}C$ values[78] and caffeine was analyzed for its $\delta^{15}N$ values.[79] The $\delta^{15}N$ values in caffeine were found to range widely (–30.31 to –1.69‰) and, hence, were found to be useful.

4.4.3.4 *Synthetic Drugs*

Natural isotope abundances of carbon and hydrogen in several production batches of the commercial drug diazepam (7-chloro-1,3-dihydro-1-methyl-5-phenyl-2H-1,4-benzodiazepin-2-on) have been studied. The results showed significant differences in $\delta^{13}C$ values between the diazepam synthesized by Hofmann–LaRoche Switzerland (mean, –35.5‰; range, 2.1‰) and Hofmann–LaRoche U.S. (mean, –30.5‰; range, 1.3‰).[80]

More recently, $^{13}C/^{12}C$ and $^{15}N/^{14}N$ isotopic ratios were used to differentiate confiscated 3,4-methylenedioxymethamphetamine (MDMA) samples.[81] In this study, MDMA was extracted from the tablet and the compound's isotopic ratios were determined using GC-IRMS analysis. Carbon isotopic ratio data allowed for classifying 16 MDMA samples into 4 groups (Figure 4.7). Samples 8 and 9 were further differentiated by the nitrogen isotopic data (Figure 4.8).

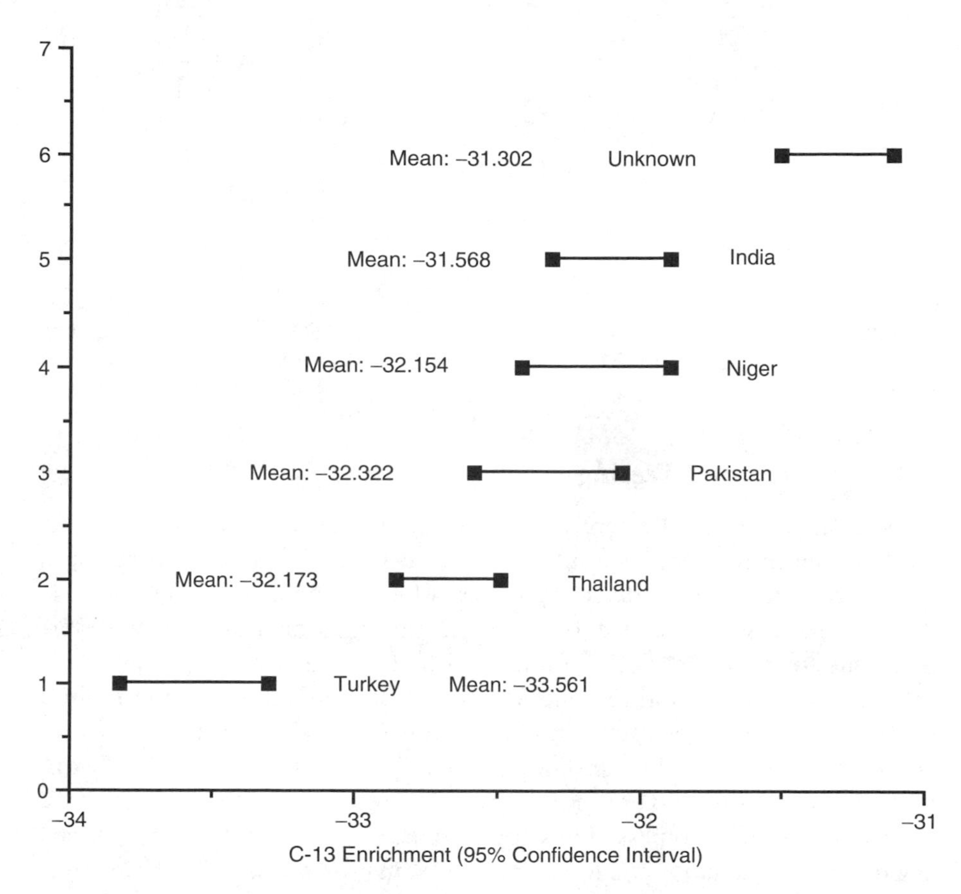

Figure 4.6 Means and ranges of $\delta^{13}C$ (‰) of heroin samples from different geographic regions. (Constructed based on data reported in Desage, M. et al., *Anal. Chim. Acta*, 247, 249, 1991.)

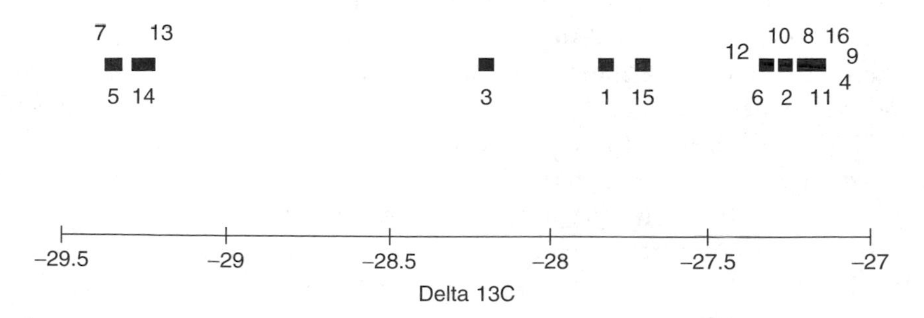

Figure 4.7 $\delta^{13}C$ values of 16 confiscated MDMA samples (From Mas, F. et al., *Forensic Sci. Int.*, 71, 225, 1995. With permission.)

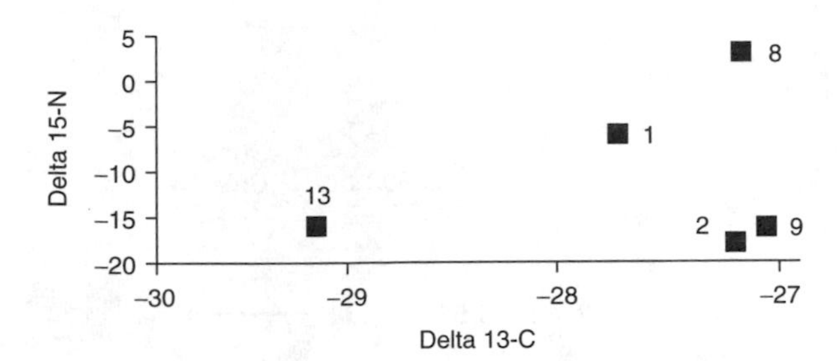

Figure 4.8 Further characterization of confiscated MDMA samples using both $\delta^{13}C$ and $\delta^{15}N$ values (From Mas, F. et al., *Forensic Sci. Int.*, 71, 225, 1995. With permission.)

4.4.4 Antidoping Testing

A major area of interest recently in the antidoping testing community is ^{13}C-content determination. This interest derived from the desire to develop a subjective parameter for detecting misuse of androgens, especially testosterone. Since testosterone concentration varies among individuals, the testosterone/epitestosterone (the 17α epimer and metabolite of testosterone) (T/E) ratio was used as an indirect marker of testosterone administration. Since the T/E ratio also varies among individuals and epitestosterone can be administered to circumvent this detection mechanism, determination of ^{13}C contents in testosterone and epitestosterone is now considered the most effective approach. The effectiveness of this approach is based on the fact that synthetic androgens are generally made from a single plant species, which have lower ^{13}C contents than those derived from endogenous sources.

Several studies have been carried out to use metabolic-pathway related ^{13}C isotope patterns to differentiate between endogenous human testosterone and exogenous testosterone.[82] In the human body, testosterone is synthesized from cholesterol via dehydroepiandrosterone and is then further metabolized to androstanediol. Becchi, Aguilera and coworkers found that the averaged $\delta^{13}C$-values for endogenous 5α- and 5β-androstanediol dropped from −26.52‰ before synthetic testosterone administration (natural background or baseline value) to −32.44‰ during testosterone administration.[83] Independent work by Shackleton et al. obtained a similar baseline $\delta^{13}C$-value of −26.87‰ and a drop down to −30.21‰ on average for androstanediol samples after a bolus administration of 250 mg exogenous testosterone.[84,85] This drop of about −4‰ lasted for up to 10 d before $\delta^{13}C$ values of androstanediol returned to their former baseline value. They also reported a narrow range of −29.15 to −30.41‰ for 5 synthetic testosterone samples manufactured in 5 different countries. Based on the results reported by Aguilera et al.[83] and their own, Shackleton et al. suggested a conservative cut-off value of −29.0‰

for androstanediol to identify unambiguously testosterone abuse for up to 7 d after administration.

In a recently published study, it has been reported that the $\delta^{13}C$ values for 4 synthetic epitestosterone samples examined were all –30.3‰, while the mean $\delta^{13}C$ value of epitestosterone obtained from 43 healthy male urine samples was –23.8‰ (SD, 0.93‰). Nine of 10 athletes' urine samples with epitestosterone concentrations >180 μg/l had their $\delta^{13}C$ values within ±3 SD of the control group. The other athlete's $\delta^{13}C$ value was found to be –32.6‰ and the sample suggested the administering of synthetic epitestosterone.[86]

4.4.5 Explosives

Samples of TNT from a known country of origin (but unknown date of manufacturing) were analyzed for carbon isotope composition.[87] The $\delta^{13}C$ values of these samples were compared with that of coal and crude petroleum. Variations of $^{13}C/^{12}C$ ratios of these samples were clearly demonstrated and may provide information for sample differentiation purposes. The exact cause of this variation is not obvious. If all toluene is manufactured from coal or petroleum sources, these variations must be due to a difference in the initial isotopic composition of the parent material or a difference introduced in the toluene manufacturing process, depending on the type of hydroforming process, the kind of catalyst, and the reaction temperature used.

4.5 Stable Isotope Coding as a Tracing Mechanism: Methamphetamine Example

Isotope ratio measurement of compounds that have been deliberately tagged with selected stable isotopes can also be used as a mechanism for sample differentiation purposes. Parallel with the mechanism of labeling medically important research drugs with radioactive and nonradioactive isotopes,[88] drugs with high potential for being channeled into illicit use could be labeled with specific amounts of a nonradioactive isotope during the manufacturing process. A subsequent isotope ratio measurement could then be used to differentiate samples from different sources and to further identify the source of the drug and/or its precursors.

To illustrate this concept, an example is made of the frequently abused methamphetamine. A study[89] from this author's group incorporates varying amounts of ^{13}C-labeled methylamine during the methamphetamine synthesis process. The ^{13}C to ^{12}C ratios in the resulting products are monitored by an electron impact/quadrupole mass spectrometer, with selected ion monitoring

Table 4.5 Variations of ^{13}C-Enrichment Level in Methamphetamine Synthesized with Different Percentage of ^{13}C-Methylamine.

Sample No.	Methamphetamine Yield (%)	^{13}C-Methylamine Used (%)	^{13}C to ^{12}C Calculated	Ratio in $[C_3H_8N]^+$ Measured	
				Average	Std. dev.
7	31.0	0.0	0.03961	0.03873	0.00068
9	22.0	0.0		0.03944	0.00045
14	7.5	0.25	0.04199	0.04140	0.00074
16	6.4	0.25		0.04156	0.00071
17	58.0	0.25		0.04166	0.00071
13	26.0	0.50	0.04439	0.04414	0.00066
12	18.0	1.0	0.04932	0.04844	0.00083
15	40.0	1.0		0.04871	0.00103
10	31.0	2.0	0.05932	0.05803	0.00096
11	23.0	4.0	0.07956	0.07623	0.00143

Source: Low, I.A. et al., Gas chromatographic/mass spectrometric determination of carbon Isotope composition in unpurified samples: methamphetamine example, *Biomed. Environ. Mass Spectrom.*, 13, 531, 1986.

mode measuring the intensity ratio of the *m/z* 59 and the *m/z* 58 ions of the $[C_3H_8N]^+$ fragment. Synthesized products are introduced into the mass spectrometer through the gas chromatograph inlet without prior clean-up. The measured ratios are in excellent agreement with the calculated ones as shown in Table 4.5.

The artificial enrichment of ^{13}C in the compound of interest allows the isotope ratio measurements to be made with a "direct" approach using a conventional mass spectrometer. Data shown in Table 4.5 indicate that a variation step of 0.25% in the ^{13}C enrichment of methylamine is sufficient for product differentiation. Calculation indicates that this variation step could be reduced by at least 50-fold with the use of an isotope ratio mass spectrometer. With the development of on-line gas chromatograph/combustion/isotope ratio mass spectrometer systems, the approach presented here may well provide a sensible mechanism for monitoring controlled substances. It should be further noted that the isotope approach exemplified here allows sample differentiations after human consumption. Analysis of appropriate metabolites may provide valuable sample source information. While scientific merits and economical feasibility are considered in this study, obvious sociological and legislative concerns remain to be addressed.

References

1. Bricout, J., Isotope ratios in environmental studies, *Int. J. Mass Spectrom. Ion Phys.*, 46, 195, 1982.

2. Wolfe, R.R., *Tracers in Metabolic Research — Radioactive and Stable Isotope/Mass Spectrometry Methods*, Alan R. Liss, New York, 1983.
3. Baillie, T.A., The use of stable isotopes in pharmacological research, *Pharmacol. Rev.*, 33, 81, 1981.
4. Haskins, N.J., The application of stable isotope in biomedical research, *Biomed. Mass Spectrom.*, 9, 269, 1982.
5. Ramesh, S. et al., Three-electron oxidations. 18. Carbon-13 and deuterium isotope effects in the cooxidation of 2-hydroxy-2-methylbutyric acid and 2-propanol. Evidence for a two-step mechanism, *J. Am. Chem. Soc.*, 103, 5172, 1981.
6. Liu, R.H. et al., Isotopically labelled analogues for drug quantitation, *Anal. Chem.*, 74, 618A, 2002.
7. Liu, R.H., Sample differentiation by stable-isotope-ratio mass spectrometry, in *Analytical Methods in Forensic Chemistry*, Ho, M.H., Ed., Ellis Horwood, New York, 1990, Chap. 4.
8. Rouhi, A.M., Nitrogen isotope tales from a tomb, *C&EN*, November 12, 2001, p. 34.
9. Gonfiantini, R., Standards for stable isotope measurements in natural compounds, *Nature*, 271, 534, 1978.
10. Urey, H.C. et al., Measurement of paleotemperatures and temperatures of the upper cretaceous of England, Denmark, and the Southeastern United States, *Bull. Geol. Soc. Am.*, 62, 399, 1951.
11. Meier-Augenstein, W., The chromatographic side of isotope ratio mass spectrometry — Pitfalls and answers, *LC/GC Int.*, 10(1), 17–25, 1997.
12. Craig, H., Standard for reporting concentrations of deuterium and oxygen-18 in natural waters, *Science*, 133, 1833, 1961.
13. Junk, G. and Svec, H., The absolute abundance of the nitrogen isotopes in the atmosphere and compressed gas from various sources, *Geochim. Cosmochim. Acta*, 14, 234, 1958.
14. Melander, L. and Saunders, W.H., *Reaction Rates of Isotopic Molecules*, Wiley & Sons, New York, 1980.
15. Rieley, G., Derivatization of organic-compounds prior to gas-chromatographic combustion-isotope ratio mass-spectrometric analysis — identification of isotope fractionation processes, *Analyst*, 119, 915, 1994.
16. Meier-Augenstein, W., Applied gas chromatography coupled to isotope ratio mass spectrometry, *J. Chromatogr. A*, 842, 351, 1999.
17. Ehleringer, J.R. et al., Tracing the geographical origin of cocaine, *Nature*, 408, 311, 2000.
18. Burlingame, A.L. et al., Mass spectrometry, *Anal. Chem.*, 70, R647, 1998.
19. Preston, T. and Slater, C., Mass-spectrometric analysis of stable-isotope-labeled amino-acid tracers, *Proc. Nutr. Soc.*, 53, 363, 1994.

20. Rennie, M.J. et al., Use of continuous-flow combustion ms in studies of human metabolism, *Biochem. Soc. Trans.*, 24, 927, 1996.
21. Preston, T. and Owens, N.J.P., Interfacing an automatic elemental analyser with an isotope ratio mass spectrometer: the potential for fully automated total nitrogen and nitrogen-15 analysis, *Analyst*, 108, 971, 1983.
22. Preston, T. and Owens, N.J.P., Preliminary ^{13}C measurements using a gas chromatograph interfaced to an isotope ratio mass spectrometer, *Biomed. Mass Spectrom.*, 12, 510, 1985.
23. Brand, W.A. et al., Compound-specific isotope analysis — extending toward N-15/N-14 and O-18/O-16, *Org. Geochem.*, 21, 585, 1994.
24. Werner, R.A. et al., Online determination of delta-O-18 values of organic-substances, *Anal. Chim. Acta*, 319, 159, 1996.
25. Begley, I.S. and Scrimgeour, C.M., High-precision delta H-2 and delta O-18 measurement for water and volatile organic compounds by continuous-flow pyrolysis isotope ratio mass spectrometry, *Anal. Chem.*, 69, 1530, 1997.
26. Farquhar, G.D. et al., A rapid on-line technique for determination of oxygen isotope composition of nitrogen-containing organic matter and water, *Rapid Commun. Mass Spectrom.*, 11, 1554, 1997.
27. Meier-Augenstein, W., Stable isotope analysis of fatty acids by gas chromatography — isotope ratio mass spectrometry, *Anal. Chim. Acta*, 465, 63, 2002.
28. Brenna, J.T., High-precision gas isotope ratio mass-spectrometry — recent advances in instrumentation and biomedical applications, *Acc. Chem. Res.*, 27, 340, 1994.
29. Brand, W.A., High precision isotope ratio monitoring techniques in mass spectrometry, *J. Mass Spectrom.*, 31, 225, 1996.
30. Ellis, L. and Fincannon, A.L., Analytical improvements in irm-GC/MS analyses: advanced techniques in tube furnace design and sample preparation, *Org. Geochem.*, 29, 1101, 1998.
31. Metges, C.C. and Petzke, K.J., The use of GC-C-IRMS for the analysis of stable isotope enrichment in nitrogenous compounds, in *Methods of Investigation of Amino Acid and Protein Metabolism*, El-Khoury, A.E., Ed., CRC Press, Boca Raton, FL, 1999, p. 121.
32. Rautenschlein, M. et al., High-precision measurement of $^{13}C/^{12}C$ ratios by on-line combustion of GC eluates and isotope ratio mass spectrometry, in *Stable Isotopes in Paediatric, Nutritional and Metabolic Research*, Chapman, T.E. et al., Eds., Intercept, Andover, 1990, p. 133.
33. Merritt, D.A. et al., Performance and optimization of a combustion interface for isotope ratio monitoring gas-chromatography mass-spectrometry, *Anal. Chem.*, 67, 2461, 1995.
34. Eakin, P.A. et al., Some instrumental effects in the determination of stable carbon isotope ratios by gas-chromatography isotope ratio mass-spectrometry, *Chem. Geol.*, 101, 71, 1992.

35. Leckrone, K.J. and Hayes, J.M., Water-induced errors in continuous-flow carbon isotope ratio mass spectrometry, *Anal. Chem.*, 70, 2737, 1998.
36. Matucha, M. et al., Isotope effect in gas-liquid-chromatography of labeled compounds, *J. Chromatogr*, 588, 251, 1991.
37. Matucha, M., Isotope effects (IEs) in gas chromatography (GC) of labelled compounds (LCs), in *Synthesis and Applications of Isotopically Labelled Compounds*, Allen, J., Ed., John Wiley & Sons, New York, 1995, p. 489.
38. Meier-Augenstein, W. et al., Influence of gas-chromatographic parameters on measurement of C-13/C-12 isotope ratios by gas-liquid-chromatography combustion isotope ratio mass-spectrometry. 1, *J. Chromatogr. A*, 752, 233, 1996.
39. Prosser, S.J. and Scrimgeour, C.M., High-precision determination of H-2/H-1 in H_2 and H_2O by continuous-flow isotope ratio mass-spectrometry, *Anal. Chem.*, 67, 1992, 1995.
40. Scrimgeour, C.M. et al., Measurement of deuterium incorporation into fatty acids by gas chromatography isotope ratio mass spectrometry, *Rapid Commun. Mass Spectrom.*, 13, 271, 1999.
41. Corso, T.N. and Brenna, J.T., High-precision position-specific isotope analysis, *Proc. Natl. Acad. Sci. U.S.A.*, 94, 1049, 1997.
42. Meier-Augenstein, W. et al., Bridging the information gap between isotope ratio mass-spectrometry and conventional mass-spectrometry, *Biol. Mass Spectrom.*, 23, 376, 1994.
43. Meier-Augenstein, W., Online recording of C-13/C-12 ratios and mass-spectra in one gas-chromatographic analysis., *J. High Res. Chromatogr.*, 18, 28, 1995.
44. Meier-Augenstein, W. et al., Determination of ^{13}C enrichment by conventional GC-MS and GC-(MS)-C-IRMS, *Isotopes Environ. Health Stud.*, 31, 261, 1995.
45. Chang, W.-T., Smith, J., and Liu, R.H., Isotopic analogs as internal standards for quantitative GC/MS analysis — molecular abundance and retention time difference as interference factors, *J. Forensic Sci.*, 47, 873, 2002.
46. Smith, B.N. and Epstein, S., Two categories of $^{13}C/^{12}C$ ratios of higher plants, *Plant Physiol.*, 47, 308, 1971.
47. Dellens, E., La discrimination du ^{12}C et des trois types de metabolisme des plantes, *Physiol. Veget.*, 14, 641, 1976.
48. Rittenberg, D. and Ponticorvo, L., A method for the determination of the ^{18}O concentration of the oxygen of organic compounds, *Int. J. Appl. Radiat. Isot.*, 1, 208, 1956.
49. Parker, P.L., The chemical basis of the use of $^{13}C/^{12}C$ ratios to detect the addition of sweeteners to fruit juice concentrates, in *Proc. Symp. Technol. Probl. Fruit Juice Concentrates*, Oregon State University, Portland, OR, 1981, p. 50.

50. White, J. and Doner, L., The $^{13}C/^{12}C$ ratio in honey, *J. Apic. Res.*, 17, 94, 1978.
51. Krueger, H.W. and Reesman, R.H., Carbon isotope analyses in food technology, *Mass Spectrom. Rev.*, 1, 205, 1982.
52. Doner, L.W., Krueger, H.W., and Reesman, R.H., Isotope composition of carbon in apple juice, *J. Agric. Food Chem.*, 28, 362, 1980.
53. Doner, L.W. and Bills, D.D., Stable carbon isotope ratios in orange juice, *J. Agric. Food Chem.*, 29, 803, 1981.
54. White, J.W. and Doner, L.W., Mass spectrometric detection of high-fructose corn syrup in honey by use of $^{13}C/^{12}C$ ratio: collaborative study, *J. Assoc. Off. Anal. Chem.*, 61, 747, 1978.
55. Doner, L.W. and Phillips, J.G., Detection of high fructose corn syrup in apple juice by mass spectrometric $^{13}C/^{12}C$ analysis: collaborative study, *J. Assoc. Off. Anal. Chem.*, 64, 85, 1981.
56. Brause, A.R. et al., Verification of authenticity of orange juice, *J. Assoc. Off. Anal. Chem.*, 67, 535, 1984.
57. Doner, L.W., Carbon isotope ratios in natural and synthetic citric acid as indicators of lemon juice adulteration, *J. Agric. Food Chem.*, 33, 770, 1985.
58. Doner, L.W., Chia, D., and White, J.W., Mass spectrometric $^{13}C/^{12}C$ determination to distinguish honey and C3 plant syrup from C4 plant syrups (sugar cane and corn) in candied pineapple and papaya, *J. Assoc. Off. Anal. Chem.*, 62, 928, 1979.
59. White, J.W. et al., Stable carbon isotope ratio analysis of honey: validation of internal standard procedure for worldwide application, *J. AOAC Int.*, 81, 610, 1998.
60. Schmid, E.R. et al., Intramolecular $^{13}C/^{12}C$ isotope ratios of acetic acid of biological and synthetic origin, *Biomed. Mass Spectrom.*, 8, 496, 1981.
61. Hoffman, P. and Salb, M.J., Isolation and stable isotope ratio analysis of vanillin, *J. Agric. Food Chem.*, 27, 352, 1979.
62. Winkler, F. and Schmidt, H.-L., Application possibilities of carbon-13 isotope mass spectrometry in food analysis, *Z. Lebensm. Unters. Forsch.*, 171, 85, 1980.
63. Angerosa, F. et al., Carbon stable isotopes and olive oil adulteration with pomace oil, *J. Agric. Food Chem.*, 45, 3044, 1997.
64. Angerosa, F. et al., Application of stable isotope ratio analysis to the characterization of the geographical origin of olive oils, *J. Agric. Food Chem.*, 47, 1013, 1999.
65. Dunbar, J. and Wilson, A.T., Determination of geographic origin of caffeine by stable isotope analysis, *Anal. Chem.*, 54, 590, 1982.
66. Rossmann, A. et al., Stable isotope oxygen isotope content of water of EU data-bank wines from Italy, France and Germany, *Z. Lebensm. Uners. Forsch. A*, 208, 400, 1999.

67. Nakamura, K. et al., Geographical variations in the carbon isotope composition of the diet and hair in contemporary man, *Biomed. Mass Spectrom.*, 9, 390, 1982.
68. Borobia, M. et al., Blubber fatty acids of finback and humpback whales from the Gulf of St. Lawrence, *Mar. Biol.*, 122, 341, 1995.
69. Horita, J. and Vass, A.A., Stable-isotope fingerprints of biological agents as forensic tool, *J. Forensic Sci.*, 48, 122, 2003.
70. Liu, J.[R.]H. et al., Possible characterization of samples of *Cannabis sativa* L. by their carbon isotopic distribution, *J. Forensic Sci.*, 24, 814, 1979.
71. Keelin, C.D., A mechanism for cyclic environment of carbon-12 by terrestrial plants, *Geochim. Cosmochim. Acta*, 24, 299, 1961.
72. Desage, M. et al., Gas chromatography with mass spectrometry or isotope-ratio mass spectrometry in studying the geographical origin of heroin, *Anal. Chim. Acta*, 247, 249, 1991.
73. Besacier, F. et al., Isotopic analysis of ^{13}C as a tool for comparison and origin assignment of seized heroin samples, *J. Forensic Sci.*, 42, 429, 1997.
74. Ihle, E. and Schmidt, H.L., Multielement isotope analysis on drugs of abuse. Possibility for their origin assignment, *Isotope Environ. Health Stud.*, 32, 226, 1996.
75. Ehleringer, J.R. et al., Geo-location of heroin and cocaine by stable isotope ratios, *Forensic Sci. Int.*, 106, 27, 1999.
76. Hays, P.A. et al., Geographic origin determination of heroin and cocaine using site-specific isotopic ratio deuterium NMR, *J. Forensic Sci.*, 45, 552, 2000.
77. Besacier, F. and Chaudron-Thozet, H., Chemical profiling of illicit heroin samples, *Forensic Sci. Rev.*, 11, 105, 1999.
78. Dautraix, S. et al., ^{13}C isotopic analysis of an acetaminophen and diacetylmorphine mixture, *J. Chromatogr. A*, 756, 203, 1996.
79. Besacier, F. et al., Application du couplage chromatographie gazeuse-spectrometrie de masse isotopique de l'azote á l'analyse d'échantillons de drogues, *Analysis*, 27, 17, 1999.
80. Bommer, P. et al., Determination of the origin of drugs by measuring natural isotope contents: D/H and $^{13}C/^{12}C$ ratio of some diazepam samples, *Zeit. Naturforsch.*, 31c, 111, 1976.
81. Mas, F. et al., Determination of "common-batch" members in a set of confiscated 3,4-(methylenedioxy)methylamphetamine samples by measuring the natural isotope abundances: a prelimiary study, *Forensic Sci. Int.*, 71, 225, 1995.
82. Becchi, M. et al., Gas chromatography combustion isotope ratio mass-spectrometry analysis of urinary steroids to detect misuse of testosterone in sport, *Rapid Commun. Mass Spectrom.*, 8, 304, 1994.

83. Aguilera, R. et al., Improved method of detection of testosterone abuse by gas chromatography/combustion/isotope ratio mass spectrometry analysis of urinary steroids, *J. Mass Spectrom.*, 31, 169, 1997.
84. Shackleton, C.H.L. et al., Confirming testosterone administration by isotope ratio mass spectrometric analysis of urinary androstanediols, *Steroids*, 62, 379, 1997.
85. Shackleton, C.H. L. et al., Androstanediol and 5-androstenediol profiling for detecting exogenously administered dihydrotestosterone, epitestosterone, and dehydroepiandrosterone: potential use in gas chromatography isotope ratio mass spectrometry, *Steroids*, 62, 665, 1997.
86. Aguilera, R., Hatton, C.K., and Catlin, D.H., Detection of epitestosterone doping by isotope ratio mass spectrometry, *Clin. Chem.*, 48, 629, 2002.
87. Nissenbaum, A., The distribution of natural stable isotopes of carbon as a possible tool for the differentiation of samples of TNT, *J. Forensic Sci.*, 20, 455, 1975.
88. Wolfe, R.R., *Tracers in Metabolic Research — Radioactive and Stable Isotope/Mass Spectrometry Methods*, Alan R. Liss, New York, 1983.
89. Low, I.A. et al., Gas chromatographic/mass spectrometric determination of carbon isotope composition in unpurified samples: methamphetamine example, *Biomed. Environ. Mass Spectrom.*, 13, 531, 1986.

Identification of Ignitable Liquid Residues in Fire Debris by GC/MS/MS

5

DALE SUTHERLAND
JEANNETTE PERR
JOSÉ R. ALMIRALL

Two different approaches to forensic analysis of accelerants in fire debris by GC/MS/MS are described in this chapter. One method is to optimize MS/MS parameters and conditions for a series of target compounds in order to obtain maximum sensitivity and specificity for ignitable liquids. A specific compound of interest is isolated by selecting a parent ion, which is passed through the MS/MS collision chamber, to be decomposed by collision-induced dissociation (CID), followed by analysis of the daughter ions for compound identification. This is being discussed as an optimized target compound GC/MS/MS analysis.

The other approach is to select parent ions characteristic of several chemical classes in order to optimize the selectivity of the GC/MS/MS analysis through class pattern recognition. MS/MS is being used to "clean up" the GC/MS chromatogram, followed by comparison of the profile of the investigated analyte with profiles of standards. This is being discussed as target class GC/MS/MS analysis.

Contents

0-8493-1522-0/04/$0.00+$1.50

5.1 The Use of Compound-Specific MS/MS for the Identification of Ignitable Liquid Residues in Fire Debris Analysis

José R. Almirall and Jeannette Perr

5.1.1 Introduction

The crime of arson is very difficult to prosecute for several reasons, including the usual lack of physical evidence at the crime scene and the difficulty in associating with a particular suspect any physical evidence found, such as the presence of an ignitable liquid residue. The detection and identification of an *ignitable liquid residue*, the term used for the compounds or mixtures of compounds often used to accelerate a fire, is one of the first steps in the investigation of a suspected arson. As often noted, the arson scene can be quite complex and chaotic. The forensic chemist considers every suspected arson scene as unique, taking into account all the many different variables that can affect the extraction and identification of any ignitable liquid residues present. In a case where there is ample evidence of an ignitable liquid residue, the analysis of the evidence is very straightforward. The American Society for Testing and Materials (ASTM) International has published several excellent standards for the extraction[1–6] of ignitable liquids and for the classification and interpretation[7–8] of the analytical results from this examination. It is when the sample contains these compounds in very low concentrations or in the presence of a large amount of background substrate that more sensitive techniques are required. In the absence of these sensitive techniques, the samples would be determined as negative for the presence of ignitable liquid residues. This chapter describes the recent application of a tandem (MS/MS) mass spectrometry experiment for the purposes of isolating and identifying ignitable liquid residues in the presence of background products that may mask the target analytes of interest. Our approach involves compound specific MS/MS experiments that can be used to first isolate a specific compound of interest by isolating the parent ion (or one of the fragment ions) in the first MS experiment, conducting a collisional-induced dissociation, and then resolving the resulting daughter ions for compound identification, all within the approximate time frame of a single mass spectrometry experiment. The approach, previously described for the identification of ignitable liquid residues from fire debris,[9–12] is widely used in analytical chemistry for the identification of many different analytes of interest, including drugs of abuse.[13] Other researchers[14] are currently investigating MS/MS methods for the analysis of fire debris evidence in order to distinguish pyrolysate markers from fuels.

5.1.2 Tandem Mass Spectrometry (MS/MS)

Mass spectrometry is now widely used in the identification of organic compounds such as the components found in ignitable liquid residues. The result of an MS experiment is an ion that could represent the molecular ion or one of the fragment ions of the target analyte. Tandem mass spectrometry involves a second MS experiment on the isolated ion from the first experiment. MS/MS can be achieved in space (spatial resolution) with magnetic sector or quadrupole instruments connected in series or in time (temporal resolution) with the use of an ion trap mass spectrometer instrument. The triple-quadrupole mass spectrometer was developed[15] and commercialized for routine use of this technique. Following the chromatographic separation of the compounds, an ion source provides the ionization step and is followed by component ion separation in the first quadrupole, which acts as a mass filter. In the typical GC/MS experiment, detection of the ions would follow this step to end the experiment. In a triple-quadrupole MS/MS experiment, a second quadrupole acts as a collision chamber to produce collisional induced dissociation (CID) of the selected ions. The third quadrupole, also a mass analyzer, can resolve the resulting daughter ions produced in the collision chamber and detect the ions. MS/MS can be coupled to gas or liquid chromatography systems.

This chapter will describe the operation of an ion trap for MS/MS analysis of fire debris evidence. The ion trap can store ions for subsequent ejection and detection. A schematic diagram of an ion trap is shown in Figure 5.1. It consists of three cylindrical electrodes, a ring, and two caps on each end of the ring. The space within the rings is carefully designed in order to provide for hyperbolic orbits of the ions formed. A radio frequency (RF) oscillating potential is applied to the ring electrode while the end cap electrodes are held at a fixed potential. Electron impact ionization can be achieved by the injection of electrons with an energy of 70 eV into the trap through an orifice in one of the end cap electrodes. The RF potential provides stable orbits for a range of ion masses so that these can be stored (typically the 30 to 450 *m/z* range is used in the analysis of fire debris evidence). The ions may be ejected from the trap, in order of increasing *m/z* values, by increasing the amplitude of the RF potential. The ions will exit the trap through the opposite end cap for detection by an electron multiplier. Commercially available ion traps can provide unit resolution. A tandem (MS/MS) experiment is achieved by selecting a single ion (or a range of ions having different *m/z* values) that may be stored for subsequent dissociation and eventual detection, as described above. A thorough treatise on the theory and operation of the ion trap can be found in the work by March and Todd.[16]

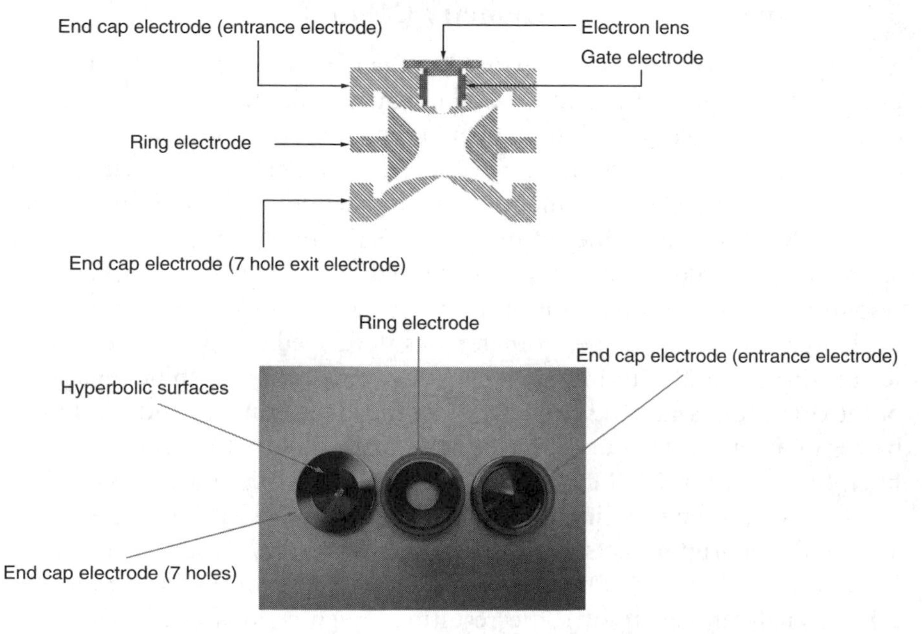

Figure 5.1 Schematic diagram of an ion trap consisting of three cylindrical electrodes.

5.1.3 Fire Debris Analysis

The investigation of a suspected arson includes the extraction, isolation, and analysis of the debris resulting from the fire, in order to determine the presence of any ignitable liquid residues (ILR) that may have been used as an "accelerant" and could serve as evidence of an arson. Typical accelerants used are automobile gasoline, diesel fuel, and other volatile mixtures that can be acquired commercially. The literature describing the extraction of these compounds from debris, the analysis of the extracts, and the interpretation of the data is very mature.[17–24] Forensic scientists, through the ASTM, have developed and published consensus standards for the analysis and interpretation of fire debris evidence.[1–8] These advances have led to what are now fairly routine and standardized analytical methodologies for the analysis of fire debris. In most cases, there is enough target compound concentration in the sample so that the extraction and analysis is straightforward. In the absence of methods for the extraction and analysis of very low concentrations of the volatile organic compounds that make up ILRs, the sample may be determined as negative, possibly ending the investigation process. Recent reports of sensitive methods for the extraction and analysis of ILR target compounds, including the use of solid phase microextraction (SPME), have described the application of SPME as an improvement in the extraction and preconcentration of the analytes of interest in a single, simple step prior to GC/MS analysis.[25–35,12] These methods have been shown to improve the sensitivity of the analysis by improving the extraction efficiency of very low concentrations of the target compounds. Further complications during the analysis of these kinds of samples include the presence of compounds from the substrate background and product compounds formed during the combustion and pyrolysis of substrate material. It has been reported that these interferences could have an effect on the analysis of the ILRs,[24] even when GC/MS is used as the analytical technique. One possible solution to the resolution of small quantities of target compounds in the presence of interfering species is the use of GC/MS/MS, which isolates the target compounds of interest in the first MS experiment (using an ion trap), followed by dissociation of the "trapped" target compound, and finally, the identification of the target compound, free of interferences. The following describes the use of MS/MS in the isolation and analysis of ILRs in the presence of interfering species when the sample contains ILRs in very low concentrations, and GC/MS alone cannot solve the problem.

5.1.3.1 Ion Trap MS/MS in Fire Debris Analysis

Electron ionization (EI) is typically used as the source of ionization in the mass spectral analysis of ignitable liquid residues. A voltage of 70 V imparts enough energy in the ionization process to form positive ion fragments.

While chemical ionization (CI) is also possible,[12,36] the following describes only EI followed by MS/MS in the analysis of ILRs. Ion traps have an inherent advantage in sensitivity over quadrupole mass spectrometers in that the target analytes can be stored in the trap during the analysis while maintaining a high efficiency of detection. The daughter-ion mass spectra obtained during an MS/MS experiment differ from EI mass spectra, requiring a new library of MS/MS spectra to be created by the user.

The sample stream exits the chromatographic column and enters the trap through a hole between one of the quartz spacers and one of the end caps (see Figure 5.1). The internal ionization ion trap experiment first creates the ions by sending a pulse of electrons into the trap. In MS/MS, a predetermined ion (called parent or precursor ion) formed in the first MS stage can be separated from all other ions, trapped, and subjected to collision induced dissociation (CID) through collision with an inert gas such as helium, argon, or nitrogen. The precursor ion can be either the molecular ion or a fragment ion but needs to be unique to the component of interest. Ions formed by CID, referred to as daughter ions, are ejected and detected. Figure 5.2 illustrates the different steps in an MS/MS experiment. The true value of an MS/MS experiment is the ability to isolate ignitable liquid residue target components from coeluting peaks of interfering compounds. This helps to improve the signal-to-noise ratios of the target compounds, thereby improving the level of detection while simultaneously improving the selectivity of the target compounds.

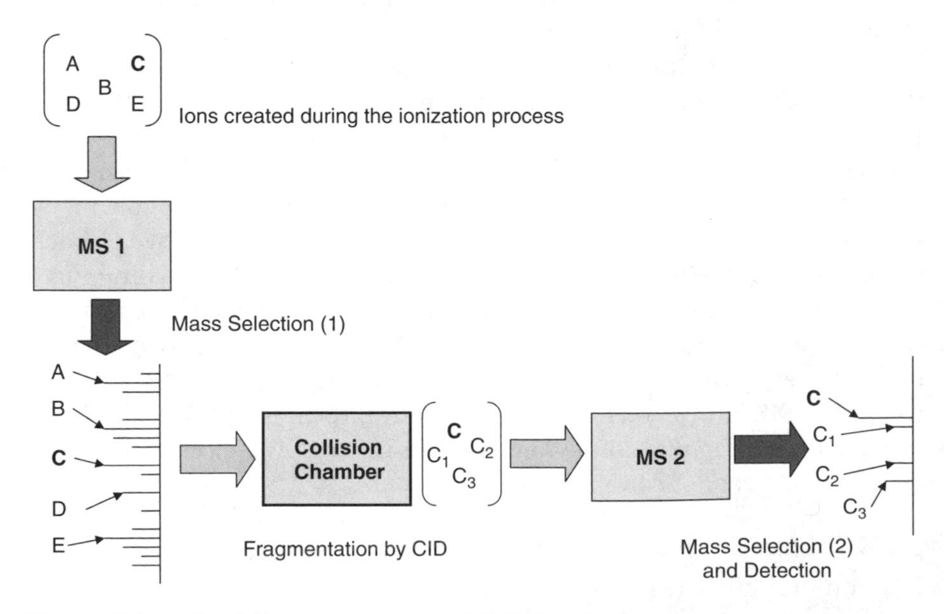

Figure 5.2 The different steps in an MS/MS experiment.

MS/MS experiments are capable of selecting the compounds of interest and isolating them from the interfering components that coelute from the GC. In aromatic compounds, the molecular ion is selected, while in alkanes, which form fragment ions a typical fragment ion such as *m/z* 85 is chosen.

5.1.3.2 MS/MS Examples

Figure 5.3 illustrates the chromatogram of pyrolysis products (top) and pyrolysis products with a number of the components in gasoline added to form a total mixture concentration of 1,000 ppm (middle) along with a 1,000 ppm gasoline standard (bottom), using GC/MS analysis. Pyrolysis products were formed by burning a 2" × 2" section of nylon carpet and extracting the headspace over the burned carpet using an activated charcoal strip (ACS) measuring 8 × 20 mm and eluting the ACS with 100 μl of carbon disulfide. Figure 5.4 shows the same samples analyzed under GC/MS/MS conditions. The GC/MS/MS chromatograms are more easily interpreted, and the gasoline components are clearly isolated from the coeluting interferants from the burned carpet. In order to conduct such an experiment, the expected retention times for the target compounds must be known and the MS/MS conditions applied for each retention time. The gasoline standard was used to determine the retention times and MS/MS conditions. The MS/MS conditions using the Varian Saturn 2000 ion trap are summarized in Table 5.1. The trap temperature is set lower than normal operation of the trap (~ 100°C to 170°C). The excitation voltages for each of the compounds of interest are first determined in a preliminary experiment and the resulting MS/MS spectra are saved in a library for comparison to spectra generated by the samples.

In a second example of the utility of GC/MS/MS analysis, a small amount (<1 μl) of gasoline was added to previously burned nylon carpet and the heated headspace above the burned debris was extracted using an ACS. The resulting chromatograms from the GC/MS analysis (Figure 5.5) and the GC/MS/MS analysis (Figure 5.6) further illustrate the improvements in selectivity of the MS/MS technique and the simplification of the interpretation of the analytical results. A gasoline standard is shown on top of both Figure 5.5 and Figure 5.6. The gasoline components in the bottom chromatogram of Figure 5.5 are masked when using only GC/MS analysis. The gasoline components isolated from the interfering pyrolysis products are clearly shown on the bottom chromatogram of Figure 5.6.

Figure 5.7 and Figure 5.8 illustrate the similarities and differences between the MS spectrum of *m*-xylene (Figure 5.7) and the MS/MS spectrum of *m*-xylene (Figure 5.8). Similarly, Figure 5.9 and Figure 5.10 illustrate the differences in mass spectra for naphthalene for the MS and MS/MS modes of operation, respectively.

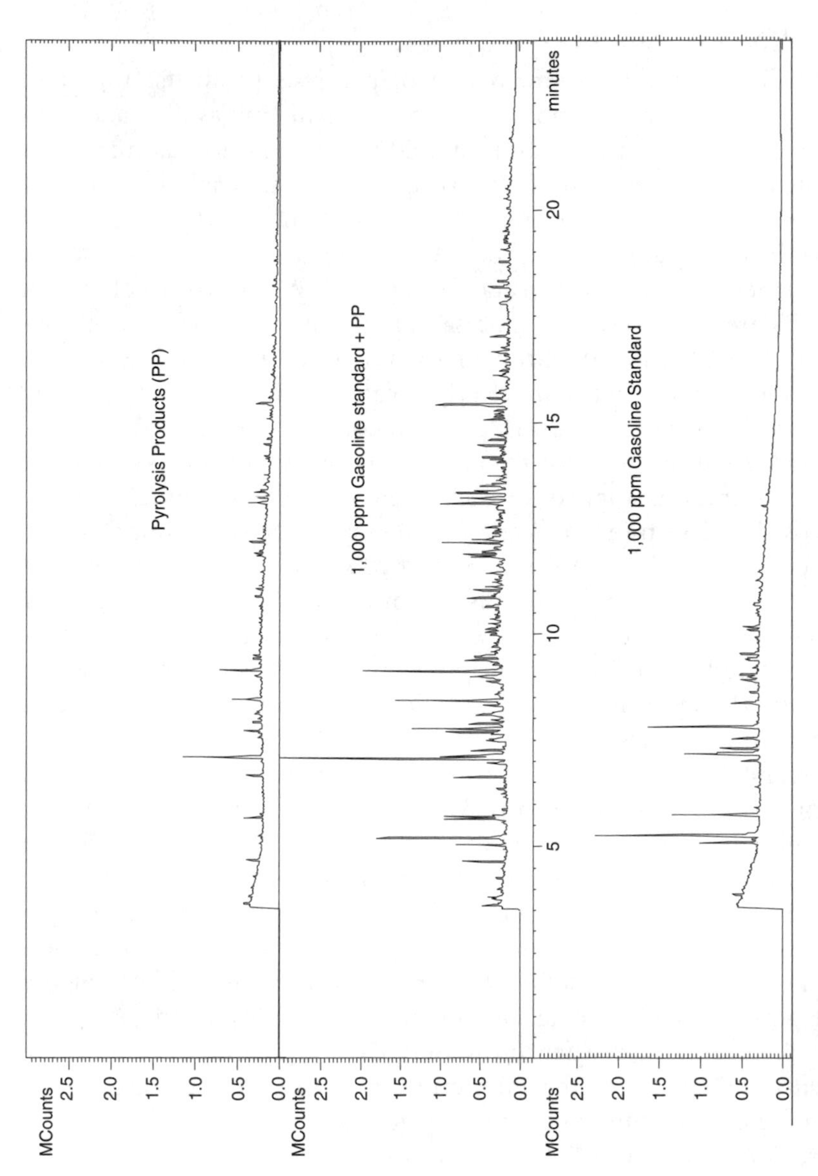

Figure 5.3 The chromatogram of pyrolysis products alone (top) and pyrolysis products with a spike of gasoline added to form a total mixture concentration of 1,000 ppm of gasoline (middle), along with a 1,000 ppm gasoline standard (bottom), using GC/MS analysis.

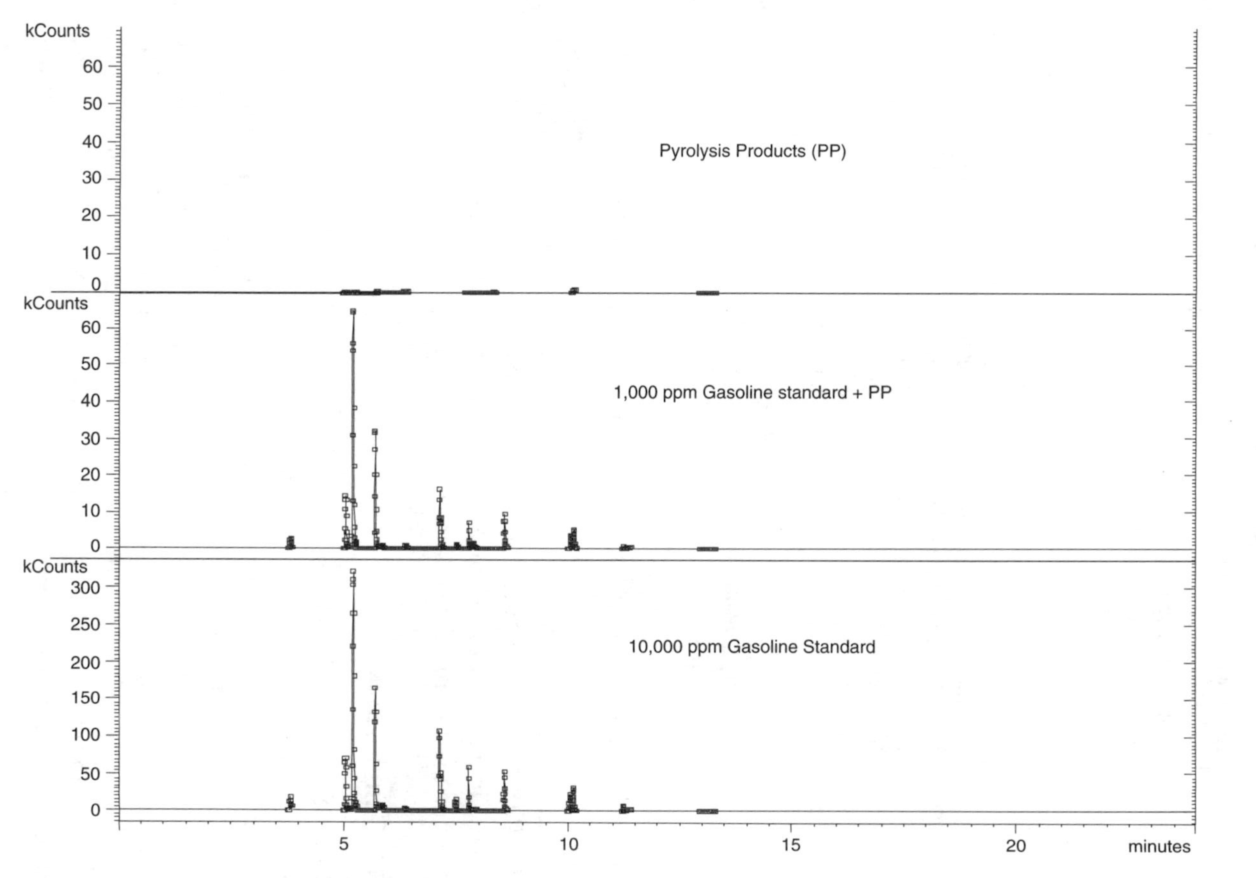

Figure 5.4 The GC/MS/MS analysis of pyrolysis products alone (top) and pyrolysis products with gasoline added to form a total mixture concentration of 1,000 ppm of gasoline (middle), along with a 10,000 ppm gasoline standard (bottom).

Table 5.1 The GC, MS, and MS/MS Conditions Using the Varian Saturn 2000 Ion Trap

GC Conditions	
He flow rate (ml/min)	1
Injection mode	Split and splitless
Split ratio in split mode	18:1
Injector temperature (°C)	280
Run time (min)	23
MS Conditions	
Scan time (μsec)	2
Target TIC (counts)	5000
Maximum ionization time (μsec)	25000
Ion trap temperature (°C)	180
Prescan ionization time (μsec)	1500
Background mass (amu)	38
RF dump mass (amu)	650
Scan range (amu)	40–400
MS/MS Conditions	
Excitation voltage (V)	40–60 (aromatics) 0 (aliphatics)

One advantage of the MS/MS mode is the substantial reduction in background ions for a particular selected ion. Figure 5.11 is an example of the increase in signal relative to the noise when *m*-xylene is detected in MS (top) and MS/MS (bottom) modes. Table 5.2 summarizes the improvements in sensitivity, in most cases, when MS/MS is used for the analysis of target compounds used in fire debris analysis.

5.1.4 Conclusions

GC/MS/MS provides improved selectivity for the identification of target compounds when coeluting compounds mask the compounds of interest. The detection limit (S/N>3) for a single component target compound in an ignitable liquid residue is ~ 10 times lower when using GC/MS/MS in the splitless mode over GC/MS in the splitless mode. The combination of improved selectivity and improved sensitivity provides GC/MS/MS with the potential for improving both the detection and the identification of target compounds in residues extracted from fire debris, especially in cases where the sample concentration is very low and/or when the sample contains interfering species. It is expected that future work will include the generation of compound specific MS/MS spectra under standardized conditions for target compounds of interest in fire debris analysis.

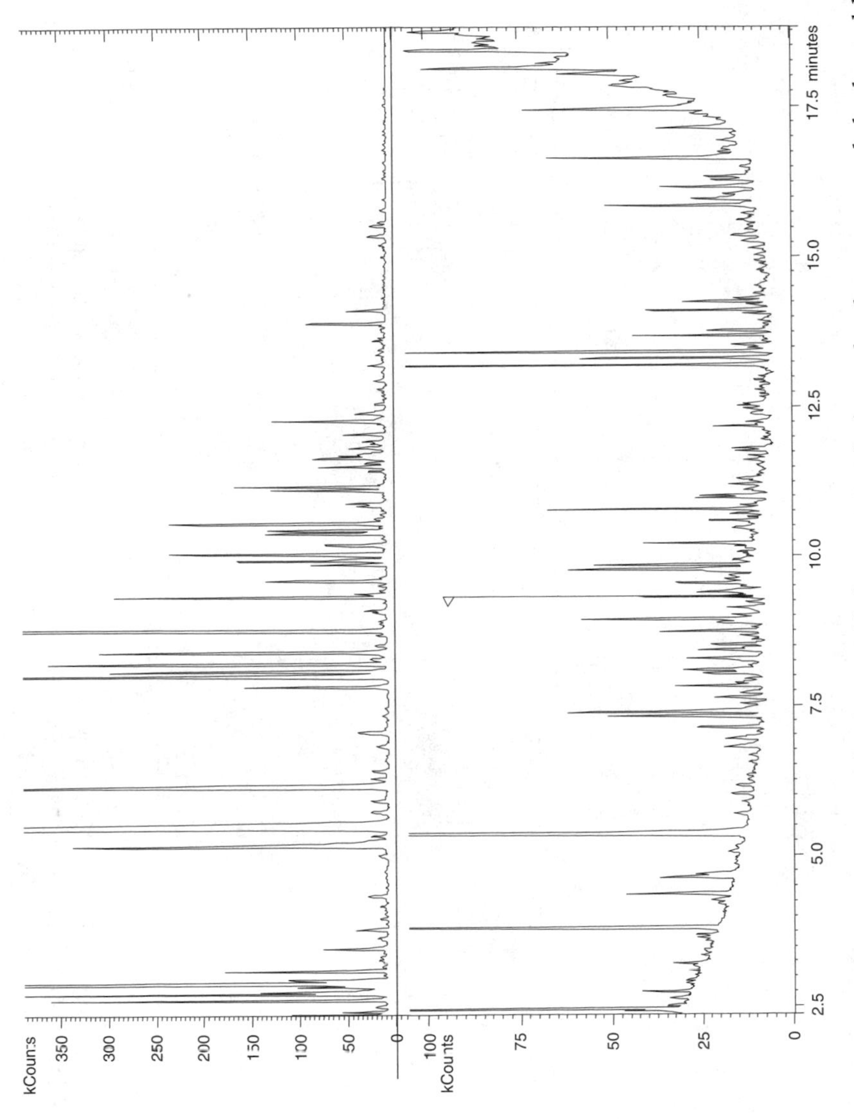

Figure 5.5 A 1-μl spike of gasoline standard (top) was added to previously burned nylon carpet and the heated headspace above the burned debris was extracted using an ACS (bottom) and analyzed using GC/MS.

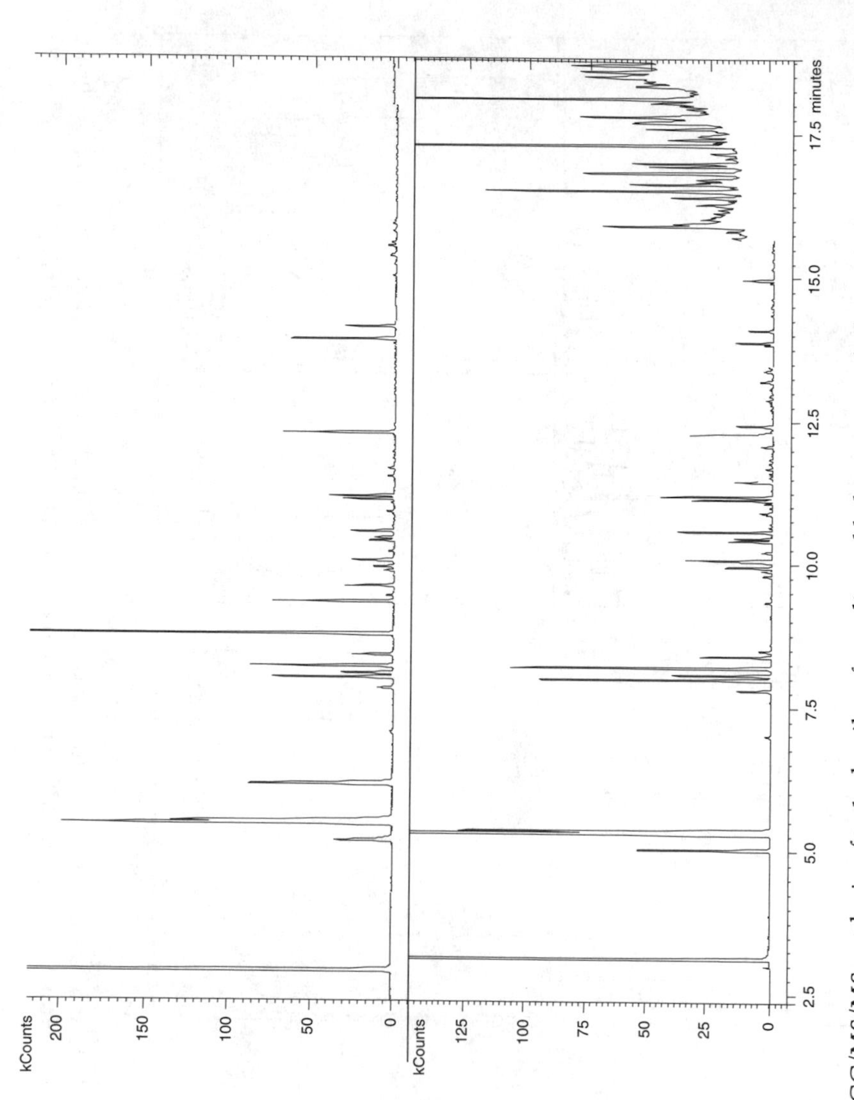

Figure 5.6 The GC/MS/MS analysis of a 1-µl spike of gasoline added to previously burned nylon carpet. The heated headspace above the burned debris was extracted using an ACS (bottom) and compared to a 1-µl spike of gasoline standard (top). (Note: The GC/MS/MS experiment was ended after ~ 16 min).

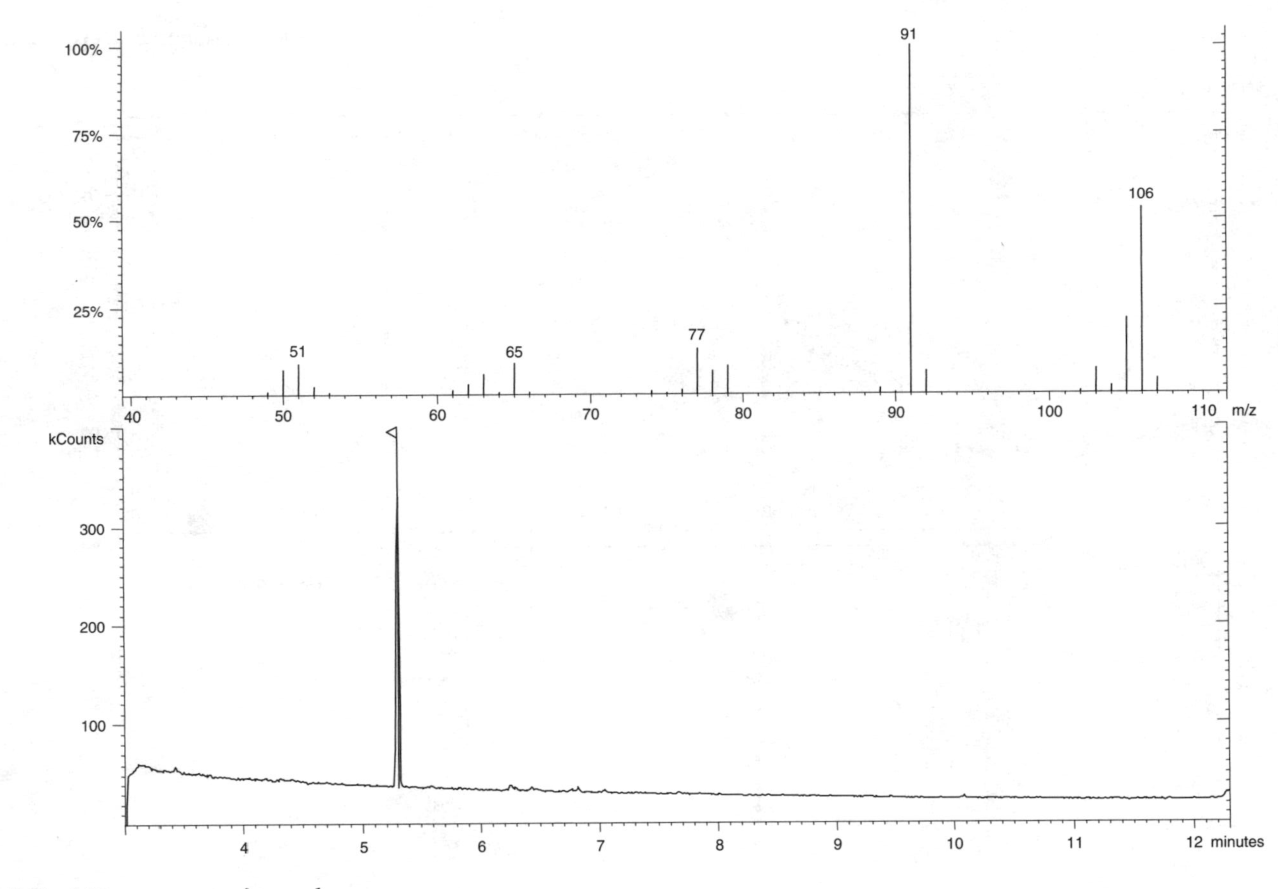

Figure 5.7 MS spectrum of *m*-xylene.

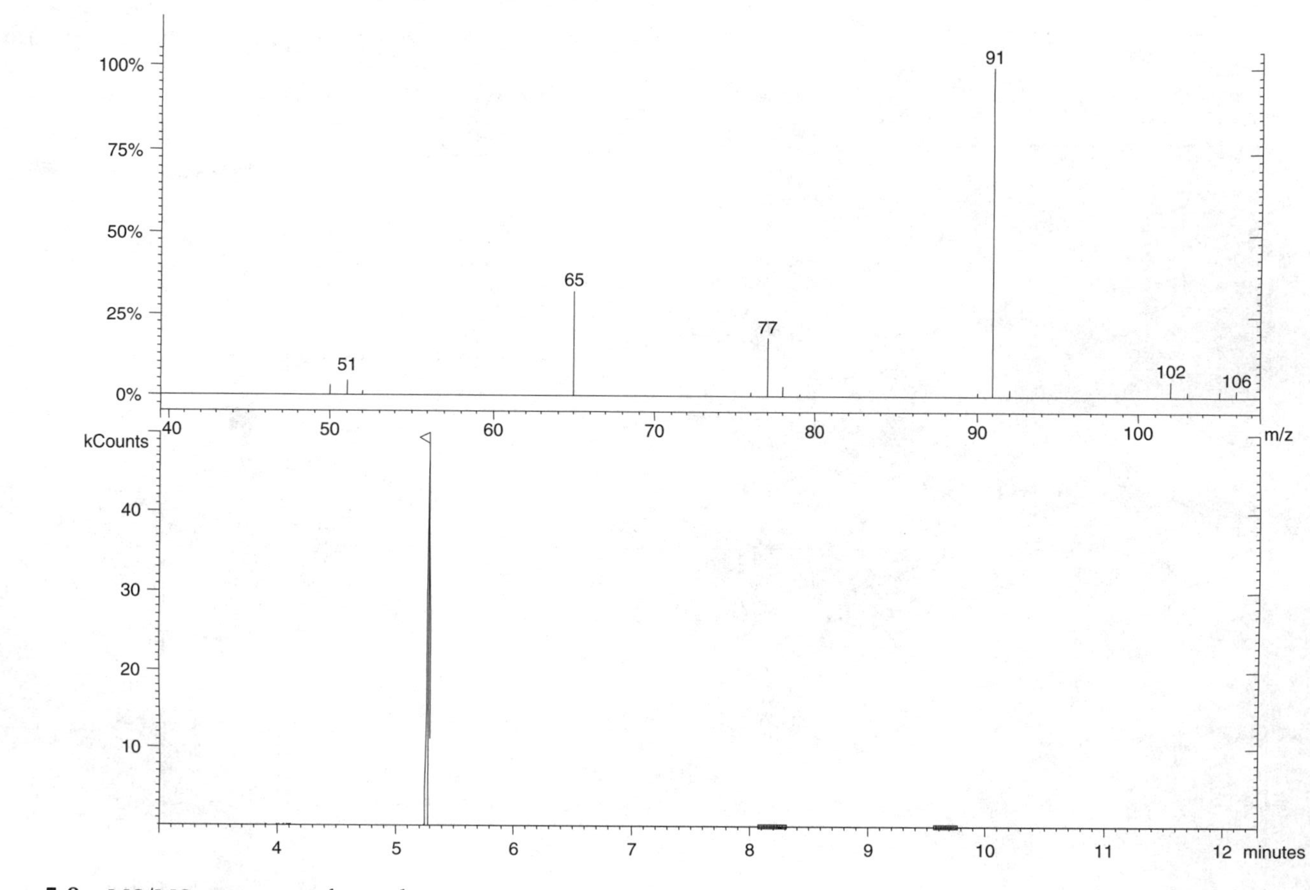

Figure 5.8 MS/MS spectrum of *m*-xylene.

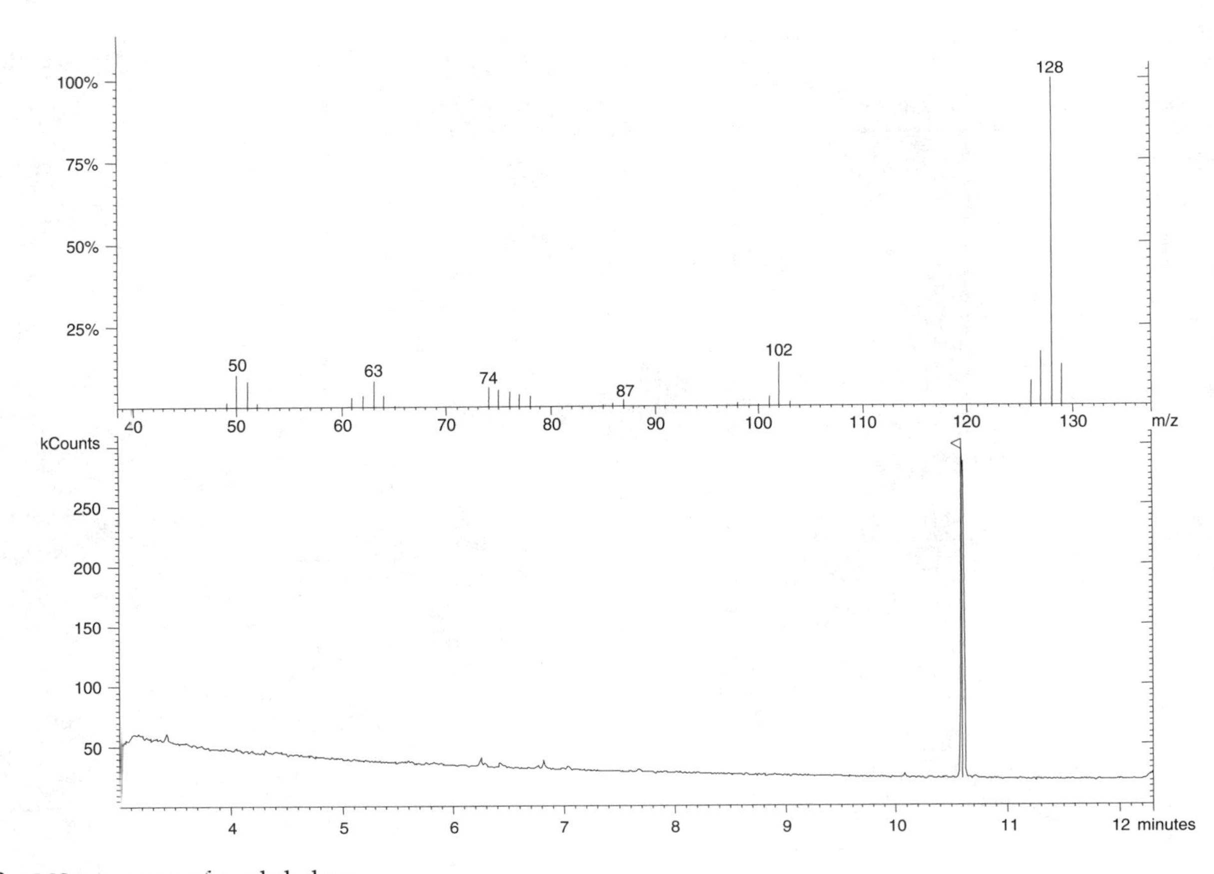

Figure 5.9 MS spectrum of naphthalene.

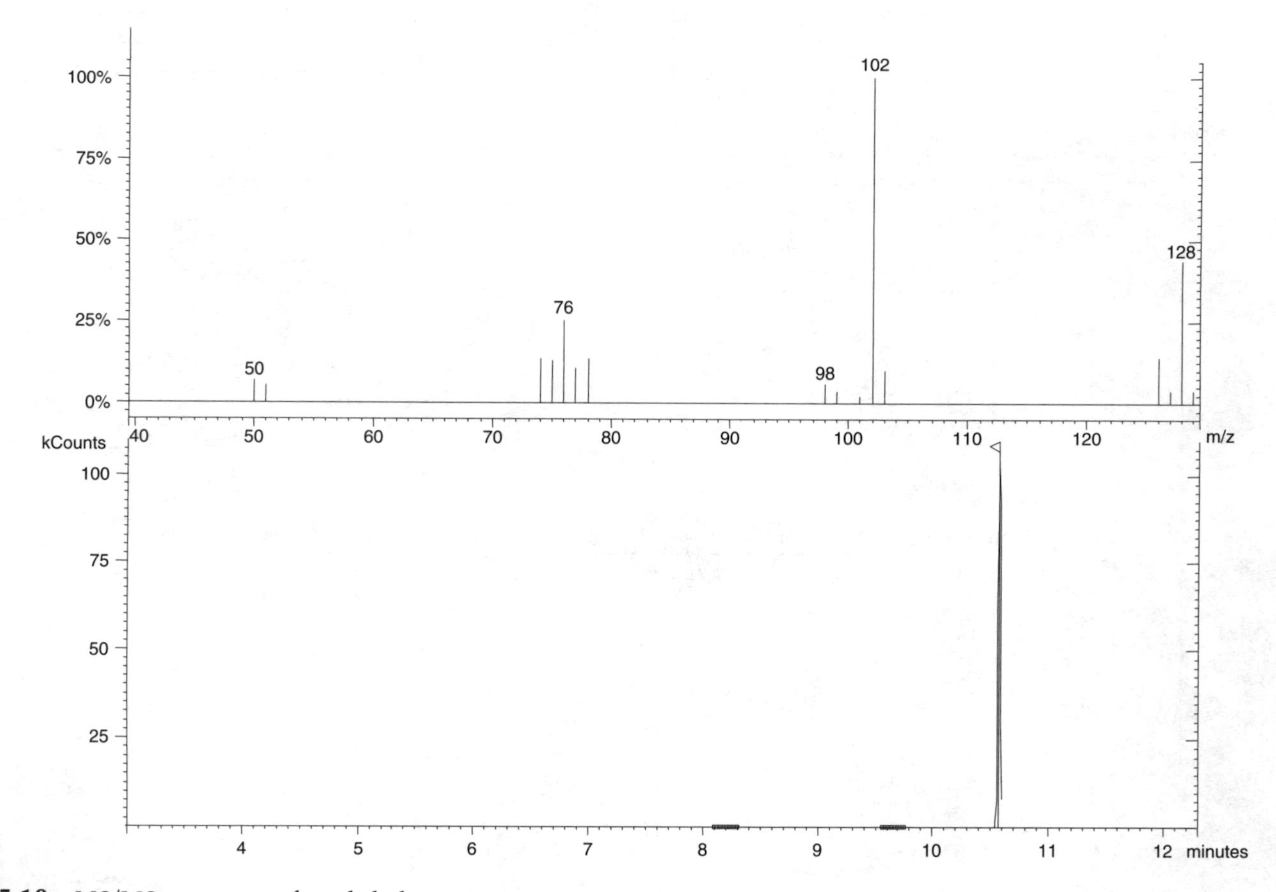

Figure 5.10 MS/MS spectrum of naphthalene.

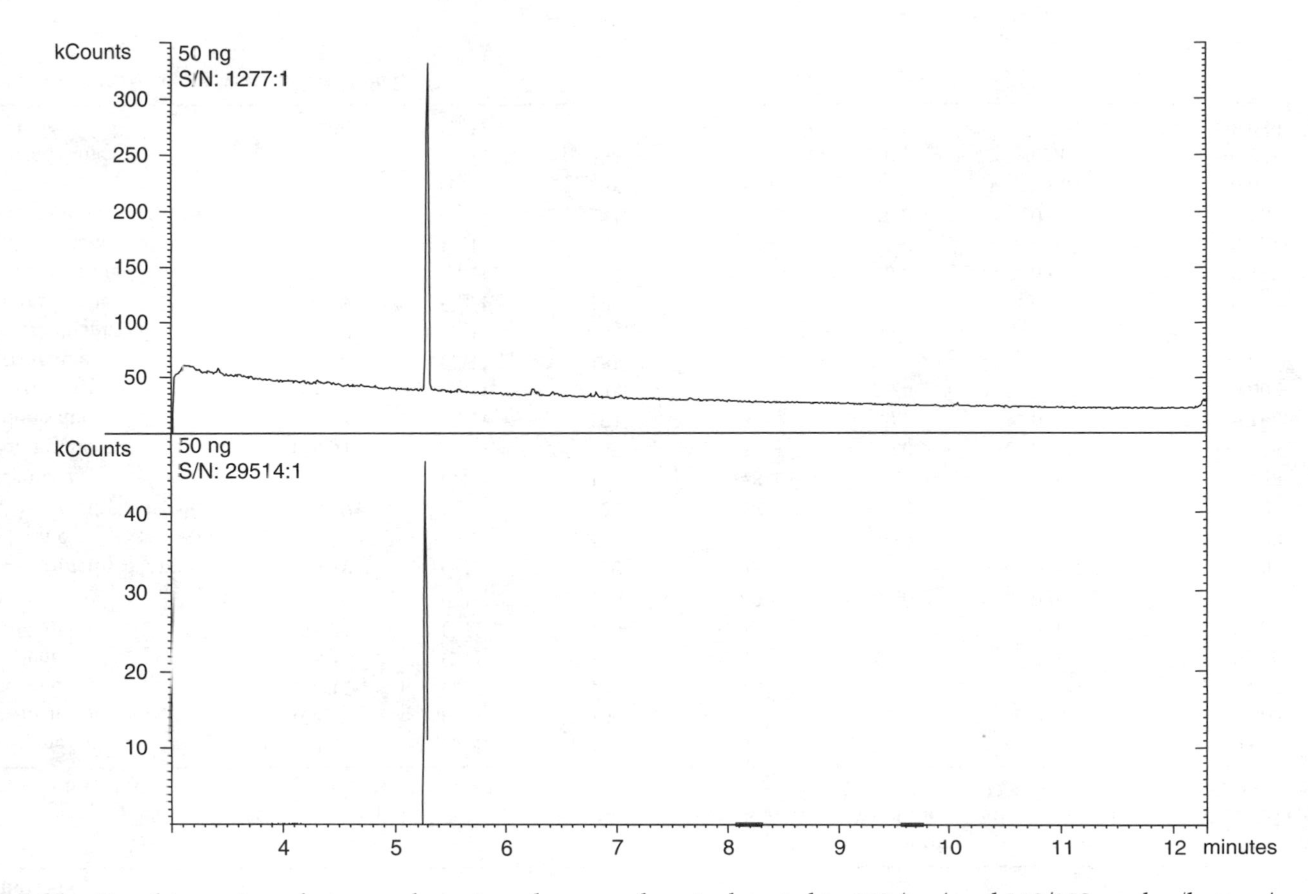

Figure 5.11 Signal intensity relative to the noise when *m*-xylene is detected in MS (top) and MS/MS modes (bottom).

Table 5.2 Comparison of the Sensitivity between MS and MS/MS for Many of the Target Compounds Used in Fire Debris Analysis

Compound	Molecular Weight (g/mol)	Formula	Parent Ion (*m/z*)	Excitation Storage Level (*m/z*)	Excitation Amplitude (V)	EI/MS Level of Detection (ng)	EI/MS/MS Level of Detection (ng)
n-octane	114.23	C_8H_{18}	115	37	0.0	0.040	0.642
ethylbenzene	106.17	C_8H_{10}	106	48	51.5	0.021	0.002
p-xylene	106.17	C_8H_{10}	106	48	52.5	0.034	0.019
o-xylene	106.17	C_8H_{10}	106	48	50.5	0.036	0.020
n-nonane	128.26	C_9H_{20}	129	37	0.0	0.020	0.568
cumene	120.19	C_9H_{12}	120	48	43.0	0.016	0.004
1,2,4-trimethylbenzene	120.19	C_9H_{12}	120	48	49.0	0.011	0.002
n-decane	142.28	$C_{10}H_{22}$	143	37	0.0	0.013	0.289
1,2,3-trimethylbenzene	120.19	C_9H_{12}	121	48	50.0	0.017	0.023
p-cymene	134.22	$C_{10}H_{14}$	134	48	43.0	0.009	0.148
n-undecane	156.31	$C_{11}H_{24}$	157	37	0.0	0.004	0.316
n-dodecane	170.33	$C_{12}H_{26}$	171	37	0.0	0.004	0.156
2-methylnaphthalene	184.36	$C_{13}H_{28}$	143	48	53.0	0.007	0.002
n-tridecane	142.20	$C_{11}H_{10}$	185	37	0.0	0.005	0.078
1-methylnaphthalene	142.20	$C_{11}H_{10}$	143	48	53.5	0.009	0.004
n-tetradecane	198.39	$C_{14}H_{30}$	199	37	0.0	0.002	0.080
n-pentadecane	212.41	$C_{15}H_{32}$	213	37	0.0	0.003	0.115
n-hexadecane	226.44	$C_{16}H_{34}$	227	37	0.0	0.002	0.024
n-heptadecane	240.47	$C_{17}H_{36}$	241	37	0.0	0.001	0.063
pristine	268.52	$C_{19}H_{40}$	268	37	0.0	0.001	0.074
n-octadecane	254.49	$C_{18}H_{38}$	255	37	0.0	0.001	0.038
phytane	283.60	$C_{20}H_{42}$	283	37	0.0	0.001	0.039

Note: Detection limits are determined as 3 times S/N.

Acknowledgments

The authors would like to acknowledge Supelco Sigma-Aldrich and the College of Arts and Sciences of Florida International University for financial support and Varian Inc. for a partial instrument grant which enabled us to conduct some of the work referred to in this chapter.

References

1. ASTM E1412-00 Standard Practice for Separation of Ignitable Liquid Residues from Fire Debris Samples by Passive Headspace Concentration With Activated Charcoal, *ASTM Annual Book of Standards, Volume 14.02*, ASTM International, West Conshohocken, PA, 2002.
2. ASTM E1413-00 Standard Practice for Separation and Concentration of Ignitable Liquid Residues from Fire Debris Samples by Dynamic Headspace Concentration, *ASTM Annual Book of Standards, Volume 14.02*, ASTM International, West Conshohocken, PA, 2002.
3. ASTM E1388-00 Standard Practice for Sampling of Headspace Vapors from Fire Debris Samples, *ASTM Annual Book of Standards, Volume 14.02*, ASTM International, West Conshohocken, PA, 2002.
4. ASTM E1385-00 Standard Practice for Separation and Concentration of Ignitable Liquid Residues from Fire Debris Samples by Steam Distillation, *ASTM Annual Book of Standards, Volume 14.02*, ASTM International, West Conshohocken, PA, 2002.
5. ASTM E1386-00 Standard Practice for Separation and Concentration of Ignitable Liquid Residues from Fire Debris Samples by Solvent Extraction, *ASTM Annual Book of Standards, Volume 14.02*, ASTM International, West Conshohocken, PA, 2002.
6. ASTM E2154-01 Standard Practice for Separation and Concentration of Ignitable Liquid Residues from Fire Debris Samples by Passive Headspace Concentration with Solid Phase Microextraction (SPME), *ASTM Annual Book of Standards, Volume 14.02*, ASTM International, West Conshohocken, PA, 2002.
7. ASTM E1387-01 Standard Test Method for Ignitable Liquid Residues in Extracts from Fire Debris Samples by Gas Chromatography, *ASTM Annual Book of Standards, Volume 14.02*, ASTM International, West Conshohocken, PA, 2002.
8. ASTM E1618-01 Standard Test Method for Ignitable Liquid Residues in Extracts from Fire Debris Samples by Gas Chromatography-Mass Spectrometry, *ASTM Annual Book of Standards, Volume 14.02*, ASTM International, West Conshohocken, PA, 2002.

9. Plasencia, M. D., Montero, S., Krivis, J., Armstrong, A., and Almirall, J., Improved sensitivity and selectivity for the detection and identification of ignitable liquid residues from fire debris by ion trap mass spectrometry I, in *Proc. of the American Academy of Forensic Sciences Meeting*, Reno, NV, 2000.
10. Perr, J. M., Diaz, C., Furton, K. G., and Almirall, J. R., A comprehensive approach for analyzing ignitable liquid residues and explosive material using SPME/GC/MS/MS, in *Proc. of the International Association of Forensic Science Meeting*, Montpellier, France, 2002.
11. Perr, J. M., Furton, K. G., and Almirall, J. R., The analysis of ignitable liquid residues and explosive material using SPME/GC/MS/MS, in *Proc. of the American Academy of Forensic Sciences Meeting*, Chicago, IL, 2003.
12. Almirall, J. R. and Perr, J., New developments and quality assurance in fire debris analysis, in *Analysis and Interpretation of Fire Scene Evidence*, Almirall, J. R. and Furton, K. G., Eds., CRC Press, Boca Raton, FL, 2003, in press.
13. Yost, R. A. and Fetterolf, D. D., MS/MS: Tandem mass spectrometry for forensic analysis, in *Analytical Methods in Forensic Chemistry*, Ho, M. H., Ed., Ellis Horwood, West Sussex, England, 1990.
14. Homeyer, S., The FBI research partnership program, *Forensic Sci. Comm.*, 1, 5, 2003.
15. Yost, R. A. and Enke, C. G., Triple quadrupole mass spectrometry for direct mixture analysis and structure elucidation, *Anal. Chem.* 51, 1251A, 1979.
16. Todd, J. F. J., Introduction to practical aspects of ion trap mass spectrometry, in *Practical Aspects of Ion Trap Mass Spectrometry*, March, R. E. and Todd, J. F. J., Eds., CRC Press, Boca Raton, FL, 1995.
17. Bertsch, W. and Zhang, Q.-W., Sample preparation for the chemical analysis of debris in suspect arson cases, *Anal. Chim. Acta*, 236, 183, 1990.
18. Keto, R. O. and Wineman, P. L., Detection of petroleum-based accelerants in fire debris by target compound gas chromatography/mass spectrometry, *Anal. Chem.*, 63, 1964, 1991.
19. Caddy, B., Smith, F. P., and Macy, J., Methods of fire debris preparation for detection of accelerants, *Forensic Sci. Rev.*, 3, 57, 1991.
20. Bertsch, W., Chemical analysis of fire debris. Was it arson? *Anal. Chem.* (News & Features), 68, 541A, 1996.
21. DeHaan, J. D., *Kirk's Fire Investigation*, 5th ed., Prentice Hall, Upper Saddle River, NJ, 2002.
22. Bertsch, W., Analysis of accelerants in fire debris — data interpretation, *Forensic Sci. Rev.*, 9, 1, 1997.
23. Newman, R., Gilbert, M., and Lothridge, K., *GC-MS Guide to Ignitable Liquids*, CRC Press, Boca Raton, FL, 1998.

24. Almirall, J. R. and Furton, K. G., Characterization of background and pyrolysis products that may interfere with the forensic analysis of fire debris, *J. Anal. Appl. Pyrolysis*, in press , 2003.
25. Furton, K. G., Almirall, J. R., and Bruna, J., A simple, inexpensive, rapid, sensitive, and solventless technique for the analysis of accelerants in fire debris based on SPME, *J. High Resolution Chromatogr.*, 18, 625, 1995.
26. Almirall, J. R. and Furton, K. G., A fast and simple method for the analysis of accelerants from fire debris using headspace solid-phase microextraction, in *Proc. of the Int. Symp. on the Forensic Aspects of Arson Investigations*, U.S. Government Printing Office, Washington, D.C., 1995, p. 337.
27. Almirall, J. R., Bruna, J., and Furton, K. G., The recovery of accelerants in aqueous samples from fire debris using solid phase microextraction (SPME), *Science and Justice* (*J. Forensic Sci. Soc.*), 36, 283, 1996.
28. Furton, K. G., Almirall, J. R., and Bruna, J., A novel method for the analysis of gasoline from fire debris using headspace solid phase microextraction, *J. Forensic Sci.*, 41, 12, 1996.
29. Almirall, J. R. and Furton, K. G., Forensic and toxicology applications, in *Solid Phase Microextraction: A Practical Guide*, Scheppers Wercinski, S.A., Ed., Marcel Dekker, New York, 1999, p. 203.
30. Furton, K. G., Wang, J., Hsu, Y.-L., Walton, J., and Almirall, J. R., The use of solid-phase microextraction/gas chromatography in forensic analysis, *J. Chromatogr. Sci.*, 38, 297, 2000.
31. Almirall, J. R., Wang, J., Lothridge, K., and Furton, K. G., The detection and analysis of flammable or combustible liquid residues on human skin using SPME/GC, *J. Forensic Sci.*, 45, 461, 2000.
32. Furton, K. G., Almirall, J. R., Bi, M., Wang, J., and Wu, L., Application of solid-phase microextraction to the recovery of explosives and ignitable liquid residues from forensic specimens, *J. Chromatogr. A*, 885, 419, 2000.
33. Almirall, J. R. and Furton, K. G., New developments in sampling and sample preparation for forensic analysis, in *Sample Preparation in Field and Laboratory*, Pawliszyn, J. B., Ed., Elsevier Science, Amsterdam, 2002, chap. 17, p. 919.
34. Ren, Q. L. and Bertsch, W., A comprehensive sample preparation scheme for accelerants in suspect arson cases, *J. Forensic Sci.*, 44, 504, 1999.
35. Harris, A. C. and Wheeler, J. F., GC-MS of ignitable liquids using solvent-desorbed SPME for automated analysis, *J. Forensic Sci.*, 48, 41, 2003.
36. Armstrong, A., Plasencia, M. D., Montero, S., Krivis, J., and Almirall, J.R., Improved sensitivity and selectivity for the detection and identification of ignitable liquid residues from fire debris by ion trap mass spectrometry II, in *Proc. of the American Academy of Forensic Sciences Meeting*, Reno, NV, 2000.

5.2 Analysis of Ignitable Liquids in Fire Debris and the Role of GC/MS/MS

Dale Sutherland

5.2.1 Introduction

Black hole, copious amounts of water, thunderstorm, hot spell, decisions, decisions…

Finally, you are able to investigate, take notes, observe burn patterns, get circumstantial evidence, and obtain samples from the fire scene. But are these samples truly negative or does this field of forensic science lack the capability to handle these highly weathered, more difficult samples? In spite of burn patterns, multiple areas of origin, K9 or other sample locators, etc., these now older samples often return from the forensic laboratory with negative results. This scenario has been seen more than once by many investigators.

As a rule, samples of debris that remain after any fire are very complex as they contain large quantities of a wide variety of burnt material. These samples obtained by investigators may be turned over directly for laboratory analysis or may be stored for long periods of time awaiting other case information that may lead to a decision to have the samples analyzed. The analysis of the burnt material or fire debris can be very complex due to the large number of organic compounds present that represents the natural components of the material, compounds (called pyrolysates) made through exposure to heat and combustion, as well as potential ignitable liquid residues that may or may not have been used to accelerate a fire. No two fire debris samples are identical. The detailed analysis of fire debris samples requires the skill and knowledge of a scientist that is focused on the challenges and objectives in this field of trace evidence analysis. There are few tests available that can review and compare such a large number of characteristic targeted organic compounds and/or organic class profiles to ultimately determine whether a sample is "positive," meaning that the sample contains residues of an ignitable liquid, or "negative." The ignitable liquid characteristics are often in very small amounts relative to the presence of large quantities of pyrolysates. The most common samples contain materials such as wood, plastics, and polymers, which have a wide range of diversity without considering the potential number of pyrolysates developed from their combustion.

Aside from the following brief statements, this chapter will not dwell on the history of the analysis, which is readily available elsewhere.[1] This chapter will focus on the latest advancements and strategies in fire debris analysis today. For many years the analysis of fire debris has been conducted by the extraction of the organic components from the fire debris sample, followed by compound separation with gas chromatography and detection by a universal carbon detector such as a flame- or photo-ionization detector. This GC/FID or GC/PID approach is still being used today but is surpassed in specificity — and thus in confidence in the result — as well as in its sensitivity, by detection with a mass spectrometer (GC/MS). The mass spectrometer is a much more flexible detector with effectively a universal response when operated in a full scan mode. The result is a total ion chromatogram very much like the chromatogram obtained from a GC/FID or GC/PID. The mass spectrometer however can offer another dimension of specificity by obtaining a mass spectrum for a response that can be used to obtain a tentative identification through computer library matching algorithms. Also, the scientist can reconstruct specific-mass ions of interest that are characteristic for a specific-class of organic compounds. For example an ion at *m/z* 120 can be reconstructed to show an ion chromatogram that describes the profile for dimethylbenzenes that may be present as a chemical class. This profile can also be confirmed by similarly reconstructing (for a secondary identification) the ion at *m/z* 105.

However, when samples are very complex, the interpretation may be more difficult, and the scientist often would like to see more clear information upon which to base the result. This is especially true when some characteristics of an ignitable liquid may be observed in small quantities but may not be comparable enough to the laboratory's library of reference analyses to document a positive result for the presence of an ignitable liquid. In some complex, weak, or highly weathered samples, the characteristic ion profiles do not compare well enough for the layman to understand, whether judge or jury. More information is then needed to conclusively show whether an ignitable liquid is present or not. Other methods of analysis may provide the additional information that confidently documents a positive or negative result in an unmistakable manner to both judge and jury.

5.2.2 Specificity — The Key to Difficult Samples

There are several ways to approach the analysis of samples that have a complex makeup or matrix. Fire debris samples may be a significant challenge when they have a complex matrix dominated by large quantities of pyrolysates, combined with compounds that are inherent to the material(s). These can be many times more in quantity than any ignitable liquid residue that may be present. To uncover the ignitable liquid characteristics, the method has to either remove the unwanted pyrolysates and matrix or improve the specificity of the test. Extract cleanup approaches have not been popular and have not been routinely effective; some have recently been removed from lists published by ASTM International.

It has been argued that more sensitivity is not required as this might lead to false positive results that may really be a natural characteristic of the background, or that such small quantities of hydrocarbons could be detected that virtually any sample would be positive. In fact, more sensitivity alone is, not by itself, the solution to obtaining the additional information the forensic scientist needs. What is needed is more specificity in the test. The outcome of enhanced specificity is increased confidence in the result and a natural by-product is some increase in sensitivity. More specific methods can often reveal the organic characteristics necessary to conclusively report whether the sample is positive or negative for the presence of an ignitable liquid. The way that a scientist can improve specificity can be in the form of new chromatographic methods, new processing software, new instrumental hardware, or a combination of these.

5.2.2.1 Specificity Improvement — Chromatographic

To improve specificity, the scientist can turn to improving chromatographic resolution, which usually requires longer analysis times, or to narrower internal diameter capillary columns that have higher resolution but less sample capacity. The scientist can also use multidimensional or 2D chromatography (GCxGC) which can be very complex, especially for this matrix. Improving chromatography alone is not enough to adequately address these samples. Instead of improving chromatography — or in addition to it — we turn to detection systems to improve specificity.

5.2.2.2 Specificity Improvement — Detection

Detection using mass spectrometry has greatly improved specificity over past detectors such as FID or PID. Today, GC/MS represents the benchmark in specificity and thus the benchmark in confidence in the result. However, for weak samples, highly complex samples, or highly weathered samples, forensic scientists may still find themselves lacking some chemical characteristics necessary to provide a clear positive or negative result. In these cases analysts find that the sample comparability is not adequate relative to their library of reference analyses. To obtain more information from GC/MS analysis, there are ways of improving specificity using new software for data reduction.

5.2.2.3 Specificity Improvement — Data Processing Software

Newly developed algorithms, to deconvolute chromatographically coeluting compounds, effectively enhance selectivity. This is a result of the resurgence of Time-of-Flight (TOF) mass spectrometers which have showcased the ability to deconvolute target compounds from very difficult artificial concoctions as well as real samples. Currently, this is not just a feature for TOF spectrometers but can also be conducted on data acquired by more common ion trap or quadrupole-based mass spectrometers. Software such as Ion Fingerprint Detection (IFD) has been used by this author and has shown similarly impressive capabilities to the deconveluting algorithms associated with TOF. IFD has also impressed the U.S. Environmental Protection Agency for the analysis of target compounds in difficult environmental samples.

Another software approach to improve specificity is to use pattern recognition to search for the presence of certain chemical compounds and classes. Macro methods[1] and identification stratagems have been published[2] that aid the scientist in the interpretation of samples primarily using mass ion profiles. A consistent approach to identification is required but it is often very time-consuming. Thermo-Finnigan's Xaminer program can be used in a stand-alone fashion or is well integrated into their Xcalibur data processing software. Xaminer can process the contents of a GC/MS sample analysis and isolate target compounds. It then formulates these target compounds into a spectragram comprised of compounds of many classes. The unknown spectragram is then compared against the spectragrams developed with the laboratory's library of reference substance analyses and provides a list of probable results. The results are listed in order of a fit score that uses the same Biller–Biemann algorithms that scientists have been using for years for NIST mass spectral library searches. The Xaminer algorithm uses an 8-peak presearch and the 50 most significant spectragram peaks with proportional mass weighting are removed. The simple intensity vs. compound identification data arrays allows for very rapid searching against the laboratory's library of spectrograms developed from reference ignitable liquid analyses. This gives confidence to the result as it adds the specificity of pattern recognition across the chromatogram. Also of significant importance is that with increasing numbers of ignitable liquid substances, it becomes more and more difficult for the scientist to remember every characteristic for all the substances in the laboratory's growing reference library. It is thus a very valuable tool for the scientist, who can then ensure a complete, reproducible, and unbiased review of the sample against the reference library. The scientist can then quickly double-check the results, knowing that the sample characteristics have also been compared to not only common ignitable liquids in the reference library but also to the more obscure substances that are rarely seen. Some scientists at the Miami–Dade County Crime Laboratory are currently exploring the use of Xaminer in conjunction with GC/TOF.[3]

5.2.2.4 *Specificity Improvement — Hardware*

Specificity, beyond full scan mass spectrometry (GC/MS), can be improved by adding additional hardware to the method. This can be done by adding another gas chromatograph with the mass spectrometer as detector (GCxGC/MS), using emerging technology such as GC/isotope ratio mass spectrometry (GC/IRMS), improving the mass resolution of the spectrometer beyond unit mass (high resolution mass spectrometers [GC/HRMS]), or by the addition of another mass spectrometer procedure (GC/MS/MS).

Very recently GCxGC/MS (or two-dimensional gas chromatography/mass spectrometry, 2D GC/MS) has been used in experimentation on fire debris.[4] The objective was to address the extra and coeluting pyrolysate peaks that complicate the interpretation of class ion profiles, thus adversely affecting some compound ratios. A TOF instrument (Leco Pegasus) was used due to its very fast scanning speed, essential for 2D GC/MS. An internal oven modulator was used to sequentially heart-cut 4-second components that are diverted to another more polar capillary column. The resulting 2D data plots, similar in appearance to 2D Gel Proteomics data, illustrated the additional separation of coeluting compounds. The data is displayed as a top–down view of a now-2D total ion chromatogram. With real samples, these 2D plots, which show a dot for the presence of each compound, are difficult to compare even to the trained eye. This data has not been used in a court of law and in this format would be incomprehensible to judge and jury alike. If the software could be visually improved to more easily compare chemical class information in a courtroom situation, this approach might have merit as it adds additional specificity to the GC/MS analysis by chromatographically resolving some additional pyrolysates from compounds of interest. The 4-second analysis on the more polar column does not appear to have a particularly high level of chromatographic resolution. It is unknown whether the analyst can display selected compounds that have been deconvoluted in a selected ion profile manner to allow easy pattern recognition and comparison of compound class characteristics.

Analysis by GC/HRMS significantly improves the analysis by improving the mass resolution; e.g., the mass of an ion can be measured to 0.001 amu at a resolution of 10,000, which often completely removes all pyrolysates that have a similar mass. However, this half-million dollar instrument is usually reserved for environmental samples that are tested for the presence of polychlorinated dibenzo-p-dioxins and -furans at ultra trace parts-per-quadrillion levels.

GC/IRMS may show promise, but it is of high cost and is also in experimental use and unproven for trace evidence work.

With the advent of the benchtop ion trap mass spectrometer, at a cost similar to quadrupole units, the price to obtain MS/MS capability and its specificity is not an obstacle and is now discussed further.

5.2.3 Analysis by GC/MS/MS

As references will show, the use of GC/MS/MS is well documented in other fields of analysis including forensics. This laboratory has conducted research and development using GC/MS/MS since April 1997. GC/MS/MS is the coupling of gas chromatography with mass ion isolation with the first mass spectrometer, followed by a second MS stage of dissociation through ion collisions and subsequent detection of the daughter ions formed. The second dissociation and new daughter ion spectrum detected by the second MS operation creates the added specificity that labels GC/MS/MS as perhaps the ultimate method today for mixture analysis.

In MS/MS, a significant ion from a targeted compound or a characteristic ion from a chemical class is chosen from a full scan GC/MS mass spectrum. This is the parent (or precursor) ion. In the MS/MS acquisition, the chosen parent ion is isolated in the first MS operation. In the second MS stage it undergoes collision induced dissociation (CID) with helium reagent gas, which results in a new daughter ion (or product ion) spectrum. The isolation and CID process makes MS/MS inherently more selective than MS alone, and this presents numerous advantages for complex samples where analyte and matrix coelution are common. An ion trap mass spectrometer that includes tandem mass spectrometry in time also offers advantages of reduced electronics, manifolds, and pumps (and price), compared to a triple linear quadrupole mass spectrometer which is based on tandem mass spectrometry in space.[5] The latest ion trap mass spectrometers include higher scan rates, external ionization, and the capability of obtaining both full scan and MS/MS information in the same analytical run. This use of GC/MS/MS is not new. This tandem mass spectrometry approach has clearly been shown to be the method of choice for biological specimens, detection of explosives, and drugs of abuse, as well as for structure elucidation in pharmaceutical analysis. Its application in forensic laboratories has been limited in the past by the cost and complexity of the instrumentation. This has changed with the relatively low cost and high efficiency of conducting MS/MS in an ion trap mass spectrometer.

GC/MS/MS has been used to address problems of analytes in complex matrices such as THC in blood,[6] trace quantities in difficult matrices such as residues of drugs of abuse in hair,[7] or to minimize the matrix interferences in the analysis of pesticides in fruits and vegetables.[8]

The high sensitivity and selectivity of GC/MS/MS have been used in the detection of trace components in explosive mixtures.[9] These are important in environmental remediation projects or for high-speed security sweeps such as those done at airports.

GC/MS/MS is one of the laboratory's "big guns" used when the result must have the highest defensibility. An example is the use of MS/MS (although with liquid chromatographic separation) in the much publicized O.J. Simpson trial. A method using LC/MS/MS was developed to determine whether the blood stained cloth swatch and socks found at the O.J. Simpson residence contained the blood preservative EDTA. The analysis successfully determined that EDTA was not present in the evidence at levels indicating its use as a preservative. It was thus determined that police authorities had not tampered with the evidence. In support of the specificity of MS/MS, scientists at Cornell University published their findings[10] for determining EDTA in blood. They found that the use of LC/MS/MS removes the chemical background or interferences from the target compound of interest; in fact, in their experiments "the blank blood plasma sample is free of all matrix peaks" or potential interferences. They stated that MS/MS "provides the highest specificity and the best detection level of any method currently published." Also, in their conclusions in reference to the O.J. Simpson trial, "The question of blood-evidence tampering in a criminal trial has led not only to improved analytical techniques for the determination of EDTA but also to the demonstration that a relatively new technique (MS/MS) is ready to be used as credible evidence in the courtroom." The high specificity of MS/MS has lead some scientists to consider that the results are irrefutable.[11]

5.2.3.1 GC/MS and GC/MS/MS Spectra Comparison

The following figures will give the reader an insight into the difference between regular GC/MS spectra and the daughter ion spectra obtained from

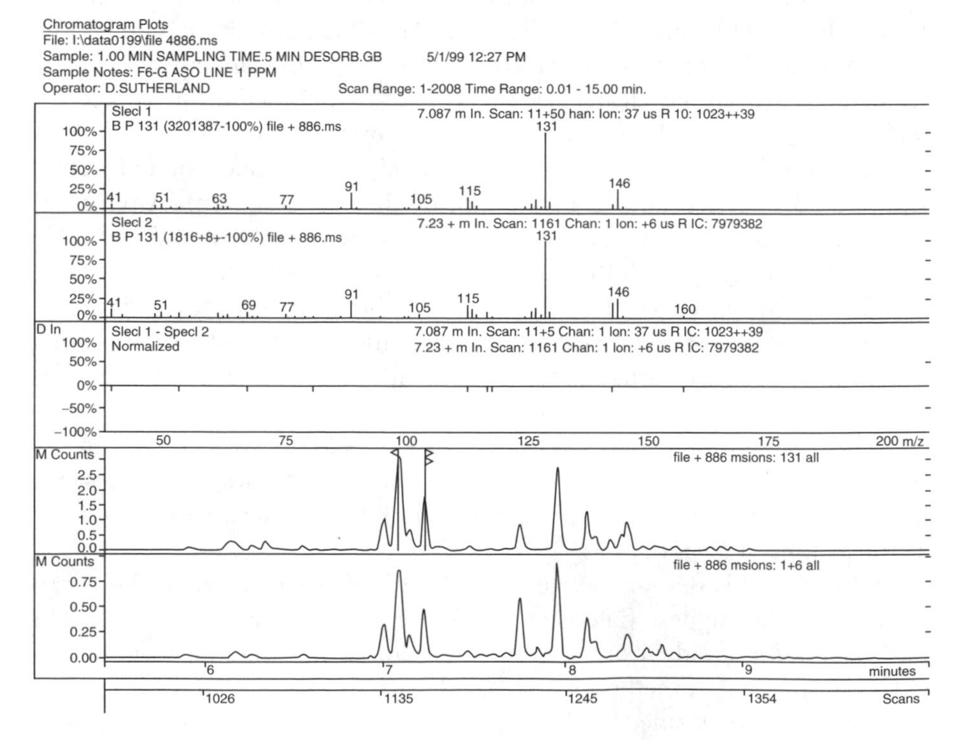

Figure 5.12 Comparison of GC/MS spectra of two indane isomers (flagged in mass ion profile): essentially identical.

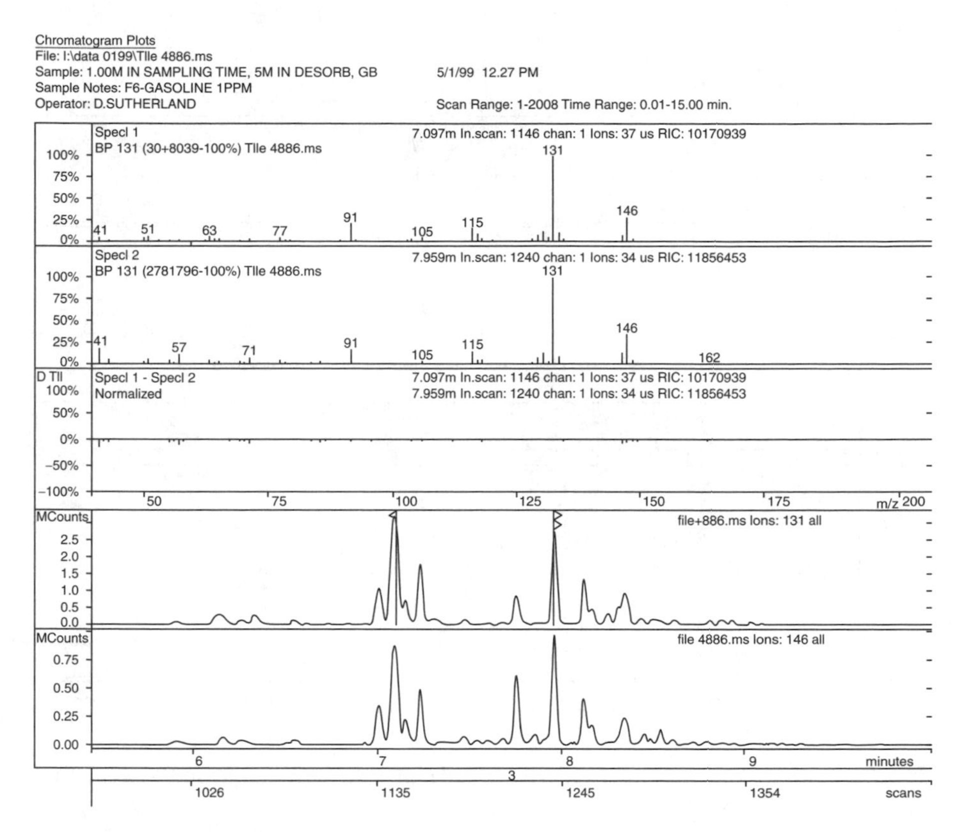

Figure 5.13 Comparison of GC/MS spectra for other indane isomers to the first isomer: essentially identical.

GC/MS/MS analyses. Figure 5.12, Figure 5.13, and Figure 5.14 show the GC/MS spectra from the series of indanes found in gasoline. The reconstructed mass ion profiles of *m/z* 131 and 146 are shown for reference. The top spectrum in each figure is for the isomer with the largest response for the ion at *m/z* 131 (cursor with single flag). The second spectrum in Figure 5.12 is for the adjacent isomer in the ion chromatogram (cursor with a double flag). The middle diagram in this figure shows the difference between these two spectra and indicates that these two isomers have virtually identical spectra, as expected, in this GC/MS analysis.

Figure 5.13 is laid out in the same way as Figure 5.12. Figure 5.13 also illustrates that another isomer (cursor with a double flag) has an identical spectrum to the first isomer. Figure 5.14 illustrates that a third isomer has a spectrum nearly identical to the first isomer. The small deviation at *m/z* 133 in the spectrum is from another compound that is not fully separated from this third isomer.

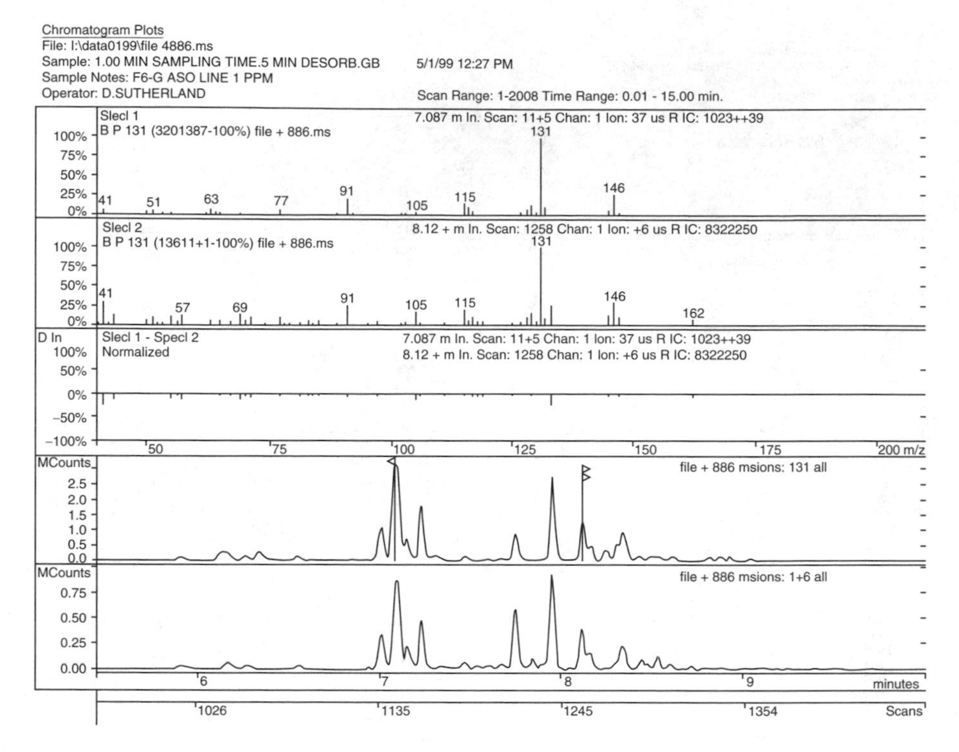

Figure 5.14 Comparison of GC/MS spectra of a third indane isomer to the first isomer: essentially identical.

These figures illustrate that, aside from chromatographic separation, single stage mass spectrometry cannot differentiate isomers based on their mass spectra.

Figure 5.15 compares the spectra obtained by GC/MS (top spectrum and top chromatogram for the ion at *m/z* 131) for the largest indane response to the daughter ion spectrum obtained by GC/MS/MS (second from top). There are two things of note in this figure. As the ion at *m/z* 131 is the largest ion in both GC/MS and GC/MS/MS analyses, the mass ion profiles for 131 are very similar for GC/MS and GC/MS/MS. However, the mass spectra are quite different. The collision induced dissociation in the second stage of the GC/MS/MS analysis, using the ion at *m/z* 146 as the parent ion, creates a unique daughter ion spectrum containing an ion at *m/z* 119. This adds a great deal of specificity. Thus, if there was a coeluting interference with the ion at *m/z* 131, only an indane would have a daughter ion spectrum that contains an ion at *m/z* 119. Also, note that the ion at *m/z* 146 is reduced in the GC/MS/MS spectrum as it was isolated as the parent ion and subsequently reduced through CID.

Figure 5.16, Figure 5.17, and Figure 5.18 show the same series of indane isomers from an analysis by GC/MS/MS.

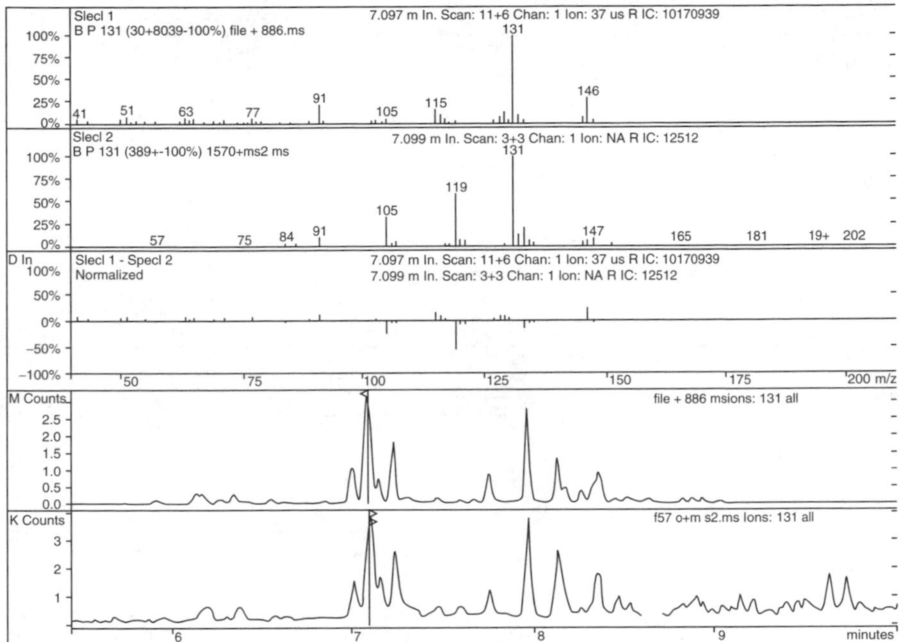

Figure 5.15 Comparison of GC/MS (upper spectra and ion chromatogram) to GC/MS/MS (lower spectra and ion chromatogram) for the same indane isomer: significantly different spectra.

In Figure 5.16 there is a distinct difference between the spectra of the indanes indicated by the flagged cursors. Ions at *m/z* 119 and 121 are present in different amounts. Note that the ion profile of the ion at *m/z* 146 is again shown, this time only to illustrate that this parent ion is significantly lost in the CID operation of the second MS stage. The ion at *m/z* 146 would not be diagnostic for indanes in the GC/MS/MS analysis.

In Figure 5.17, the ion at *m/z* 119 is also present but in a different ratio to the ion at *m/z* 131.

In Figure 5.18 this isomer differs greatly in the ratio of ions at *m/z* 105, 119, and 133 to the ion at *m/z* 131.

These series of figures illustrate that the ion at *m/z* 119 has been created in the CID process. The ratio of the ion at *m/z* 119, as well as the changing ratios of ions *m/z* 105 and 133 to the ion at *m/z* 131, can even provide isomeric specificity where GC/MS could not differentiate between isomers. Thus, the result of the CID process in the second MS stage provides excellent structural specificity in the GC/MS/MS analysis and thus much greater confidence in individual compound identification.

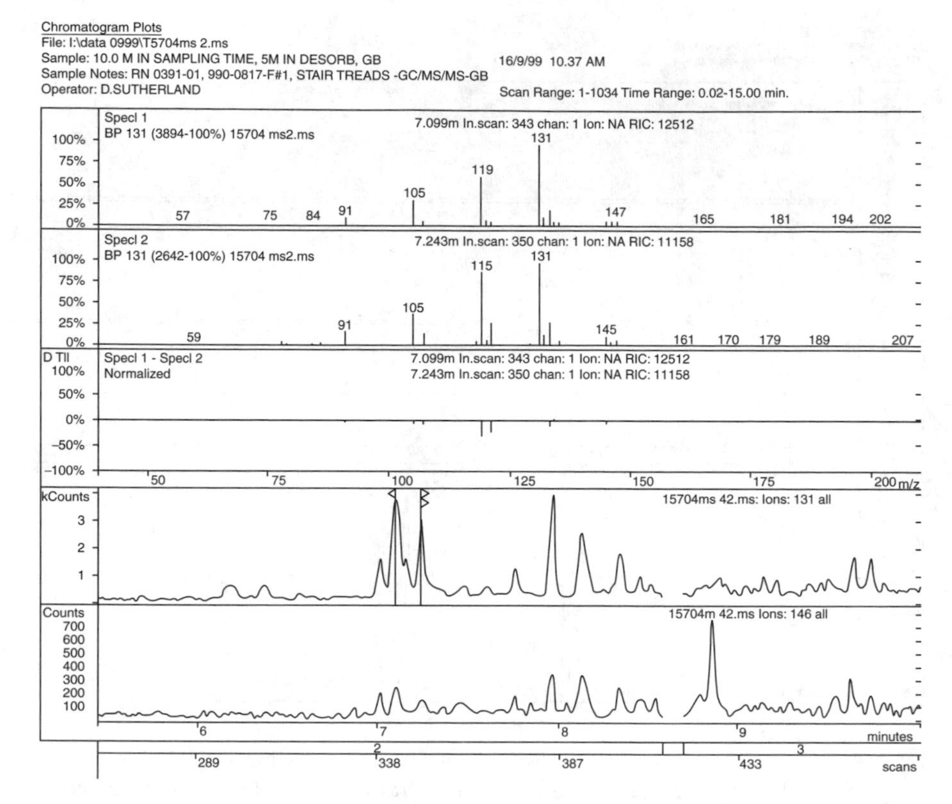

Figure 5.16 Comparison of two GC/MS/MS spectra for two indane isomers: ion ratios different. Note that the parent or precursor ion 146 has been reduced due to CID.

5.2.4 GC/MS/MS — Target Class Analysis

In the trace evidence field of forensics, the author's laboratory has used GC/MS/MS as a second confirmatory analysis to GC/MS on fire debris samples. This method was reviewed by forensic scientists, accepted, and subsequently published.[12]

GC/MS/MS is the closest thing available to an ignitable liquid detector as it may be programmed to isolate the parent ions of specific compounds of the specific chemical classes characteristic to ignitable liquids. This author uses a target chemical class approach (e.g., alkanes, dimethyl (C2)-benzenes, indanes, etc.) rather than a target compound approach (e.g., such as using the 51 compounds listed in ASTM E-1618-01[13]). The method of chemical class specificity, rather than enhancing a target compound list, allows for pattern comparison over a wide chromatographic range. This additional specificity from pattern identification is realized but with some loss of sensitivity when compared to an analysis of individual specific compounds that have each been optimized for the MS/MS analysis.

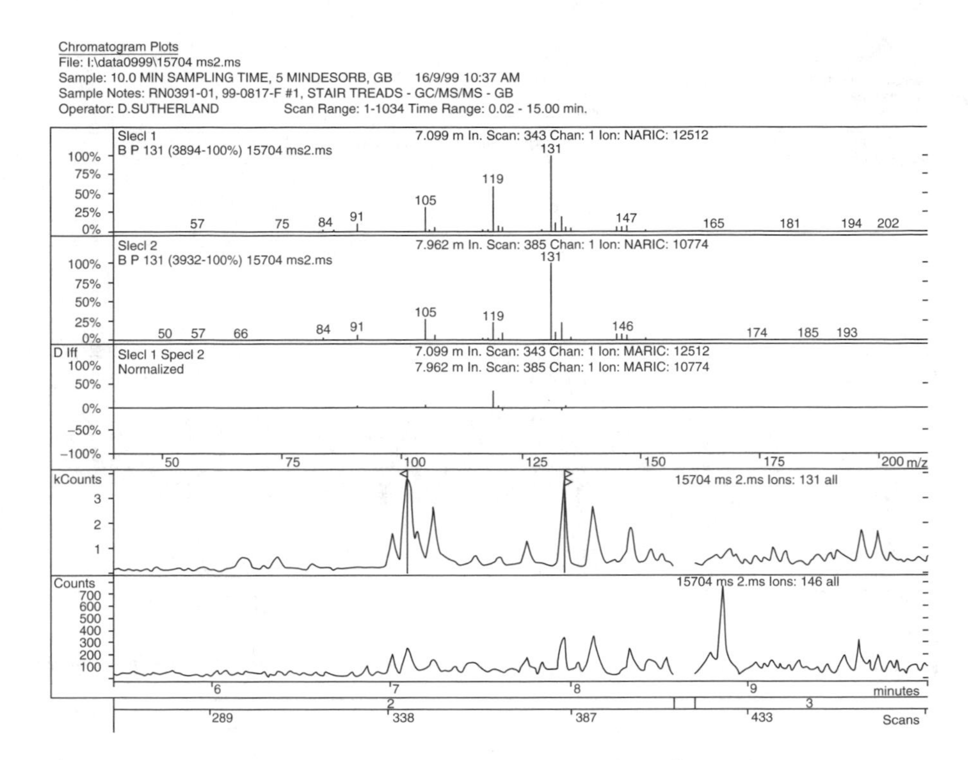

Figure 5.17 Comparison of GC/MS/MS spectra for other indane isomers: ion ratios different.

The instrumental conditions in a target class analysis are more general and not compound specific. The following conditions in Table 5.3 are presented as an example only and should be regarded as a minimum number of precursor and product ions that may be chosen to define the organic chemical classes that make up the majority of ignitable liquids. Energy isolation windows, etc. may be different, depending on the instrument. MS/MS is different in that the spectra and the class ion profile will be different from instrument to instrument, as the energies programmed will result in some difference in profiles. The important aspect is that the MS/MS acquisition parameters should be identical for both sample and reference material analysis. Thus, sample-to-reference data comparisons originating on the same instrument will be excellent. Precursor and product ion class profiles may be different from one instrument to another due to different product ion efficiencies and energies.

5.2.4.1 GC/MS/MS — Chemical Analysis — Examples

We have utilized GC/MS/MS as a second confirmation analysis for samples that were weak, complex, or highly weathered in the GC/MS analysis for over 6 years. This two-analysis approach has been in response to the fact

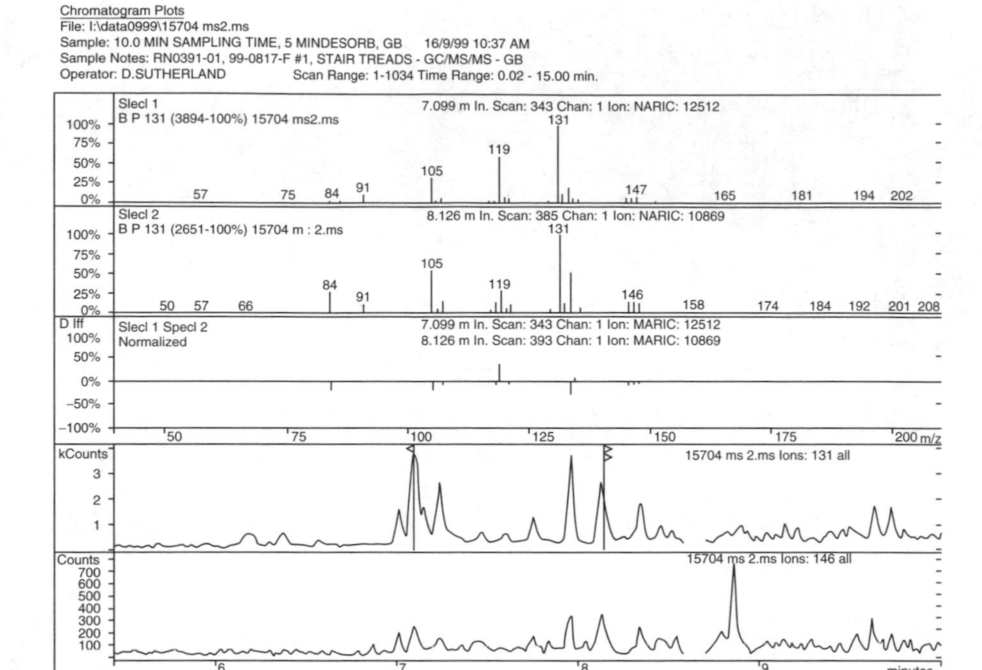

Figure 5.18 Comparison of GC/MS/MS spectra for a third indane isomer: ion ratios different.

Table 5.3 GC/MS/MS — Target Class Approach

Center of parent ion window (*m/z*)	Window Range (+/– *m/z*)	CID Energy (V)	Example parent or daughter ion (*m/z*)	Target chemical classes
71	3	8	71, 69	Alkanes, alkylcyclohexanes
106	5	35	91, 105	Alkylbenzenes
118	5	35	117, 118	Indanes, alkylbenzenes
133	5	35	119, 133	Alkylnaphthalenes, alkylbenzenes
145	7	40	141, 142, 145	Alkylbenzenes, indanes
158	5	35	154, 156	Alkylnaphthalenes, alkylbiphenyls
169	5	45	155, 170, 168	Indanes, alkylbiphenyls
183	5	40	156, 171	Alkylnaphthalenes, alkylanthracenes

that an ASTM method does not yet exist for GC/MS/MS as investigators and reviewers are not familiar with this method, and that there needs to be an adequate time of comparison of the method with the current benchmark of GC/MS.

A single GC/MS/MS method can be used to analyze fire debris as is being discussed in a sub-committee of ASTM E30.01 that is charged with the review and potential publication of a GC/MS/MS method for fire debris analysis. A single method approach with GC/MS/MS is actually an initial acquisition segment, using a standard full-scan approach (GC/MS) to detect relatively low molecular weight, and single-component, ignitable liquids, followed by a series of MS/MS acquisition segments using the target class approach with GC/MS/MS to address petroleum distillates and other mixtures.

The primary objective of GC/MS/MS analysis is to remove responses due to pyrolysates that may complicate the interpretation of class ion chromatograms (e.g., ion at *m/z* 119 for dimethyl-benzenes/C2-benzenes). As seen in Figure 5.19, the class profile for C2-benzenes is very comparable in the

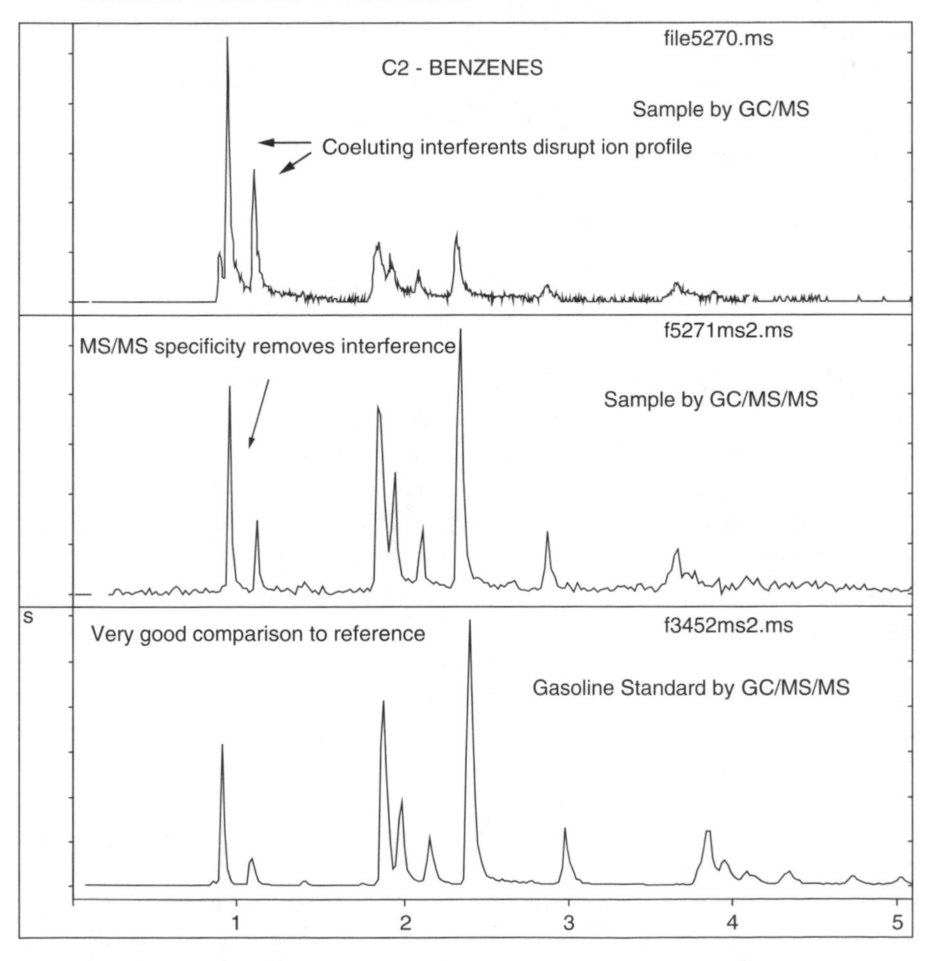

Figure 5.19 Interferents in GC/MS analysis (top ion profile) are removed by GC/MS/MS providing excellent comparability to reference material (bottom ion profile).

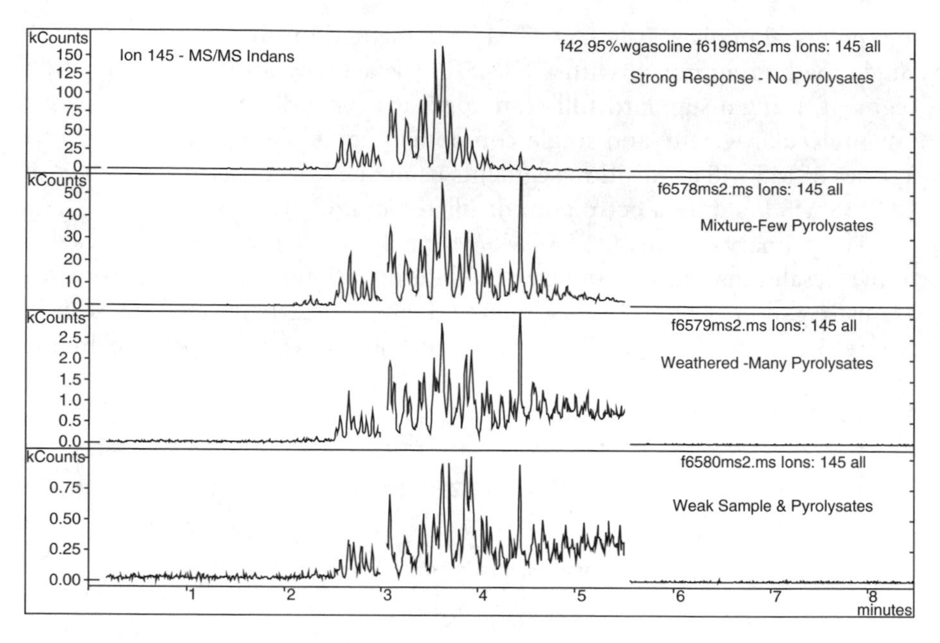

Figure 5.20 GC/MS/MS mass ion chromatograms are robust to the amount of pyrolysates present. Top-to-bottom indane ion profiles: 95% weathered gasoline, gasoline residue in sample of concrete and floor debris, gasoline residue in sample from car footwell, gasoline residue from under car passenger seat (bottom profile).

GC/MS/MS analysis compared to the reference standard of gasoline, illustrating the removal of the pyrolysate response that affects the class ion profile in the GC/MS trace. Work has also been published with a focus on the removal of pyrolysates to improve the detection of gasoline.[14]

As mentioned previously for indanes, GC/MS/MS also adds robustness to the method in the presence of a variety of pyrolysates. This is shown in Figure 5.20. In this diagram of four profiles, the top profile is 95% weathered gasoline, followed by weak pyrolysates from a sample with a mixture of concrete and floor debris. The third profile is a very complex sample of charred debris from the footwell of a car, and the bottom profile is a complex but weak sample taken from beneath the passenger seat of another vehicle.

As shown, the indane class profile, such as in gasoline, does not vary significantly with the type or quantity of pyrolysates present, illustrating the robust nature of MS/MS data. When an ignitable liquid is present, GC/MS/MS is more capable of determining a positive identification, regardless of the amount of weathering or amount of pyrolysates present.

The review of indanes and the use of GC/MS/MS can in some cases enhance the identification characteristics of incidental material, thus further reducing the possibility of reporting a false positive result. A recent study[15] has shown that there are specific indane patterns for a variety of inks found

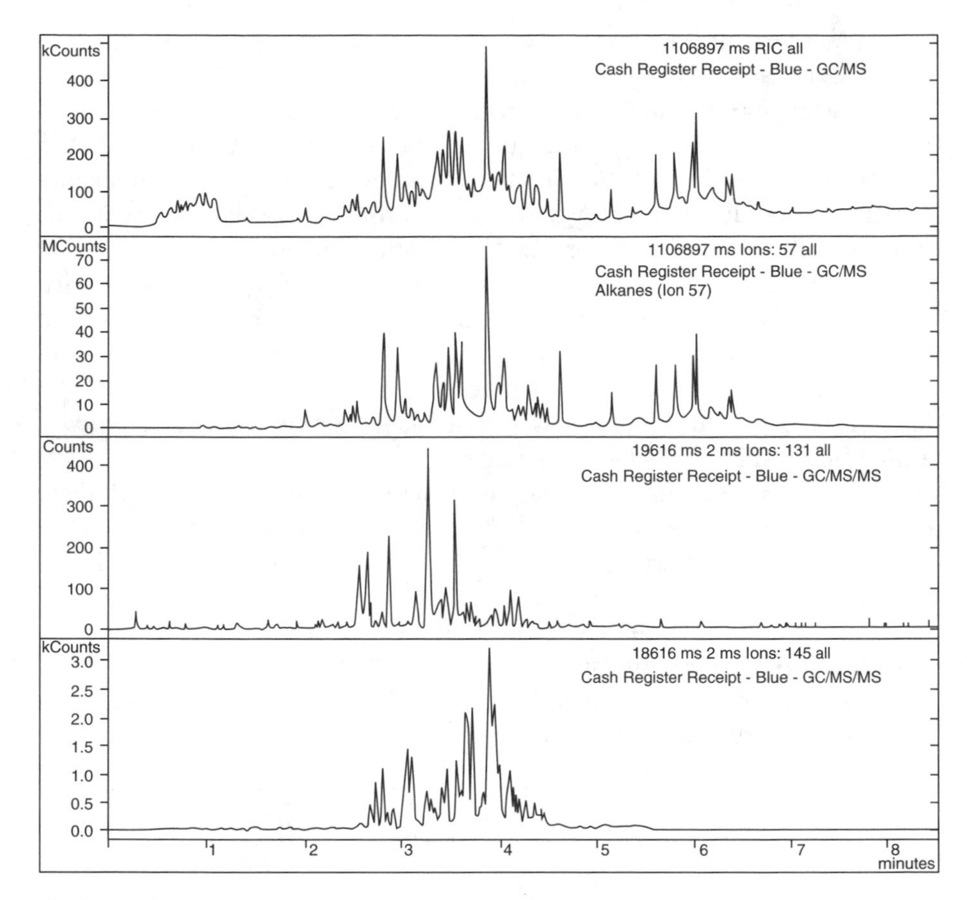

Figure 5.21 GC/MS/MS can display unique profiles of some incidental materials.

in newspapers, courier slips, magazines, etc. that are enhanced in GC/MS/MS, providing a well-recognized marker that the forensic scientist can use in interpretation of the potential presence of ignitable liquid characteristics. An example of indanes in a cash register receipt is shown as the profiles of ions at m/z 131 and 145 in Figure 5.21.

This lab has defined criteria for situations where reanalysis by GC/MS/MS for confirmation is considered. These criteria are based on the minimum presence of branched alkanes and dimethylbenzenes in an initial GC/MS analysis. The detection of the straight chain (n-alkanes) and branched alkanes requires particular attention. The detection of alkanes is important. They are, however, already so fragmented that there is no molecular ion or other significant diagnostic ions to isolate as a precursor ion for CID fragmentation in the second stage MS. MS/MS is very sensitive for aromatic components, but to obtain the branched and straight-chain alkane pattern at all, this class is analyzed by acquiring a precursor ion with no collision energy added. In a chemical-class specific analysis, an ion at *m/z* 71 or 85 is chosen for monitoring alkanes and is subsequently reconstructed for data review and assessment as in GC/MS. This ion is acquired in a very similar fashion to that in selected ion monitoring (SIM) in GC/MS. As evidence of branched alkanes is required for distillate or gasoline identification, this class represents the limiting factor in sensitivity for MS/MS. Beyond zero energy addition, other instrumental parameters can be modified to enhance alkane class sensitivity. Some researchers have worked on obtaining a response for the molecular ion to use as a precursor ion in order to obtain an associated daughter ion in MS/MS. These parameters have been recently reported by researchers[16] who have also recognized that the alkane class is a limitation to the sensitivity of a target-class specific GC/MS/MS analysis.

Using this approach, the analysis of a stair tread had some of these criteria, but there were insufficiently clear characteristics to define a positive result to present in court (see Figure 5.22). In review of the TIC in the initial analysis, this sample would have been reported as negative by GC/MS, whereas the TIC from the GC/MS/MS analysis shows that the sample has many additional compounds not seen in the GC/MS analysis.

A part of the examination of the stair tread sample, Figure 5.23 illustrates the improved comparability of this sample to a gasoline reference for these

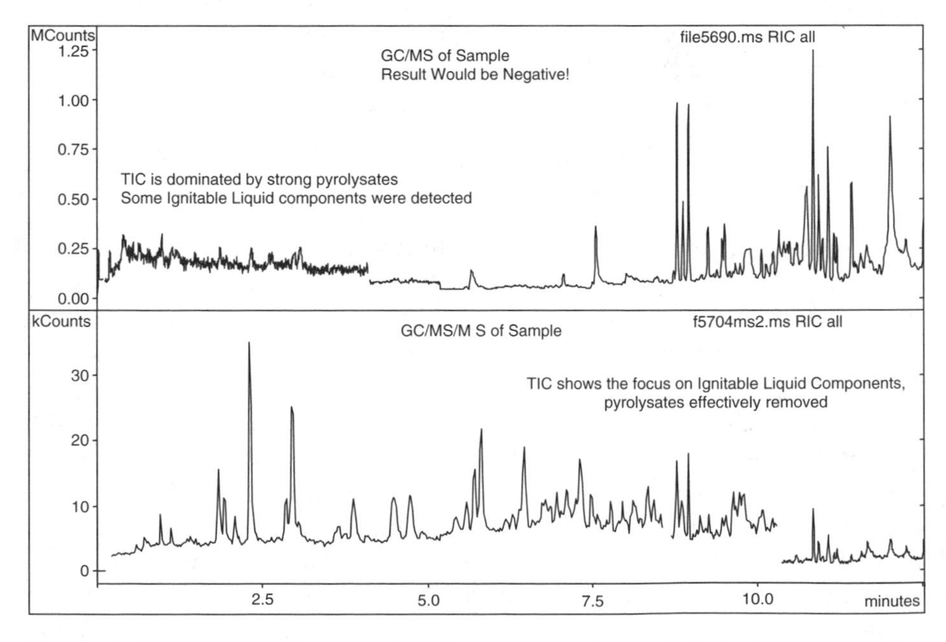

Figure 5.22 Top profile: Sample of a stair tread by GC/MS; bottom profile: same sample by GC/MS/MS illustrating increased detection of ignitable liquid components and reduced effect by pyrolysates.

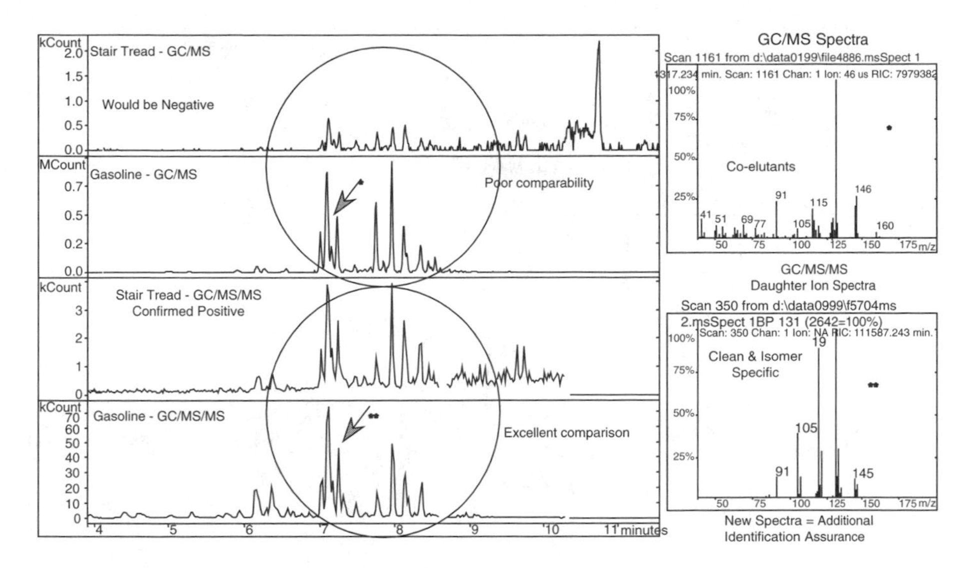

Figure 5.23 Stair tread sample: study of indanes.

indane profiles. In addition, the spectra retrieved from the selected compounds illustrate that MS/MS spectra are cleaner and are most often significantly different, thus improving selectivity and confidence in the identification. Upon further review, it was found that the compounds more easily detected in the GC/MS/MS analysis were ignitable liquid components. Thus, the analysis by GC/MS/MS determined that this sample was actually positive for the presence of gasoline.

In answer to the fear that this technology, when applied to trace evidence, may exceed an appropriate level of sensitivity, the sensitivity to alkanes is in effect a self-limiting parameter and alkanes are characteristic for the determination of ignitable liquids. In GC/MS, alkanes are the primary characteristic components for the identification of distillates, but some distillates have little else to rely on but straight-chain and branched alkanes for identification. This may be critical in some samples with a large quantity of pyrolysates — many of which are alkenes — which significantly alter the alkane profile, thus making identification difficult. GC/MS/MS can significantly improve this situation due to its high sensitivity for aromatic components. Trace aromatic components that may be present are enhanced in GC/MS/MS, allowing additional identification criteria which are not readily apparent in the GC/MS analysis. This is shown in Figure 5.24. GC/MS/MS thus provides more information, reducing the sole reliance on alkanes.

In the class-specific approach, this analysis is not what could be called too sensitive. It is not able to detect positive results in ambient air conditions or environmental general deposition, as determined in a specific research.[15] GC/MS/MS can, however, be important in extending the useful

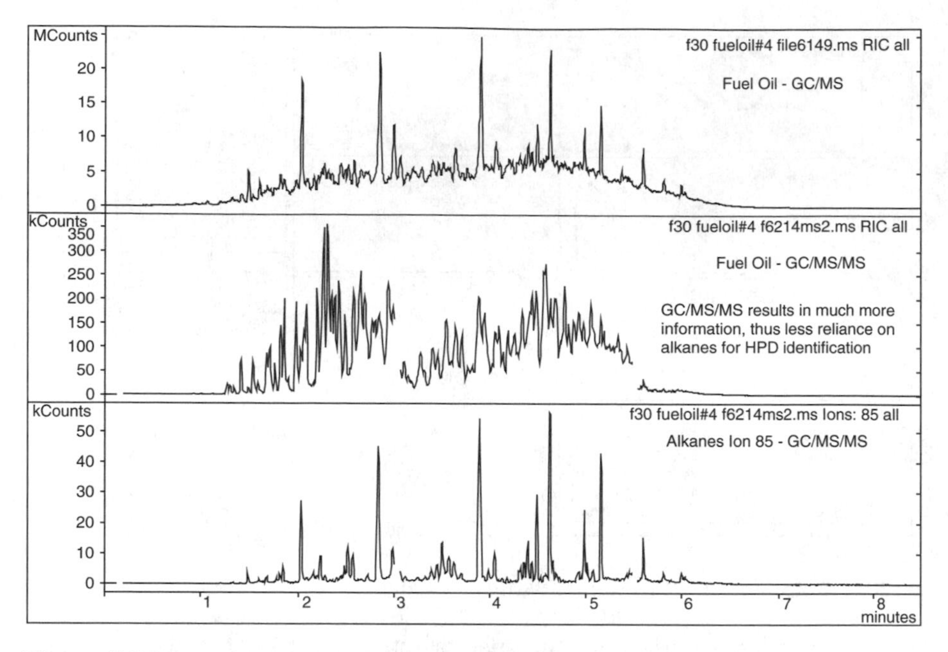

Figure 5.24 GC/MS vs. GC/MS/MS for home fuel oil reference material.

life of a sample. Samples that have been at the fire scene, or stored, for a long period of time may still have ignitable liquid residues that can be definitively detected by GC/MS/MS. A sample of a gallon-size tin can is shown in Figure 5.25. The top half of the figure is the total ion chromatogram (TIC) of this sample. If this were the chromatogram from a GC/FID analysis for screening purposes, this sample would certainly be dismissed as negative and not be further analyzed due to the lack of detection of organic components. The GC/MS chromatographic baseline changes at various segments are due to programming in the acquisition to improve sensitivity for compounds eluting in that range. This tin can had been retrieved from a fire scene and initially analyzed as negative by a laboratory that used GC/MS. The tin can had seen a considerable amount of heat as the screw cap spout was lost when the solder melted in the fire. A few months later it was submitted and analyzed by GC/MS and again found to be negative. There were a sufficient number of ignitable liquid characteristics to warrant reanalysis by GC/MS/MS. This tin can was reextracted directly in its nylon evidence bag using SPME and reanalyzed by GC/MS/MS. The TIC from the GC/MS/MS analysis illustrates the additional number of compounds that were detected.

One of the best ion profiles in GC/MS analysis of the tin can was for C3-benzenes as shown in Figure 5.26. There was some comparability to the same profile in gasoline but it was not very distinctive. The profile comparison in

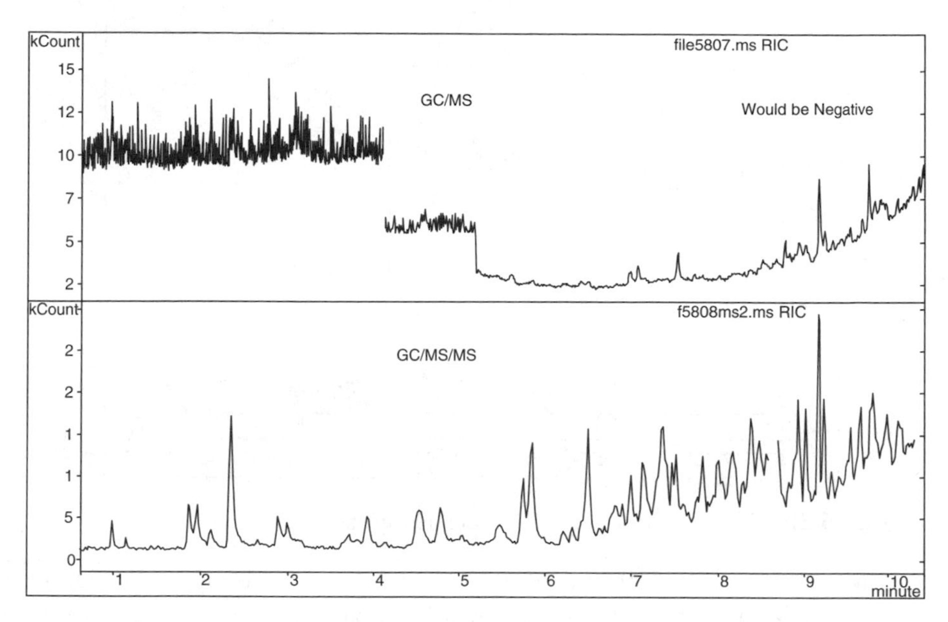

Figure 5.25 Upper TIC: tin can sample by GC/MS; bottom TIC: tin can sample by GC/MS/MS.

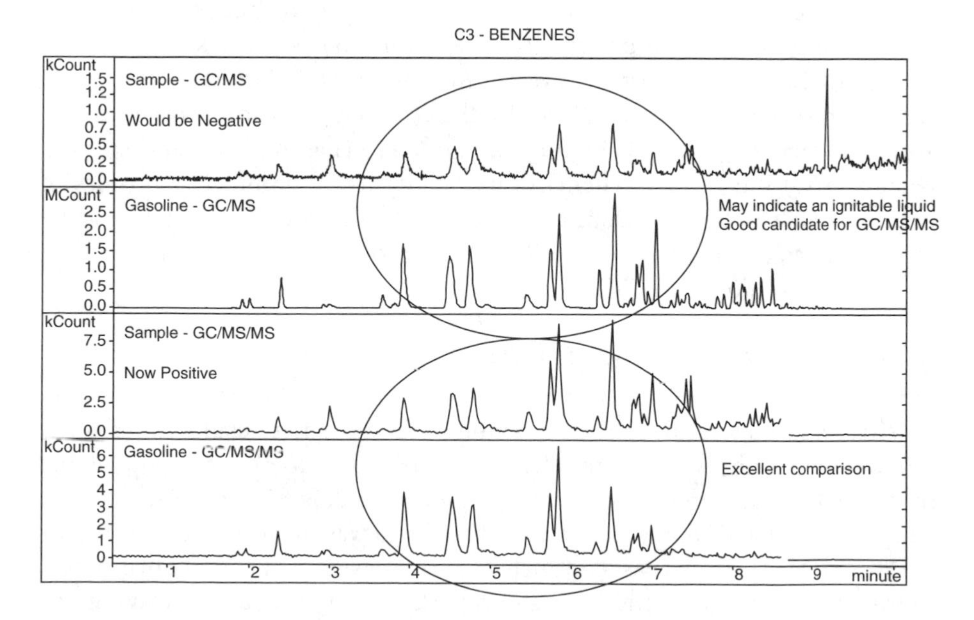

Figure 5.26 GC/MS/MS improves comparability to the reference gasoline analysis over GC/MS.

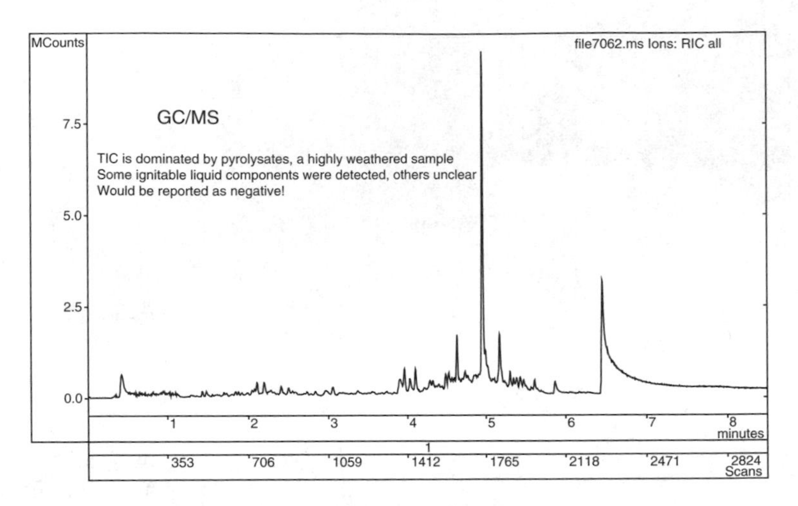

Figure 5.27 Floor sample from Porsche: TIC from GC/MS analysis.

GC/MS/MS was excellent and would be easily observable by both scientist and courtroom layperson. The GC/MS/MS analysis resulted in a positive determination for the presence of gasoline residues, showing that this technique is perhaps the closest technique to an ignitable liquid detector.

Another case example is for a floor sample taken from a fire-ravaged Porsche. The TIC in Figure 5.27 shows that few compounds were detected in GC/MS and a review of mass ion profiles indicated that some ignitable liquid components were present. Based on this analysis, the sample would be reported as negative for the presence of an ignitable liquid.

Upon reanalysis with GC/MS/MS, the class daughter ion profiles of *m/z* 106 (Figure 5.28), *m/z* 118 (Figure 5.29), *m/z* 119 (Figure 5.30), and *m/z* 133 (Figure 5.31) illustrate excellent comparability to the same class profiles in gasoline. This is another example of a situation where a sample would be falsely reported as negative by GC/MS, while proven positive for the presence of an ignitable liquid by GC/MS/MS.

5.2.4.2 Appropriate Sensitivity

Some concern has been reported that analysis by GC/MS/MS may be too sensitive and may result in the reporting of false positive results for samples that contain only material incidental to the scene, to the makeup of the material, or from some environmental background. Similar arguments were heard when GC/MS analysis was introduced and was suggested to be superior to GC/FID.[1] A study was conducted that showed that this concern was unfounded.[18] In 1998 this laboratory participated in a test involving the burning of two automobiles. The opportunity was available to take a large variety and number of passenger compartment samples from these old,

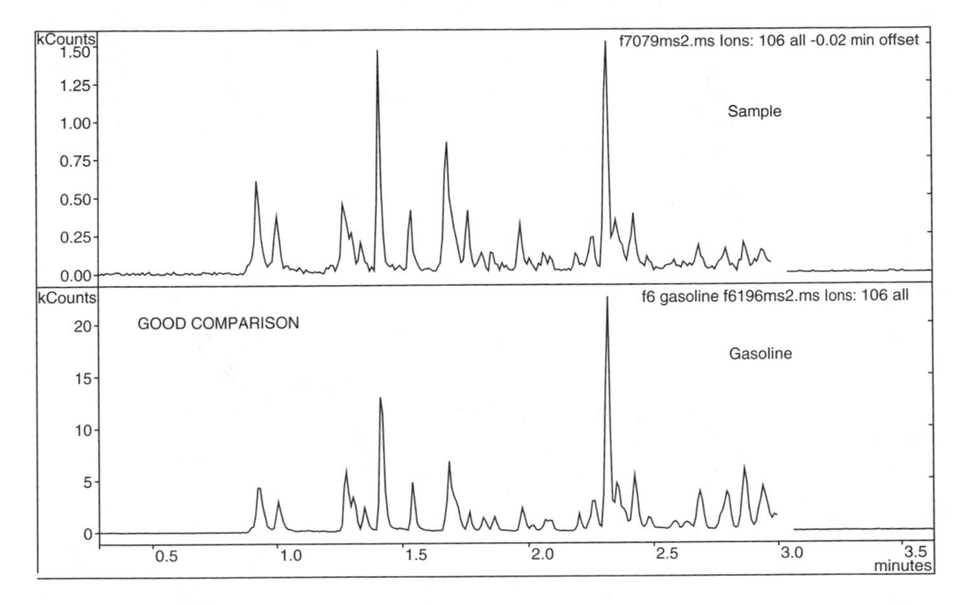

Figure 5.28 GC/MS/MS analysis of sample from Porsche: mass ion profiles for 106.

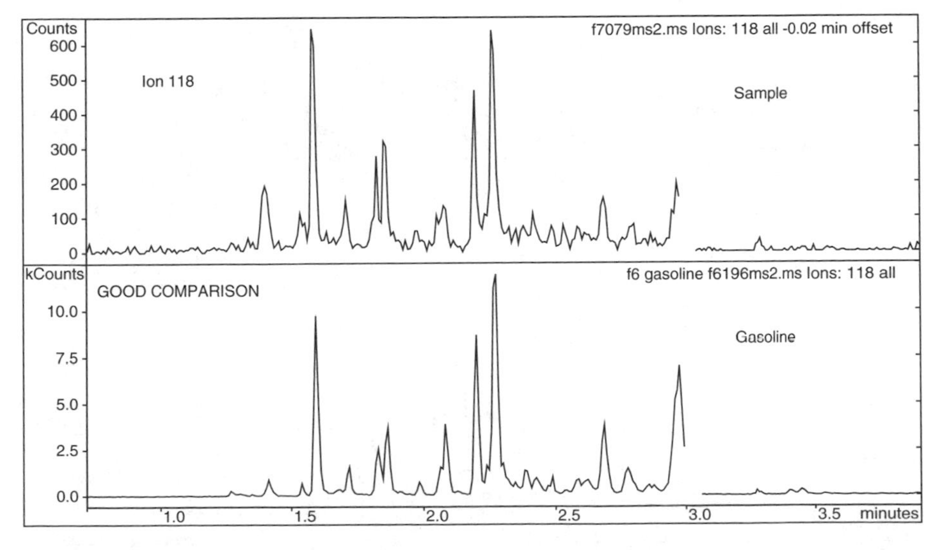

Figure 5.29 GC/MS/MS analysis of sample from Porsche: mass ion profiles for 118.

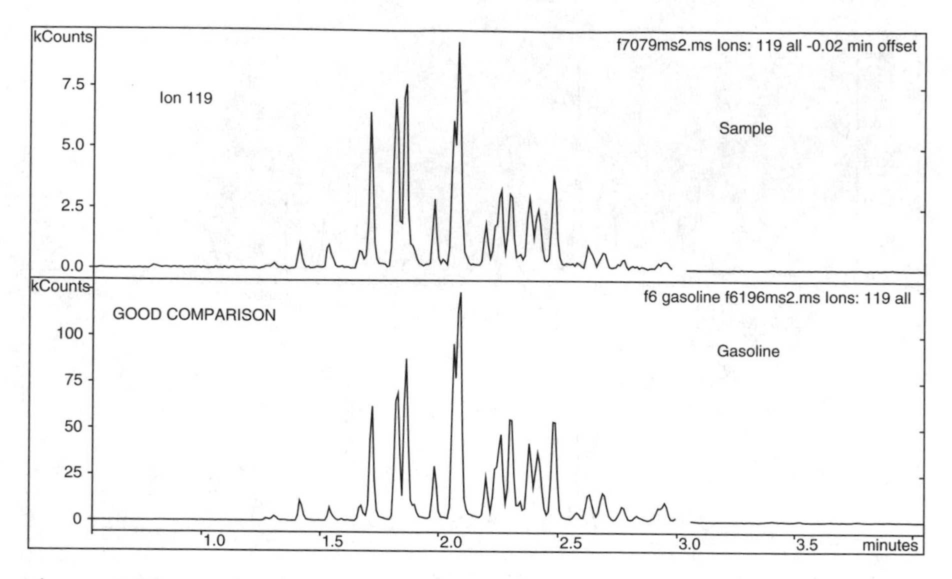

Figure 5.30 GC/MS/MS analysis of sample from Porsche: mass ion profiles for 119.

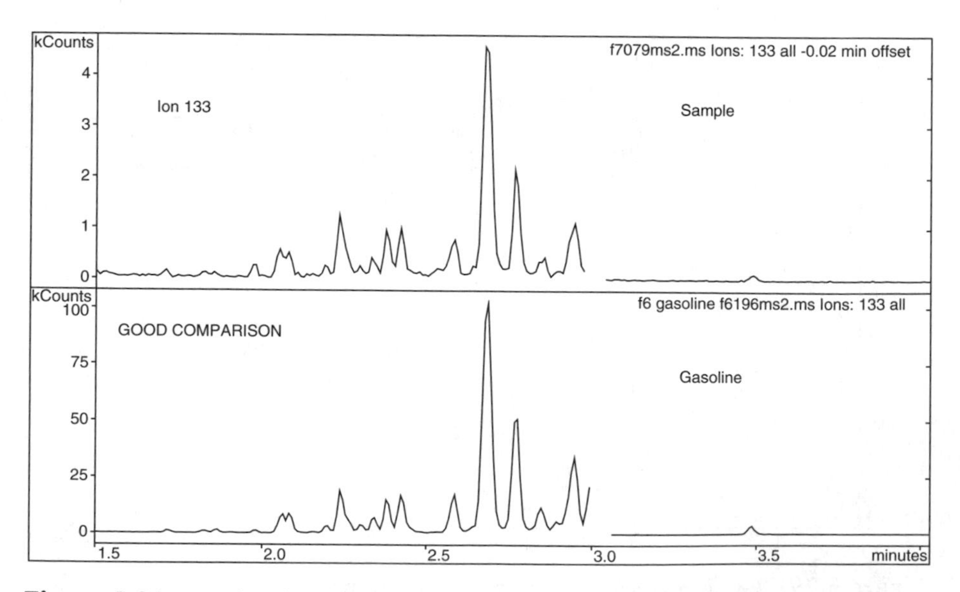

Figure 5.31 GC/MS/MS analysis of sample from Porsche: mass ion profiles for 133.

high-mileage automobiles selected from a local wrecking yard. Although their interiors had a distinct odor of a garage from oil, grease, etc., there were no positive results by either GC/MS or GC/MS/MS prior to the test burns. Thus, GC/MS/MS tests were not too sensitive. After the introduction of gasoline or diesel fuel and the subsequent extinguishing of the blaze by firemen in this training program, additional samples were taken. Some of the results of the postburn samples by GC/MS were not determined to be positive. These samples were reanalyzed by GC/MS/MS, yielding some positive results, yet some samples were still negative. This again illustrated that GC/MS/MS was more sensitive than GC/MS but not too sensitive, as some samples were determined to be negative even though an ignitable liquid was known to be present in the vicinity of the sample prior to the burn.

The objective in a "target class" type of GC/MS/MS analysis is not to develop the most sensitive analysis but to further improve upon the specificity of ASTM 1618-01 (GC/MS) by providing more diagnostic information and, hopefully, to more closely compare it with the capabilities of the K9 (canine) trained to locate ignitable liquid residues. As sensitivity is an important issue, a study was conducted by this laboratory to make a cursory comparison of our GC/MS/MS method with the capabilities of a K9.[19] This study used samples of pinewood and polypropylene carpet that were spiked with gasoline and a low-odor medium petroleum distillate (MPD, paint thinner). Some of these samples were spiked with these liquids and then burnt with a propane torch to provide highly weathered samples. All samples that had a K9 hit in this study were placed in a nylon evidence bag after the K9 inspection. The bag was then put in an oven at 90°C for 10 min and subsequently extracted by Solid Phase Micro-Extraction (SPME) using ASTM E-2154.[20] The SPME fiber was exposed to the headspace of the sample directly by poking a hole into the bag with the SPME needle and then exposing the fiber for 10 min prior to injection into the GC/MS/MS. Of particular interest were the samples of pinewood. This was an aged 2″ × 4″ pine stud where the spike of the ignitable liquid did not absorb well into the pore structure. At the lowest level of K9 sensitivity (very weak hit, K9 showing some interest) for the highly weathered, low-odor MPD spiked onto pinewood, neither GC/MS nor GC/MS/MS could confirm the K9 hit. It was surprising that the K9 still responded, although weakly, to the residues of the MPD on the heavily burned wood. This illustrated the extraordinary sensitivity of the K9. The K9 had no false hits and thus exhibited a high level of specificity. More than GC/MS, the GC/MS/MS target class, analysis appeared to resemble the K9 in terms of specificity and sensitivity. However, due to the heavy reliance on the alkane class determination of this medium petroleum distillate that had less aromatic content (MPD low odor variety), the GC/MS/MS could not improve the standard GC/MS analysis conducted in this study in 2001. Newer instrument parameters determined since that time may provide further improvement. Thus, neither GC/MS nor GC/MS/MS could confirm some of the weak K9 hits when a low odor medium petroleum distillate was present.

Although conducted with only one laboratory and one K9, this study served as a check on where this method resides in sensitivity relative to the K9. In this study, the GC/MS/MS method was easily able to detect ignitable liquids that have a significant number of aromatic components, as expected, and resulted in an improvement in the confirmation of the K9 hits over the analysis conducted by GC/MS analysis on the same instrument.

Thus, GC/MS/MS is not too sensitive and does not appear to exceed the capabilities of the K9. The moderate increase in sensitivity by GC/MS/MS is due to the increase in specificity only and not in signal strength.

5.2.4.3 Disadvantages

The disadvantage of the target class approach is the somewhat lower sensitivity than the target compound approach for one specific compound, as the MS/MS acquisition parameters are more general and not optimized for any specific compound. The analysis by MS/MS also does not allow for the review of the general quantity or type of pyrolysates that may be present in the sample. These pyrolysates have been purposefully filtered out as we program the MS/MS parameters to focus as much on ignitable liquids as possible. Similarly, an analysis by MS/MS does not allow the study of any unusual matrix components that may be of interest or be incidental. As sometimes is the case, the matrix profile may be useful to link samples to a location within a case (e.g., a case file may contain samples 2, 3, and 6 that are all carpet as found in the living room, and samples 1, 4, and 5 are wood from the floor joist in the basement). This link can be shown by a study of the polypropylene or wood pyrolysates to confirm the link of the sample to the results in the GC/MS analysis.

5.2.4.4 Benefits

The primary objective is the detection and identification of ignitable liquids. The advantage of a target class GC/MS/MS analysis is the greater specificity realized by mixture identification through comparison by pattern recognition of the reconstructed daughter ion chromatograms. GC/MS/MS is still more sensitive than GC/MS and it has been proven in one study that GC/MS/MS can lengthen the period of time that a residue of an ignitable liquid can be identified by 2 additional weeks.[15] The more sensitive and specific analysis by GC/MS/MS may also help the investigator. In theory, samples that contain an ignitable liquid can still be proven positive at a greater distance from the initial location of an ignitable liquid as it can detect more minute residues of ignitable liquids and thus be more forgiving as to the sample location. It is also important in the analysis of samples from hot or arid climates as it has more capability of detecting an ignitable liquid, if present, in very highly weathered samples.

As discussed previously,[19] the target class analysis appeared to more closely resemble the K9 in terms of specificity and sensitivity, and it has been shown that GC/MS/MS is not overly sensitive for the purposes of this test for ignitable liquids.

GC/MS/MS can significantly remove pyrolysates and thus make ion chromatograms that are representative of a particular class clearer and more comparable to reference analyses than with GC/MS data. GC/MS/MS results will easily enable a very clear decision as to whether an ignitable liquid is present or absent.

GC/MS/MS also improves the specificity and confidence in the result by providing new chromatographic patterns, with new daughter ion chromatograms and more aromatic information, which lessens the dependence on the alkanes for the identification of petroleum distillates. This is a significant benefit as alkanes are the class that is most susceptible to significant weathering. Thus, GC/MS/MS improves the analysis for highly weathered samples by providing additional characteristics for identification.

GC/MS/MS in the target class approach provides an appropriate improvement in sensitivity while adding robustness to the method. This sensitivity improvement is more probable at confirming K9 hits than GC/MS and, up to this point, has not been observed as going beyond the sensitivity of the K9. GC/MS/MS is the closest method to an ignitable liquid detector that is both sensitive and specific and provides identity information of the residue.

5.2.5 Conclusion

In 1999, this laboratory has shown that in its case analyses by GC/MS/MS, approximately 10% of the samples that were reported as negative by GC/MS were actually positive after GC/MS/MS analysis.[21] An average of 30% of the submitted samples of carpet, wood, soil, and charred debris were found positive by GC/MS. As a rule, GC/MS/MS is invaluable when a sample exhibits some ignitable liquid characteristics in a GC/MS analysis. This is especially true when the interpretation shows that there are insufficient identification characteristics, which is due to incomplete compound class patterns, heavy sample weathering, weak samples (due to low quantity), and complex samples due to a large quantity of pyrolysates. The samples that show some ignitable liquid characteristics were reanalyzed by GC/MS/MS. Of these samples that would have been reported as negative, 28% were found to be positive by GC/MS/MS. Reanalysis of samples by GC/MS/MS has very often provided the additional information to confirm the presence of an ignitable liquid. Thus, GC/MS/MS has shown that some GC/MS results can be falsely reported as negative.

In general, scientists must gravitate towards more specific methods that provide more confident results that will represent the true contents of forensic samples. There is a concern among forensic scientists that any improvement in sensitivity will result in positive results for nearly all samples and will weigh heavily on the legal process, thus automatically invoking the word "arson." Due to the specificity of GC/MS/MS the apprehension of finding an ignitable liquid in virtually every sample due to an increase in sensitivity is unfounded.

In court, the forensic scientist can only attest to the contents of the sample, analyzed to his/her best capabilities. However, the forensic scientist should also be responsible for helping investigators, lawyers, and the court by making them aware of potential incidental sources, and thus helping ensure that the data is reviewed appropriately in litigation proceedings. As a minimum, ASTM recommends that a report should state in a footnote the meaning of negative results such as "A negative result does not preclude the possibility that an ignitable liquid is present in the sample at a concentration lower than the method can detect, or that an ignitable liquid may have been present in the sample at some point in time." Perhaps forensic scientists should also be incorporating a statement on positive results such as "Positive results should also be reviewed in terms of the sampling location and history to ensure that any incidental material or external sources that may be responsible for the positive result are reviewed and assessed."

Some comments regarding a new, more sensitive technique revolve around the danger of identifying an ignitable liquid that may be present due to environmental conditions or some previous incidental material. This concern has been addressed.[15,22–24] Of course, the positive identification of a sample with a minute quantity of gasoline from an area of a garage must be carefully reviewed by the investigator vs. the presence of the same substance in a sample of pillow foam from a couch in the living room, if we use a house for an example. The forensic scientist is responsible for interpreting the sample and not the circumstances. However, it would be prudent to make a statement to ensure that the investigator will review information regarding the location and history of the sample.

GC/MS/MS is a proven technique and has previously been presented in court. Specifically, the analysis of fire debris samples by GC/MS and GC/MS/MS has been used in a case and was described in a testimony of a first-degree murder trial in January of 2002 in Ontario, Canada.

The advantages of GC/MS/MS significantly outweigh any disadvantages, and it is a most valuable tool for helping the forensic scientist interpret complex samples such as ignitable liquids in arson cases.

References

1. Bertsch, W., Holzer, G., and Sellers, C.S., *Chemical Analysis for the Arson Investigator and Attorney*, Huthig, Heidelberg, Germany, 1993.
2. Jackowski, J.P., Incidence of ignitable liquid residues in fire debris as determined by a sensitive and comprehensive analytical scheme, *J. Forensic Sci.*, 42, 828, 1997.
3. Fleisher, K., Harrison, S.J., Bukowski N., and Blackburn M., Rapid identification of accelerants in arson analysis using GC/TOF-MS and automated chromatography matching, Thermo-Finnigan Application Note AN 2004T.
4. Frysinger, G.S. and Gaines, R., CT Forensic analysis of ignitable liquids in fire debris with comprehensive 2D GC/MS, Presented at the Pittsburgh Conference, Orlando, FL, 2003.
5. March, R.E., Quadrupole ion trap mass spectrometer, in *Encyclopedia of Analytical Chemistry*, John Wiley & Sons, New York, 2000, p. 11848.
6. Collins, M., Easson, J., Hansen, G., Hodda, A. and Lewis, K., GC-MS-MS confirmation of unusually high Δ^9-tetrahydrocannabinol levels in two postmortem blood samples, *J. Anal. Toxicol.*, 21, 538, 1997.
7. Baumgartner, W.A., Cheng, C.-C., Donahue, T.D., Hayes, G.F., Hill, V.A. and Scholtz, H., Forensic drug testing by mass spectrometric analysis of hair, in *Forensic Applications of Mass Spectrometry*, Yinon, J., Ed., CRC Press, Boca Raton, FL, 1995.
8. Butler, J., Minimize matrix interferences in the analysis of residual pesticides in fruits and vegetables using PPINICI and GC-MS/MS with a large volume injection, Presented at the Pittsburg Conference, Orlando, FL, 2003.
9. Fetterolf, D.D., Detection and identification of explosives by mass spectrometer, in *Forensic Applications of Mass Spectrometry*, Yinon, J., Ed., CRC Press, Boca Raton, FL, 1995.
10. Sheppard, R.L., Henion, J., Determination of EDTA in blood, *Anal. Chem.*, 69, 477A, 1997.
11. Munoz-Guerra, J., Carreras, D., Soriano, C., Rodriguez, C., Rodriguez, A.F., Cortes, R., GC/MS/MS Analysis for anabolic steroids in urine for athletic testing, Varian Application Note No. 60.
12. Sutherland, D., The analysis of fire debris by GC/MS/MS, *J. Canad. Soc. Forensic Sci.*, 30, 185, 1997.
13. ASTM International, E-1618-01 Standard Test Method for Ignitable Liquid Residues in Extracts from Fire Debris Samples by Gas Chromatography–Mass Spectrometry, 2001.
14. de Vos, B.J., Froneman, M., Rohwer, E., Sutherland, D., Detection of petrol (gasoline) in fire debris by gas chromatography/mass spectrometry/mass spectrometry, (GC/MS/MS), *J. Forensic Sci.*, 47, 736, 2002.

15. Sutherland, D. and Penderell, K., GC/MS/MS — An important development in fire debris analysis, *Fire and Arson Investigator*, International Association of Arson Investigators, October 2000, p. 21.
16. Sittidech, M., Optimization of ion trap GC/MS and GC/MS/MS for the analysis of potential accelerants in fire debris, Presented at the Pittsburg Conference, Orlando, FL, 2003.
17. Sutherland, D. and Penderell, K., Exploring the limitations of GC/MS and GC/MS/MS: The potential interference of printed inks in fire debris, *J. Forensic Sci.* (in review).
18. Sutherland, D. and Byers, K., Vehicle test burns, *Fire and Arson Investigator*, International Association of Arson Investigators, 1998. p. 23.
19. Sutherland, D. and Lam, S., Fire debris analysis research: limitations and comparability of laboratory analysis using GC/MS and GC/MS/MS and the K9, *J. Canad. Soc. Forensic Sci.* (in review).
20. ASTM International, E2154-01 Standard Practice for Separation and Concentration of Ignitable Liquid Residues from Fire Debris Samples by Passive Headspace Concentration with Solid Phase Microextraction (SPME), 2001.
21. Sutherland, D. and Hamilton, A., Fire debris analysis statistics and the use of the latest analytical tools, *J. Canad. Assoc. of Fire Investigators*, September 1999, p. 11.
22. Sutherland, D., Fire debris analysis by GC/MS/MS: The latest in specificity and sensitivity, Presented at The Canadian Society of Forensic Science (44th AGM), September 1997, Regina, Saskatchewan, Canada.
23. Sutherland, D., Byers, K. and Sutherland, A., Fire debris analysis results from the IAAI Fire Investigation and Arson Task Force Seminar, Guelph (June 98), Simcoe (June 99), *International Association of Arson Investigators*, Ontario Chapter, Fall Issue 1999.
24. Sutherland, D., The latest in fire debris analysis — GC/MS/MS, Presented at the FRENZY Conference (Forensic Sciences and Crime Scene Technology), Washington, D.C., 2000.

Analysis of Explosives by LC/MS

6

JEHUDA YINON

Contents

6.1 Introduction

One of the challenging areas in the field of explosives is the identification of postblast explosive residues. The purpose of such analyses is twofold:

1. To serve as evidence in court. Although there is usually no need to prove that an explosion has occurred, it is necessary to explain the explosion, which includes the reconstruction of the explosive device and the identification of explosive or explosives used.
2. To help the investigation in connecting a suspect to the scene of the explosion. In addition to identifying the type of explosive, characterization of the explosive, such as country of origin and manufacturer, would be of great help to the investigation.

0-8493-1522-0/04/$0.00+$1.50

The analysis of postexplosion residues is very difficult because it is based on finding and identifying residues from the original explosive, rather than products formed during the explosion. The amount of residual unexploded explosive is usually very small and has to be isolated from large amounts of debris which could cover wide areas. A scientifically sound method has not yet been established to determine the debris on which there is a greater probability to find residual explosives.

Several comprehensive schemes for screening of the debris, clean-up procedures, extraction, and analysis have been described.[1] The combination of the separation power of high performance liquid chromatography (HPLC) with the identification capability of mass spectrometry (MS) has resulted in a powerful analytical system, LC/MS, which is being widely used in a large variety of analytical applications,[2–4] including forensic.[5–7]

A great deal of progress has been made in the technology of LC/MS during the last 20 years. The first LC/MS systems used for the analysis of explosives were the direct liquid introduction (DLI)-LC/MS systems,[8,9] which had a low sensitivity because of effluent splitting. Thermospray (TS)-LC/MS interface was the first to combine interface and ionization capabilities, in which ions in the gas phase were produced directly from solution without the need of an additional means of ionization.[10] Analysis of explosives by TS-LC/MS with a quadrupole mass spectrometer was found to have best sensitivity in the negative-ion mode with filament on.[11] Detection limits, in the full scan mode, were 200 pg for TNT, and 1 ng for RDX, HMX, and PETN. Results obtained with a triple-stage quadrupole mass spectrometer, in the single-ion monitoring (SIM) mode, yielded a detection limit of 2.5 pg for PETN.[12]

Particle beam (PB)-LC/MS, a solute transport-enrichment technique, was also used for the analysis of explosives.[13] The mode of ionization was negative-ion chemical ionization, with methane as moderator gas, at a pressure of 1.8 Torr in the ion source. Detection limits in the SIM mode were 60 pg for NG, 120 pg for TNT, and 200 pg for RDX and PETN.

The main drawbacks of thermospray and particle beam ionization for analysis of explosives were the difficulty in reproducing results and in using these techniques on a routine basis.

6.2 Electrospray Ionization (ESI) and Atmospheric Pressure Chemical Ionization (APCI)

6.2.1 Principle of Operation of ESI

Electrospray ionization (ESI) is a soft ionization technique that allows the analysis of large biomolecules, as well as a wide range of smaller polar molecules.[14–17] For molecules up to 1000 Da in molecular mass, either an [M +

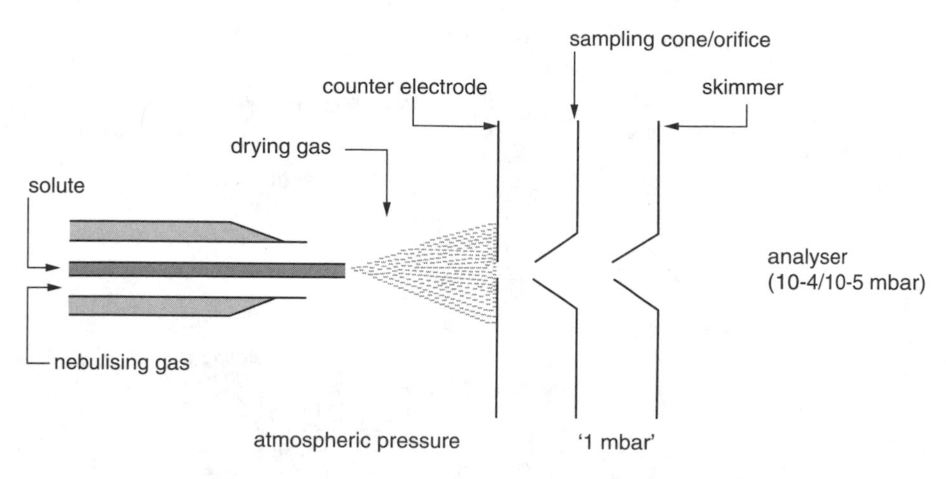

Figure 6.1 Electrospray ion source. (From www.astbury.leeds.ac.uk/Facil/Mstut/mstutorial.htm.With permission.)

H]$^+$ or an [M – H]$^-$ ion is generally detected depending on whether positive or negative ion detection has been selected. Some fragmentation may be apparent, which will provide structural information. Adduct ions are also common in ESI and are strongly dependent on the analyte and the additives or impurities in the solution. Molecules with higher molecular masses up to 200,000 Da usually produce a series of multiple charged ions which can be processed by the mass spectrometer data system to give a molecular weight profile with a mass accuracy of ±0.01%. The main features of ESI are sensitivity, ruggedness, quantitative capability, and universality. The precise molecular mechanism by which ions are desorbed from droplets remains debatable but a sufficiently consistent model is in place to understand the behavior of electrospray when applied to practical analytical problems.

Electrospray ionization operates by the process of the emission of ions from a droplet into the gas phase. In ESI operation, a solution of the analyte introduced into an ion source at atmospheric pressure is pumped through a stainless steel capillary which carries a high potential, typically 3 to 5 kV (Figure 6.1). The strong electric field generated by this potential causes the solvent to be sprayed from the end of the capillary (hence, electrospray). The charged droplets pass down a potential gradient towards the mass analyzer. During that transition the droplets reduce in size by evaporation of the solvent or by droplet subdivision resulting from the high charge density. Ultimately, fully desolvated ions result from complete evaporation of the solvent or by field desorption from the charged droplets. This process is known as "ion evaporation," and is the primary mechanism for gas-phase ion formation in electrospray. A flow of nitrogen gas through the source helps the evaporation process and removal of the solvent.

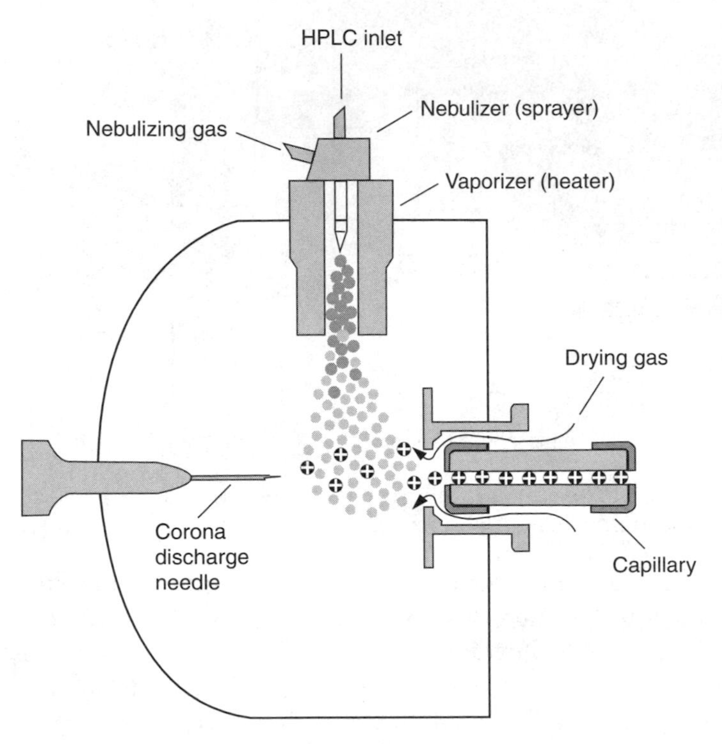

Figure 6.2 APCI ion source. (Copyright Agilent Technologies. Used with permission.)

Ions are transported into the high vacuum system of the mass spectrometer by use of a nozzle-skimmer arrangement. This acts as a momentum separator, and heavier sample molecules tend to pass through, while lighter solvent and drying gas molecules can be more readily pumped away in this differentially pumped intermediate vacuum stage. A potential gradient is also applied to the skimmers or sampling cones, which has the effect of pulling the ions into the mass spectrometer.

Polar compounds of low molecular weight (<1000 amu) will generally form singly charged ions by the loss or gain of a proton. Basic compounds, e.g., amines, can form a protonated molecule ($[M + H]^+$) which can be analyzed in the positive ion mode to give a peak at m/z (M + 1). Acidic compounds, e.g., sulphonic acids, can form a deprotonated molecule ($[M - H]^-$) which can be analyzed in the negative ion mode to give a peak at m/z (M – 1). Since ESI is a soft ionization technique, there is usually little or no fragmentation, and the spectrum contains only the $(M + H)^+$ or $(M - H)^-$ ion. The presence of additives or contaminants such as ammonium or sodium ions could cause adduct formation with ions present in solution. Common adducts occur with NH_4^+ (M + 18), Na^+ (M + 23), and K^+ (M + 39).

The main advantages of ESI are molecular weight information, good sensitivity, and suitability for large bio- or synthetic polymers as well as for polar and even ionic compounds.

6.2.2 Principle of Operation of APCI

The APCI source is based on the original API source developed by Horning's group in the mid-1970s.[18] Its commercial viability was only realized as an adjunct option to ESI when it became commercially available in 1989. The technique is very sensitive and simple to operate. It operates just as well at high mobile phase flow rates (>1 ml/min and 4.6 mm ID columns) as it does with lower flows and smaller ID columns. Thus, the full advantage of the robustness of large ID HPLC columns and ease of transfer of already established HPLC methods based on standard chromatography, is realized with APCI.

Ionization takes place at atmospheric pressure and the ions are extracted into the mass spectrometer by exactly the same set of skimmers used for electrospray (Figure 6.2). In APCI there is no voltage applied to the capillary. The liquid elutes from the capillary probe, which is surrounded by a coaxial flow of N_2 nebulizing gas into a heated region. The combination of nebulizer gas and heat forms an aerosol which begins to evaporate rapidly. At the end of the APCI probe is a high-voltage (2.5 to 3 kV) metal needle to produce a corona discharge, causing solvent molecules eluting into the source to be ionized.[2] Sample molecules which elute and pass through this region of solvent ions can be ionized by gas-phase ion molecule reactions.[3]

Chemical ionization of sample molecules is very efficient at atmospheric pressure due to the high collision frequency. Proton transfer, forming $[M + H]^+$ ions, occurs in the positive-ion mode, and either electron transfer or proton transfer, forming $[M - H]^-$ ions, occurs in the negative-ion mode. The moderating influence of the solvent clusters on the reagent ions and the influence of the high gas pressure reduce fragmentation during ionization and results in primarily $[M + H]^+$, $[M - H]^-$, and/or adduct ions. As in ESI, ions are transported into the high vacuum system of the mass spectrometer by use of a nozzle-skimmer arrangement. The probe can be situated axially or orthogonally (as in Figure 6.1) to the sampling orifice.

The main advantages of APCI are molecular weight information on volatile molecules, typical flow rates up to 2 ml/min, and good reproducibility for quantitation.

6.2.3 Comparison between ESI and APCI

Flows in the range of 0.2 to 2 ml/min can be used with APCI. This permits direct coupling of 2.1 mm and 4.6 mm ID HPLC columns to the APCI interface. ESI sources can be used up to 1 ml/min, although this is at the top of the flow range. ESI sources can also be used at flow rates down to <5 μl/min, thus allowing the interfacing of capillary columns.

Although a high temperature is applied to the APCI probe, most of the heat is used in evaporating the solvent and heating the N_2 gas, and the thermal effect on the sample is much less than may be expected. However, with very thermally labile compounds the heated probe may cause some thermal fragmentation. Since ESI uses a colder probe, it is sometimes possible to produce $[M + H]^+$ or $[M - H]^-$ ions by ESI in compounds which fragment in APCI.

Many compounds can be analyzed by either APCI or ESI. There are no definitive rules on when to use APCI over ESI, but a general observation is that APCI tends to give better sensitivity than ESI for less polar compounds.

Unlike ESI, APCI does not produce multiple charged series and is therefore unsuitable for the analysis of high molecular weight compounds such as proteins.

6.3 Analysis of Explosives

6.3.1 Nitroaromatic Explosives and Their Degradation Products

The most widely used military explosive is 2,4,6-trinitrotoluene (TNT). Its main features include low melting point, stability, low sensitivity to impact, friction, shock and electrostatic energy, and its relatively safe methods of manufacture.[1,19,20] Its melting point is 80.65°C. TNT is highly soluble in acetone (132 g per 100 g acetone at 25°C) but still soluble enough in water (10 mg per 100 g water at 25°C). Its vapor pressure at 25°C is 5.8×10^{-6} Torr. The addition of oxygen-rich products to TNT can form mixtures with enhanced explosive power. Various explosive compounds have been added to TNT to form binary explosives. Other compounds have been added to decrease sensitivity and increase the mechanical strength of the cast.

Electrospray ionization coupled to tandem mass spectrometry was used for the analysis of a series of explosives, including TNT.[21] Analyses were performed with a Perkin–Elmer Sciex API III-PLUS triple quadrupole mass spectrometer with electrospray ionization. A standard solution of TNT was made in acetonitrile–water (50:50), containing 2 m*M* of ammonium acetate.

The solution was introduced directly into the ion source by means of a syringe pump at a flow rate of 5 μl/min. The major ion in the negative-ion spectrum of TNT, at a concentration of 100 pg/μl, was the $[M - H]^-$ ion at *m/z* 226.

TNT, DNT, and 4-amino-2,6-dinitrotoluene (4-ADNT) were analyzed by electrospray-MS with various additives in order to obtain ion intensity enhancement.[22] A Finnigan TSQ 700 mass spectrometer with an electrospray ion source was used. Additives tested included ammonium nitrate, trifluoroacetate (TFA), and chloroacetonitrile. Nitrate adduct ions were more intense than TFA or chloride adduct ions by a factor of 6–40. The base peak in the negative-ion mass spectrum of TNT, with 1 m*M* ammonium nitrate in the mobile phase, was at *m/z* 226 due to the $[M - H]^-$ ion. For samples as large as 500 ng, a nitrate adduct ion $[M + NO_3]^-$ at *m/z* 283 with a relative intensity of 5% was observed. The base peak in the mass spectrum of 4-ADNT was at *m/z* 259, $[M + NO_3]^-$, while in DNT no nitrate adduct ion was detected. Additional experiments in ESI with the additives ammonium fluoride, bifluoride, chloride, bromide, and iodide[23] showed no formation of adduct ions in nitroaromatic compounds, probably due to the high electron density in their rings.

ESI and APCI in the negative-ion mode were compared for the analysis of explosives, using a Hewlett–Packard 5989B LC/MS mass spectrometer.[24] Standard solutions of the investigated explosives, which included 1,3,5-TNB, 1,3-DNB, TNT, 2-ADNT, 4-ADNT, 2,4-DNT, and 2,6-DNT, were prepared in a methanol–water (50:50) solution. Figure 6.3 shows the APCI and ESI mass spectra in the negative-ion mode of TNT. In 1,3-DNB, 2,4-, and 2,6-DNT, no ions were detected in the ESI mode. The APCI mass spectrum of 2,6-DNT is shown in Figure 6.4. In 2,4-DNT the ion at *m/z* 181 has a higher abundance than the one at *m/z* 182.

The major ions in the APCI mass spectrum of 1,3,5-TNB were at *m/z* 213,$[M]^-$ and at *m/z* 183, $[M - NO]^-$. The major ions in the APCI spectrum of 1,3-DNB were at *m/z* 168, $[M]^-$, at *m/z* 138, $[M - NO]^-$, and at *m/z* 108, $[M - 2NO]^-$. Both the ESI and APCI mass spectra of 2- and 4-ADNT contain a major ion at *m/z* 196, $[M - H]^-$.

The ESI mass spectrum of TNT without any additive was recorded with a Finnigan LCQ ion trap mass spectrometer.[25] The mass spectrum contained a larger number of fragment ions than ESI mass spectra with additives. Lowest detection limit in the SIM mode, at a signal-to-noise ratio of 3, was 5 pg.

A typical group of byproducts of industrial 2,4,6-trinitrotoluene (TNT) include isomers of trinitrotoluene, dinitrotoluene, trinitrobenzene, and dinitrobenzene. These compounds were investigated using APCI-LC/MS in the negative-ion mode and MS/MS using a Thermo–Finnigan LCQ_{DUO} ion trap mass spectrometer.[26]. HPLC separations were done with a Restek reversed-phase Allure C18 column (150 × 3.2 mm, 5 μm particle size) at a flow rate of 0.4 ml/min with sample injection volume of 10 μl. Three mobile phase systems consisting of methanol–isopropanol–water or methanol–water were used for different groups of standard mixtures. The separation of the six standard TNT isomers (2,3,4-, 2,3,5-, 2,3,6-, 2,4,5-, 3,4,5-, and 2,4,6-TNT) (Scheme 1) was carried out with a linear gradient elution of methanol–iso-

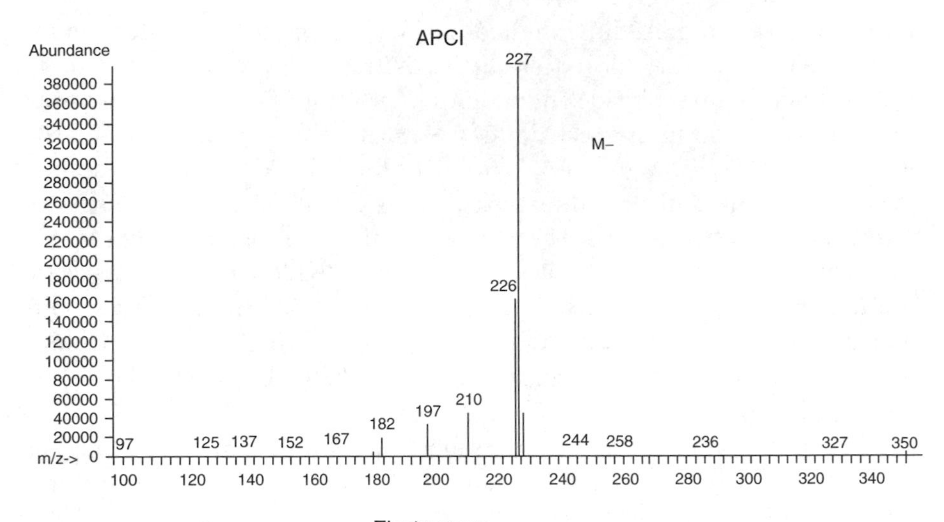

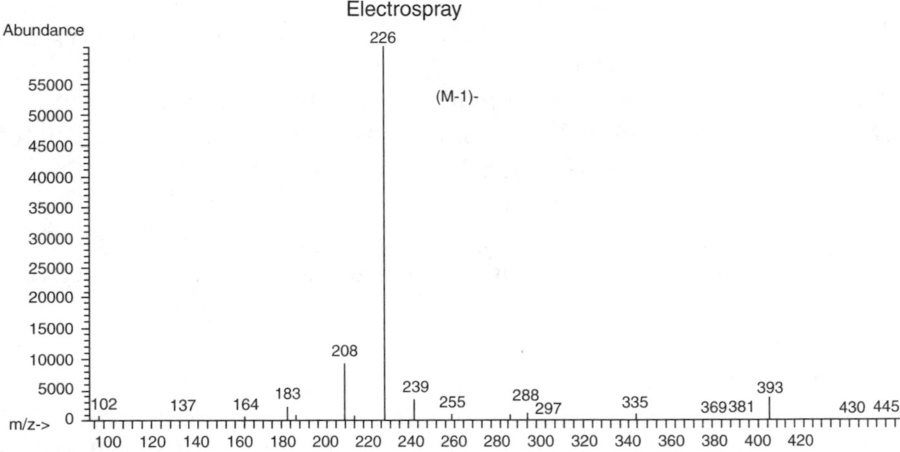

Figure 6.3 APCI and ESI mass spectra of TNT. (From Schilling, A. B., The analysis of explosives constituents and metabolites with electrospray and APCI LC/MS, presented at 44th ASMS Conference on Mass Spectrometry and Allied Topics, Portland, OR, 1996. With permission).

propanol–water. The isopropanol was kept constant at 10% through the run; methanol was held at 30% for the first 25 min, before starting a 2-min linear gradient from 30 to 85%, at which it was held for 3 min. The mobile phase was returned to its initial condition within 2 min and held for 3 min before starting a new run. For the standard mixture of four of the DNT isomers, three DNB isomers and TNB (Schemes 2 and 3), the separation was carried out by an isocratic run of methanol–water (57:43). Figure 6.5 shows the mass chromatograms of the TNT, DNB, DNT, and TNB isomers. When 2,5-DNT was included in the standard mixture of TNB, DNB, and DNT, a linear gradient elution with methanol–isopropanol–water was used, keeping a con-

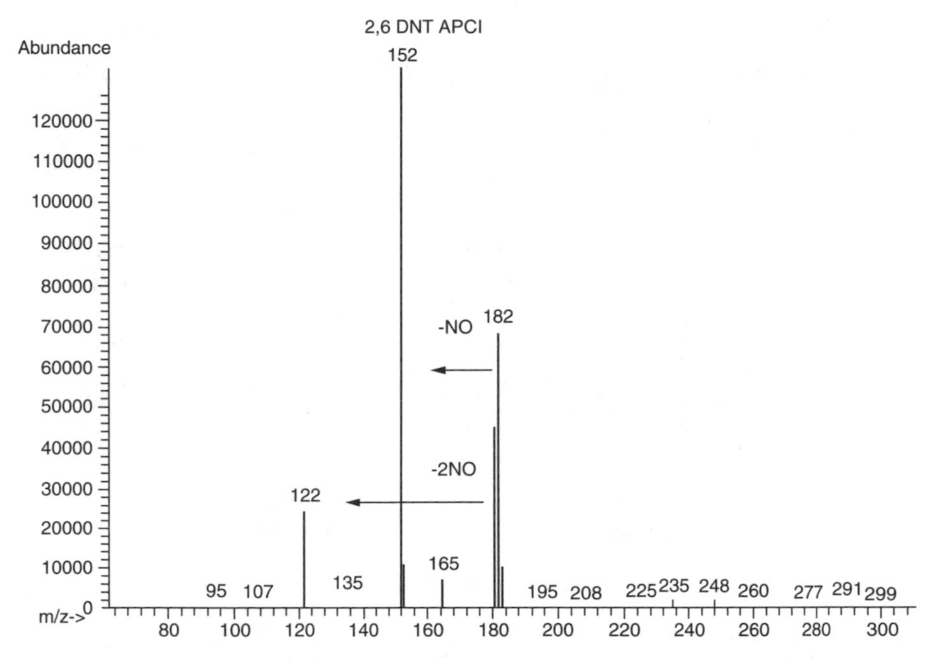

Figure 6.4 APCI mass spectrum of 2,6-DNT. (From Schilling, A.B., The analysis of explosives constituents and metabolites with electrospray and APCI LC/MS, presented at 44th ASMS Conference on Mass Spectrometry and Allied Topics, Portland, OR, 1996. With permission).

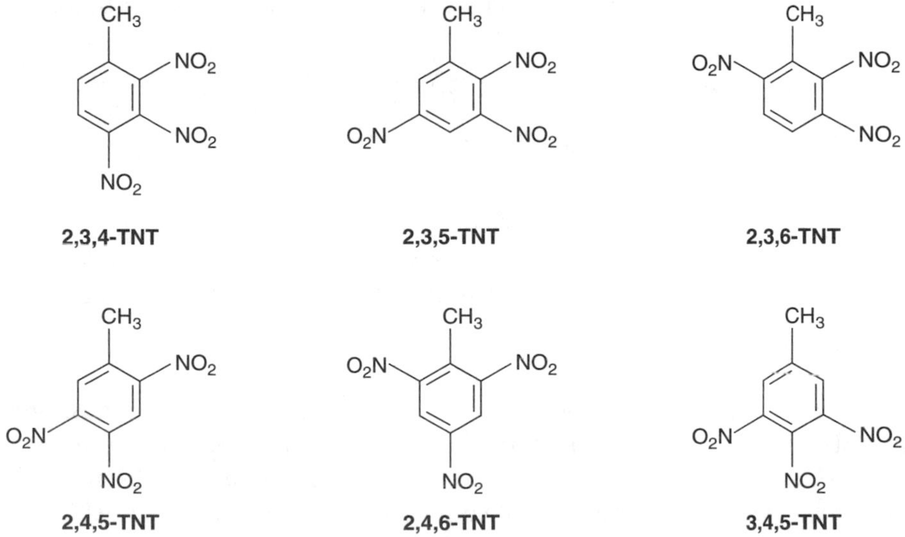

2,3,4-TNT **2,3,5-TNT** **2,3,6-TNT**

2,4,5-TNT **2,4,6-TNT** **3,4,5-TNT**

Scheme 1

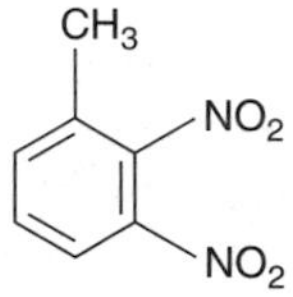

2,3-DNT 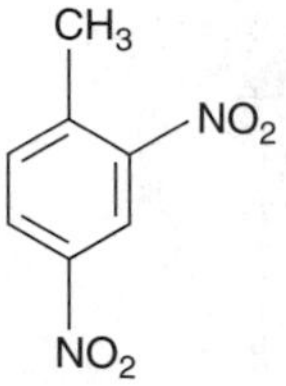 2,4-DNT 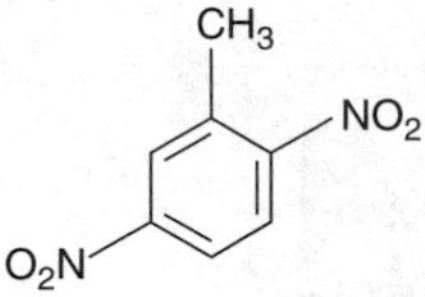 2,5-DNT

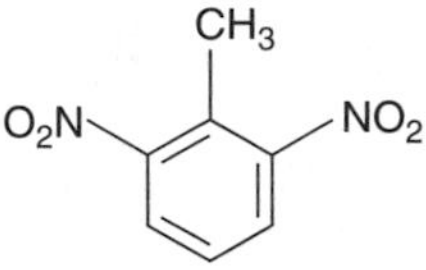

2,6-DNT 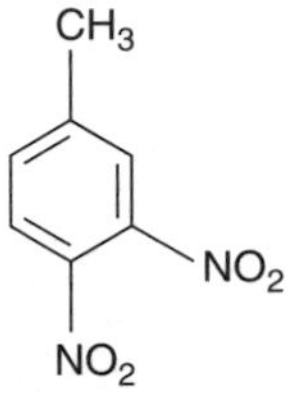 3,4-DNT

Scheme 2

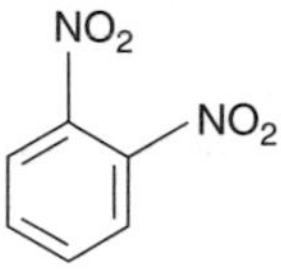

1,2-DNB 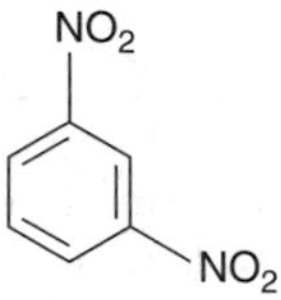 1,3-DNB  1,4-DNB

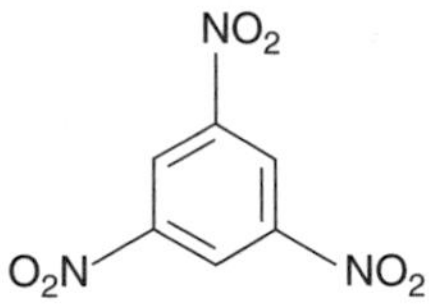

1,3,5-TNB 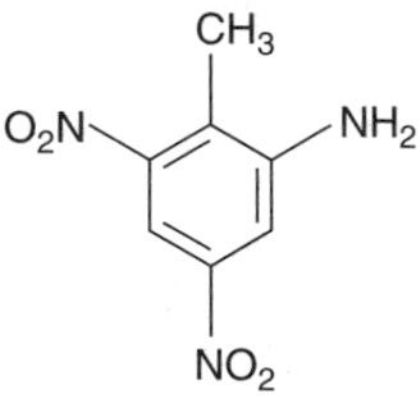 2-ADNT 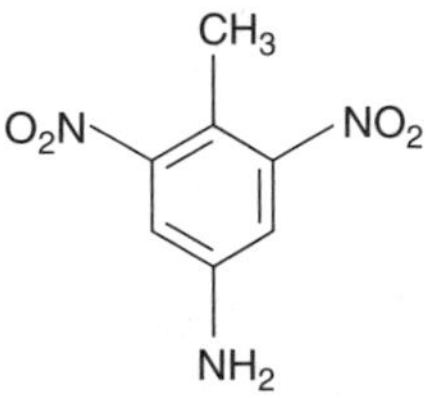 4-ADNT

Scheme 3

stant water percentage at 55%. The mobile phase starts at methanol–isopropanol–water (35:15:55), is held for 1 min, then is changed from 30 to 40% methanol over the next 14 min, followed by 2 min hold time. The mobile phase was returned to to its initial condition within one min and held for 2 min. Figure 6.6 shows the mass chromatograms under these conditions: 2,5-DNT is separated, while 2,4- and 2,6-DNT are not.

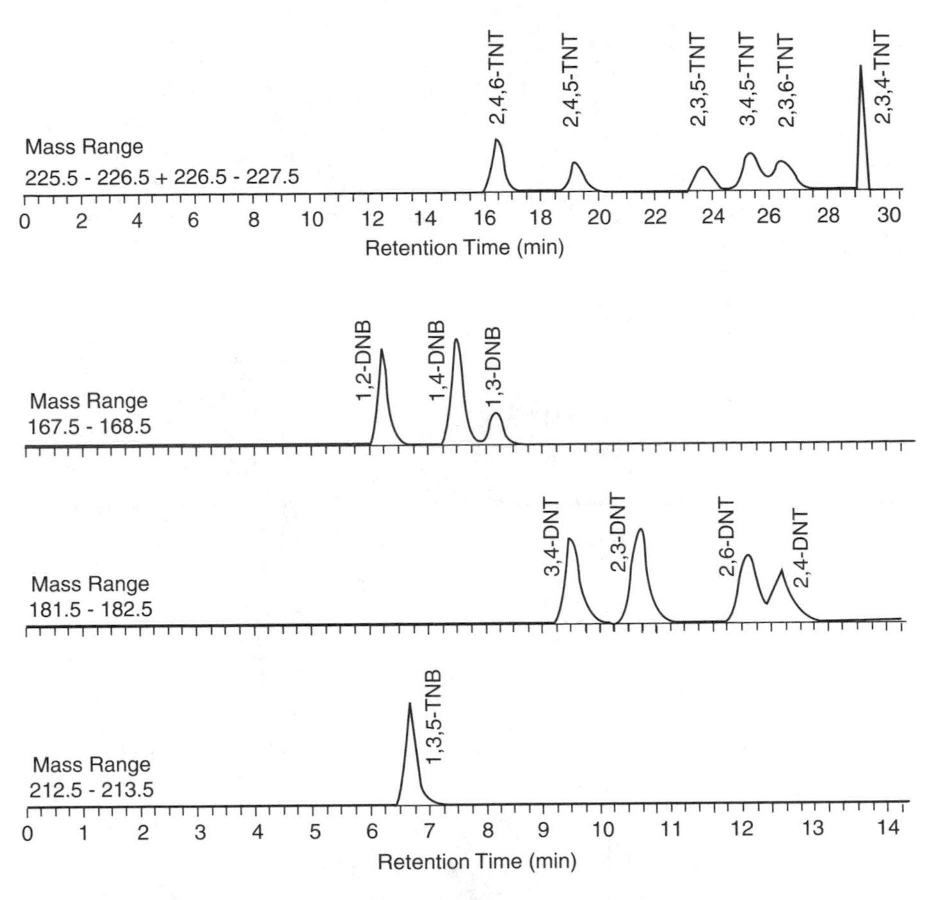

Figure 6.5 Mass chromatograms of a standard mixture of six TNT isomers and three DNB, four DNT and one TNB isomers at a concentration of 1 μg/mL each. (Reprinted from Zhao, X. and Yinon, J., Characterization and origin identification of 2,4,6-trinitrotoluene through its by-product isomers by liquid chromatography – atmospheric pressure chemical ionization mass spectrometry, *J. Chromatogr. A*, 946, 125, 2002. Copyright 2002, with permission from Elsevier Science).

The full-scan negative-ion APCI mass spectra of the TNT isomers were very similar and thus provided very little information on their characterization. MS/MS-CID provided that information. Table 6.1 shows the CID ions of the various TNT isomers. Results show that TNT isomers can be characterized by HPLC separation, followed by MS/MS-CID. In the MS/MS-CID mass spectrum of the symmetric configuration, 2,4,6-TNT, an *ortho* effect,[27] was observed, resulting in the ion at *m/z* 210, due to loss of OH from the molecular ion. In the MS/MS-CID mass spectra of the isomers 2,4,5-, 2,3,5-, 2,3,6-, and 2,3,4-TNT, who have only one nitro group ortho to the methyl group, no loss of OH was observed.

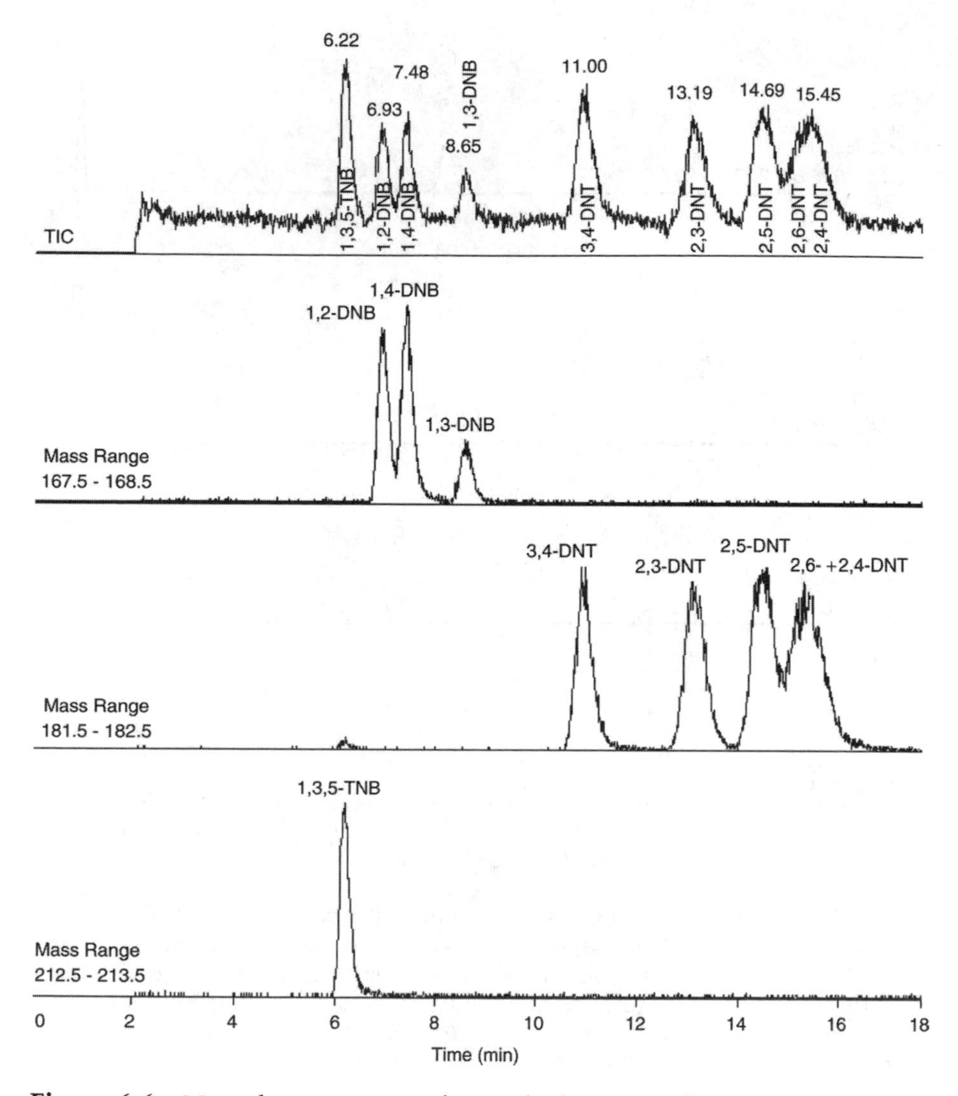

Figure 6.6 Mass chromatograms of a standard mixture of six TNT isomers and three DNB, five DNT and one TNB isomers at a concentration of 1 μg/mL each.

The mass spectra of DNB, DNT, and TNB isomers were characterized by molecular ions, namely, at *m/z* 168 for DNB, at *m/z* 182 for DNT, and at *m/z* 213 for TNB. All three DNB isomers demonstrated the same CID fragmentation pattern: loss of NO from the molecular ion. Only 2,4-DNT could be differentiated in MS/MS-CID from the other DNT isomers by producing a fragment ion at *m/z* 165 due to loss of OH. Additional fragment ions in the CID mass spectrum of 2,4-DNT are $[M - H_2O]^-$, $[M - NO]^-$, and $[M - NO - CH_3]^-$. All the other DNT isomers produced one fragment ion at *m/z* 138 due to $[M - NO]^-$.

Table 6.1 MS/MS-CID Data of TNT Isomers

Parent Ion		Daughter Ions		Tentative
m/z	ion	*m/z*	%	Identification
		2,4,6-TNT		
227	M^-	210	100	[M – OH]–
		197	48	$[M - NO]^-$
		181	5	$[M - NO_2]^-$
		167	4	$[M - 2NO]^-$
		151	5	$[M - NO_2 - NO]^-$
		137	13	$[M - 3NO]^-$
		2,4,5-TNT		
227	M^-	197	100	$[M - NO]^-$
		181	2	$[M - NO_2]^-$
		2,3,5-TNT		
227	M^-	197	100	$[M - NO]^-$
		3,4,5-TNT		
227	M^-	197	49	$[M - NO]^-$
		181	100	$[M - NO_2]^-$
		2,3,6-TNT		
227	M^-	197	100	$[M - NO]^-$
		181	10	$[M - NO_2]^-$
		2,3,4-TNT		
227	M^-	197	60	$[M - NO]^-$
		181	100	$[M - NO]^-$
		151	3	$[M - 2NO]^-$

Source: Reprinted from Zhao, X. and Yinon, J., Characterization and origin identification of 2,4,6-trinitrotoluene through its by-product isomers by liquid chromatography-atmospheric pressure chemical ionization mass spectrometry, *J. Chromatogr. A*, 946, 125, 2002. Copyright 2002, with permission from Elsevier Science.

Characterization and origin identification of explosives is important in forensic analysis of postexplosion residues. In addition to the type of explosive used in a bombing, the investigators would like to know its country of origin and, especially, its manufacturer. Each manufacturer produces explosives with characteristic differences in the type and amount of byproducts, impurities, and additives, depending on the purity of the raw materials and solvents used and the type of manufacturing process, thus resulting in a characteristic profile of byproducts, organic impurities, and additives.

The production process of TNT[28] includes toluene formation from benzene and methanol, followed by three steps of nitration: from toluene to mononitrotoluene (MNT), to dinitrotoluene (DNT), and to 2,4,6-trinitrotoluene (TNT). Nitration is carried out in the presence of nitric and sulfuric acid, followed by crystallization in alcohol or water, and washings with sodium sulfite. The three steps of nitration can be carried out as batch processes or as a continuous process.

Characterization of TNT was made possible by building an LC/MS-APCI profile of byproducts, containing the isomers of TNT, DNT, DNB, and TNB.[26] MS/MS-CID was used as an additional dimension of identification for some of the nitroaromatic isomers. Figure 6.7 to Figure 6.10 show the LC/MS-

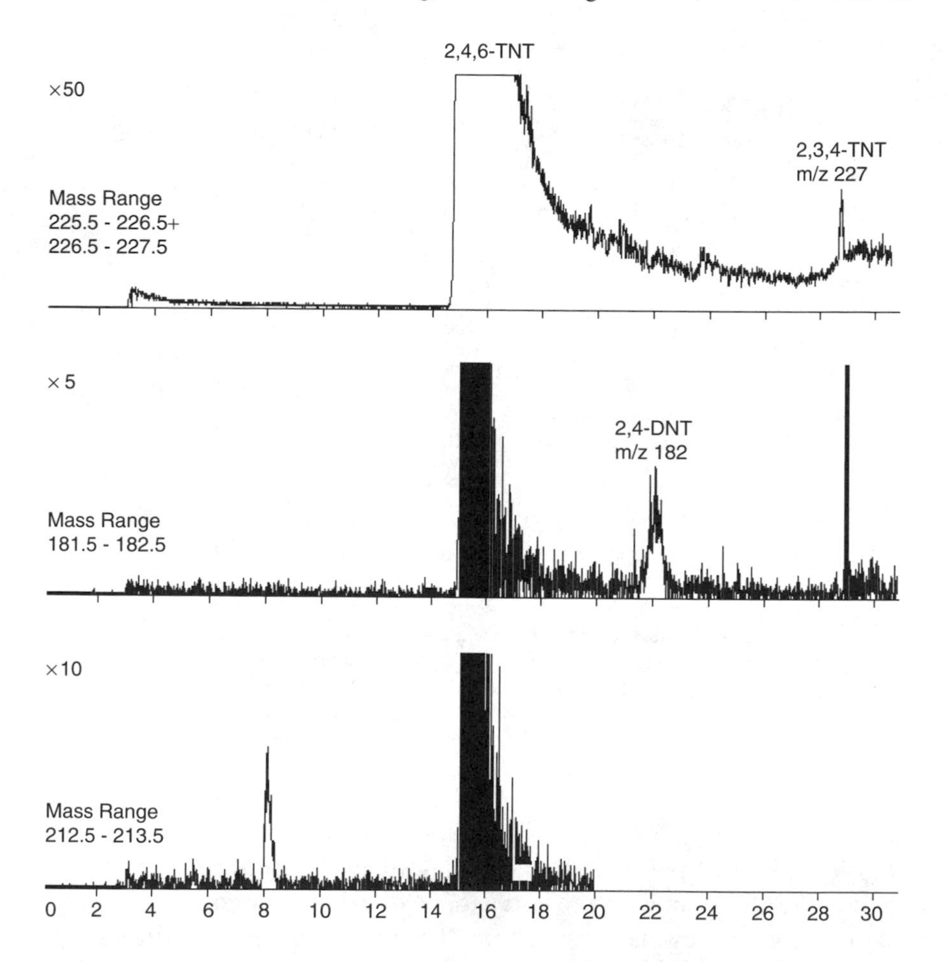

Figure 6.7 LC/MS-APCI mass chromatograms of a TNT sample from Russia. (Reprinted from Zhao, X. and Yinon, J., Characterization and origin identification of 2,4,6-trinitrotoluene through its by-product isomers by liquid chromatography – atmospheric pressure chemical ionization mass spectrometry, *J. Chromatogr. A*, 946, 125, 2002. Copyright 2002, with permission from Elsevier Science.)

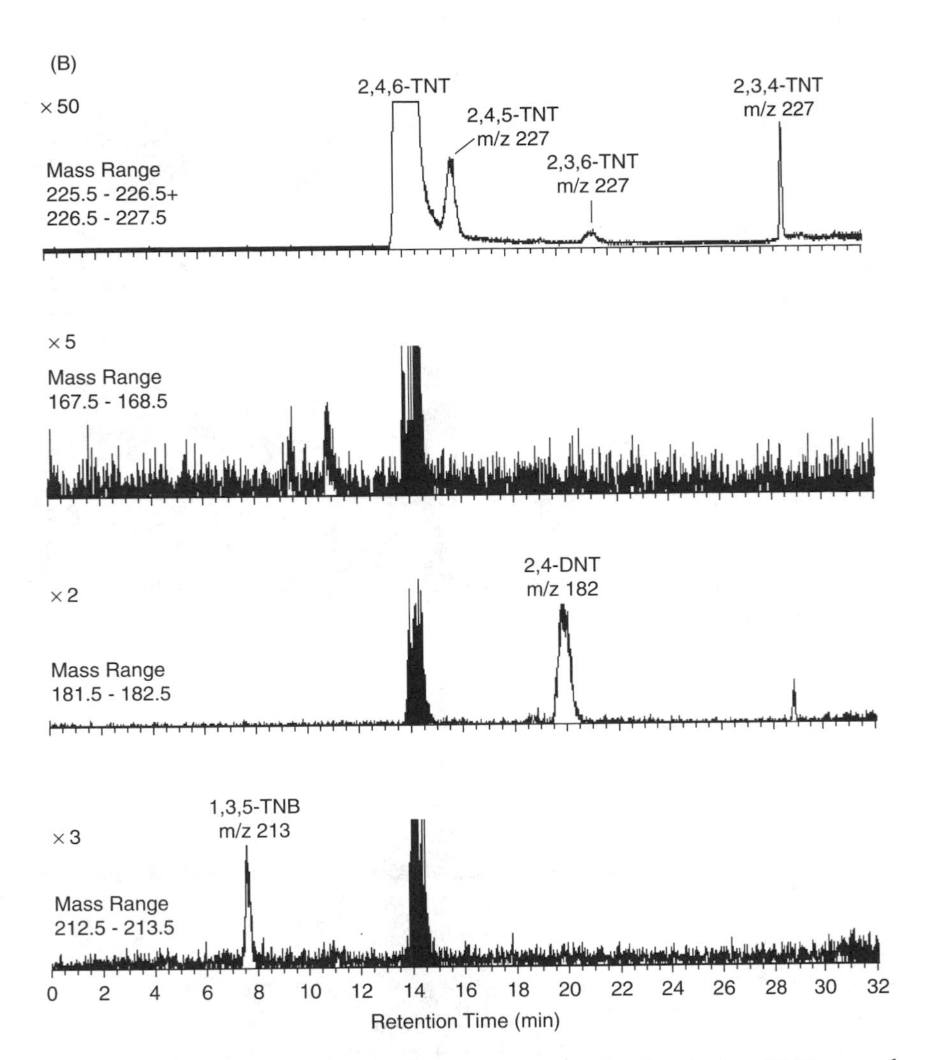

Figure 6.8 LC/MS-APCI mass chromatograms of a landmine TNT sample. (Reprinted from Zhao, X. and Yinon, J., Characterization and origin identification of 2,4,6-trinitrotoluene through its by-product isomers by liquid chromatography – atmospheric pressure chemical ionization mass spectrometry, *J. Chromatogr. A*, 946, 125, 2002. Copyright 2002, with permission from Elsevier Science.)

APCI mass chromatograms of TNT samples from Russia, from an unknown landmine, from the U.K., and from Canada, respectively. Identification of the various byproducts was confirmed by MS/MS-CID. The differences in the profiles were used to characterize the various TNT samples.

Figure 6.11 gives the LC/MS-APCI mass chromatogram of a 4-times crystallized TNT sample, showing that purification of TNT can eliminate all the byproducts. Such purification is not being done in the manufacturing

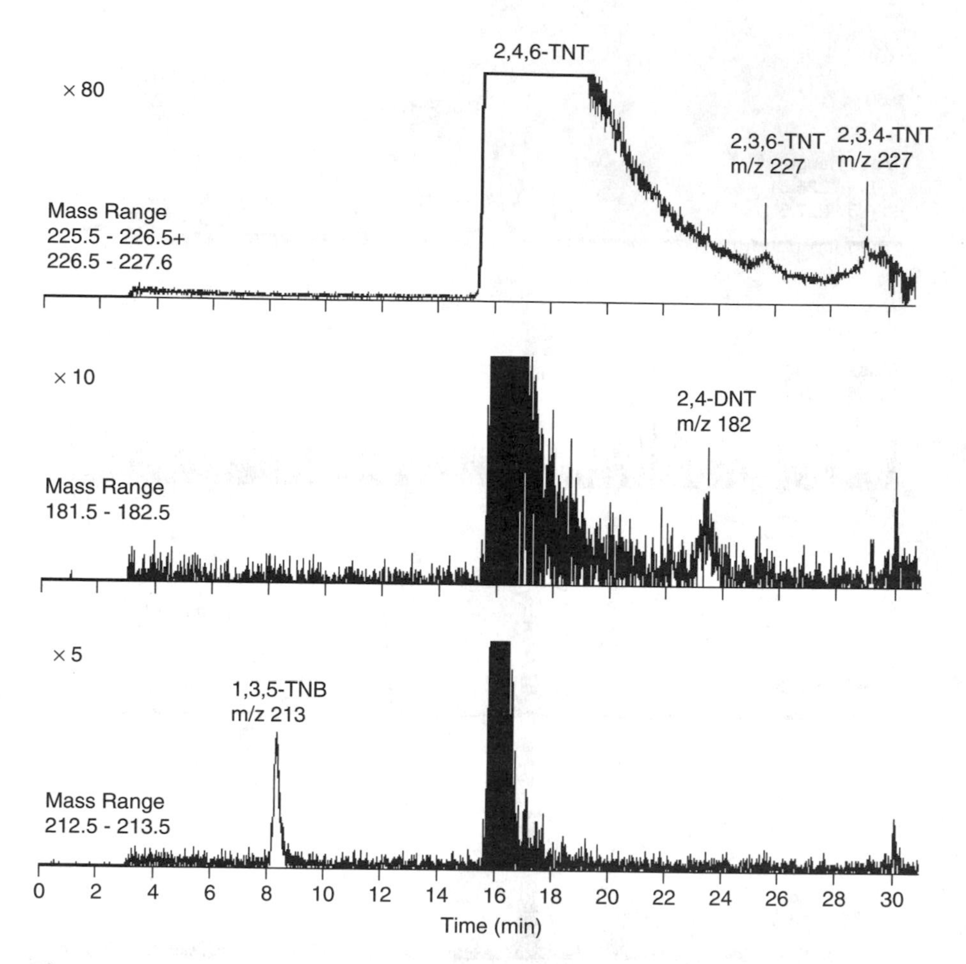

Figure 6.9 LC/MS-APCI mass chromatograms of a TNT sample from the U.K. (From Zhao, X. and Yinon, J., Characterization and origin identification of 2,4,6-trinitrotoluene through its by-product isomers by liquid chromatography – atmospheric pressure chemical ionization mass spectrometry, *J. Chromatogr. A*, 946, 125, 2002. With permission.)

process of TNT because of economic reasons. An additional reason is that the isomer impurities affect the melting point and the crystallization properties and hence the casting properties of TNT.[29] Pure 2,4,6-TNT has a tendency to crystallize in a needlelike fashion which could cause voids and cracks in shell and grenade fillings.

A series of additional nitroaromatic compounds (Scheme 4), 2,4,6-trinitro-*m*-cresol (TNC), 2,4,6-trinitroaniline (picramide), and tetryl (2,4,6,*N*-tetranitro-*N*-methylaniline) were analyzed by negative-ion ESI-MS,[30] using a Finnigan LCQ ion trap mass spectrometer. Samples were infused into the mass spectrometer at a flow rate of 5 μl/min using a syringe pump. Samples

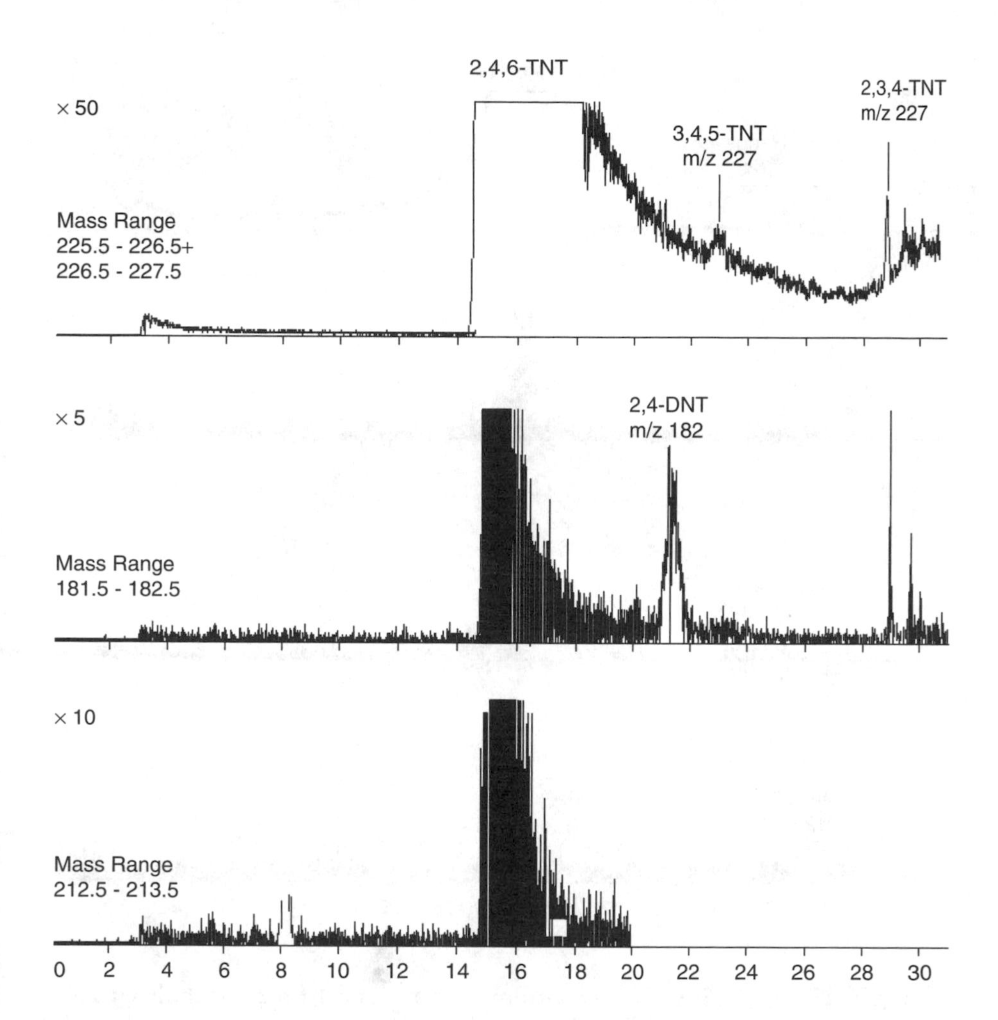

Figure 6.10 LC/MS-APCI mass chromatograms of a TNT sample from Canada (Reprinted from Zhao, X. and Yinon, J., Characterization and origin identification of 2,4,6-trinitrotoluene through its by-product isomers by liquid chromatography – atmospheric pressure chemical ionization mass spectrometry, *J. Chromatogr. A*, 946, 125, 2002. Copyright 2002, with permission from Elsevier.)

were dissolved in acetone and diluted in an isopropanol–water (50:50) solution with 0.1% ammonium hydroxide.

The major ions in the ESI mass spectrum of picramide are the $[M - H]^-$ and M^- ions, at *m/z* 227 and 228, respectively, and the adduct ion at *m/z* 285, $[M + CH_3COCH_3 - H]^-$, due to the acetone solvent. The ESI mass spectrum of TNC consists mainly of one ion, the $[M - H]^-$ ion at *m/z* 242. The ESI mass spectrum of tetryl consisted of the adduct ion $[M + CH_3COCH_3 - H]^-$ at *m/z* 344 and the ion at *m/z* 241, which is the $[M - H]^-$ ion of *N*-

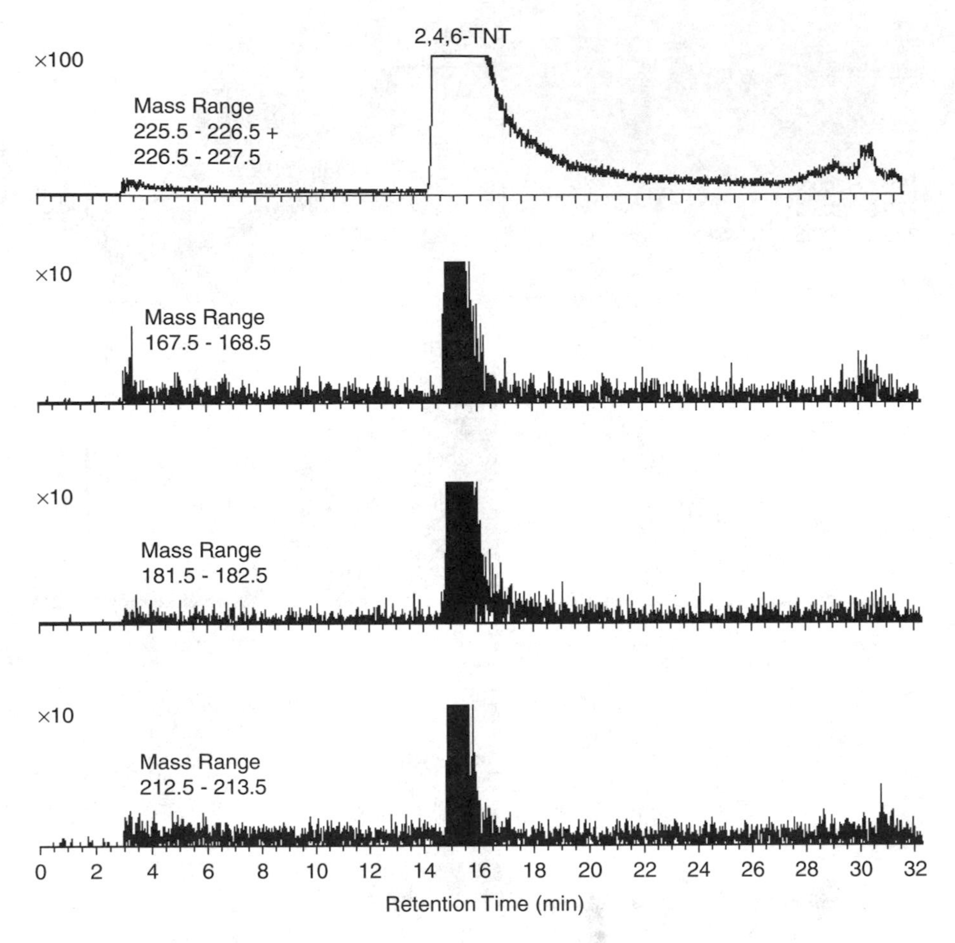

Figure 6.11 LC/MS-APCI mass chromatograms of a 4-times crystallized TNT sample. (Reprinted from Zhao, X. and Yinon, J., Characterization and origin identification of 2,4,6-trinitrotoluene through its by-product isomers by liquid chromatography – atmospheric pressure chemical ionization mass spectrometry, *J. Chromatogr. A*, 946, 125, 2002. Copyright 2002, with permission from Elsevier Science.)

CH_3, O_2N, NH_2, NO_2 — **2-ADNT**

CH_3, O_2N, NO_2, NH_2 — **4-ADNT**

Scheme 4

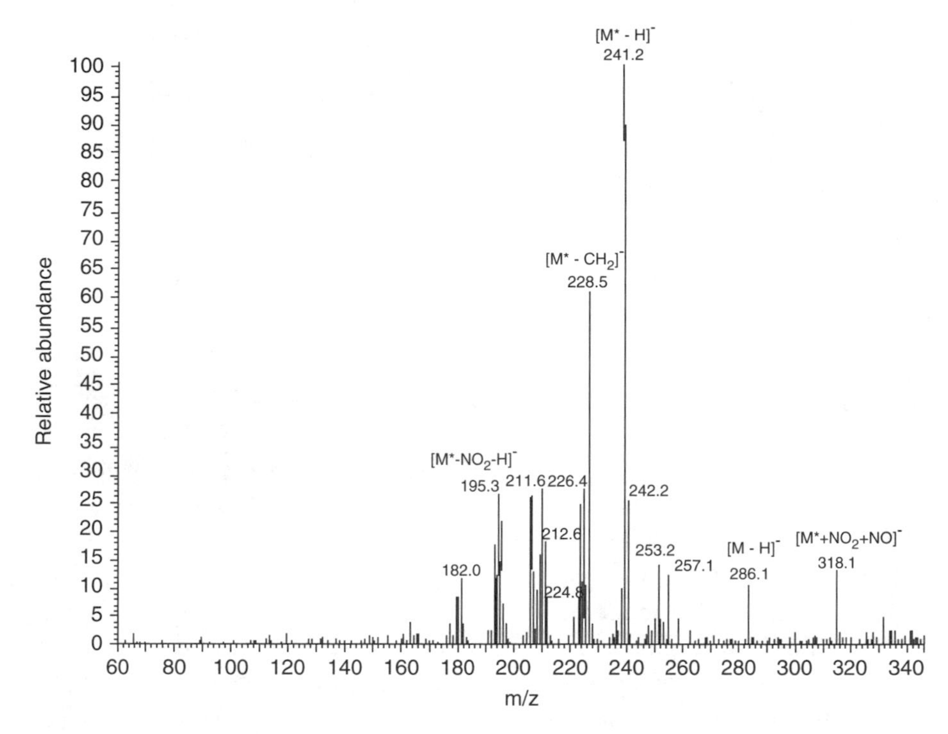

Figure 6.12 ESI mass spectrum of tetryl. M: molecular ion of tetryl; M*: molecular ion of *N*-methylpicramide (From Yinon, J., McClellan, J.E., and Yost, R.A., Electrospray ionization tandem mass spectrometry collision-induced dissociation study of explosives in an ion trap mass spectrometer, *Rap. Comm. Mass Spectrom.*, 11, 1961, 1997. © John Wiley & Sons Ltd. Reproduced with permission).

methylpicramide, a hydrolysis product of tetryl, in which the NO_2 on the amine nitrogen is replaced by an H.[31]

When introducing tetryl as a pure methanol solution,[25] the ESI spectrum (Figure 6.12) contains two sets of ions: those belonging to tetryl, which is represented by the $[M - H]^-$ ion at *m/z* 286, and those belonging to the hydrolysis product, *N*-methylpicramide, which is represented by its $[M^* - H]^-$ ion at *m/z* 241. When using a methanol–water (1:1) solution, instead of a methanol solution, no $[M - H]^-$ ion is produced in the ESI mass spectrum, but a highly abundant $[M^* + NO_2]^-$ ion at *m/z* 288 is observed. When using trifluoroacetic acid (TFA) as an additive, the ion at *m/z* 400, $[M + TFA]^-$ constituted the base peak in the ESI mass spectrum of tetryl.[24]

6.3.2 Nitramine Explosives

One of the most important military explosives is 1,3,5-trinitro-1,3,5-triazacyclohexane (RDX) (Scheme 5). RDX is relatively insensitive; it has a high chemical stability, although lower than that of TNT, and an explosive power

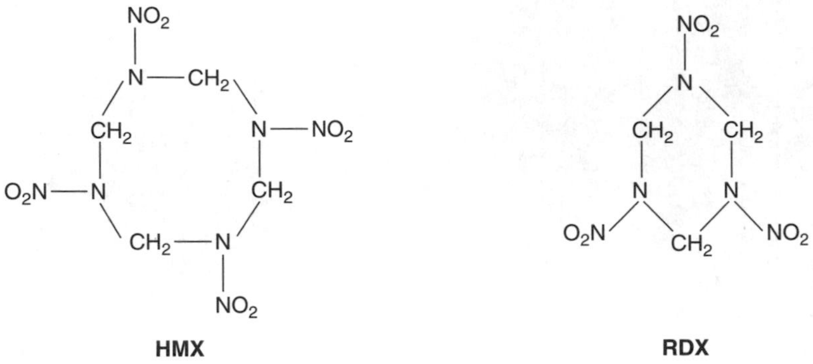

HMX

RDX

Scheme 5

much greater than that of TNT.[1,20, 28] Its melting point is 204°C. It is most soluble in dimethyl sulfoxide (41 g/100 g at 25° C), in dimethyl formamide (37 g/100 g at 25°C), and in acetone (8.3 g/100 g at 25°C), and it is slightly soluble in water (6 mg/100 g at 25°C). Its vapor pressure at 25°C is 4.6×10^{-9} Torr. The industrial product may contain traces of HMX, up to 9% depending on the manufacturing process. RDX is never handled pure and dry because of the danger of accidental explosion. Following production, it is immediately incorporated into formulations or desensitized with an additive. RDX is a component in explosive mixtures, especially plastic explosives (such as C-4 and Semtex).

Electrospray ionization coupled to tandem mass spectrometry was used for the analysis of a series of explosives, including RDX and HMX (Scheme 5).[21,32] Analyses were performed with a Perkin–Elmer Sciex API III-PLUS triple quadrupole mass spectrometer with electrospray ionization. Standard solutions of RDX and HMX were made in acetonitrile–water (50:50), containing 2 m*M* of ammonium acetate. The solutions were introduced directly into the ion source by means of a syringe pump at a flow rate of 5 μl/min.

The major ion in the mass spectrum of RDX was at *m/z* 281, interpreted as $[M + CH_3COO]^-$, the additive originating from the ammonium acetate. The MS/MS spectrum of this ion produced ions at *m/z* 46 $[NO_2]^-$ (base peak) and at *m/z* 59, due to the acetate.

Another negative-ion ESI-MS study of explosives used a PE-Sciex API 100 LC/MS system.[33] RDX and HMX were injected into an eluent consisting of methanol–water (75:25) and 5 m*M* ammonium acetate, producing major $[M + CH_3COO]^-$ ions.The mass spectrum of HMX also showed a major ion $[M + CH_3COO]^-$ at *m/z* 355.

In a study to enhance ESI intensities,[22,23] several additives were tested with a series of explosives, including RDX and HMX. Mobile phase of methanol–water (50:50) with 1 m*M* ammonium nitrate as additive produced best results, forming major $[M + NO_3]^-$ ions. Nitrate adduct ions were 6 to 40 times more intense than trifluoroacetate (TFA) or chloride (chloroacetonitrile) adduct ions. Experiments using labeled $^{15}NO_3$ in the mobile phase did not produce unlabeled nitrate adduct ions, indicating that self adduct ion formation is excluded. When using ammonium fluoride and bifluoride, no adduct ions were formed but $[M - H]^-$ ions were promoted. Ammonium chloride, bromide, and iodide formed adduct ions, the chloride and bromide adduct ions being more stable than the iodide adduct ions. Chloride and bromide adduct ions have the advantage of isotopic patterns, making identification easier.

ESI and APCI in the negative-ion mode were compared for the analysis of explosives, using a Hewlett–Packard 5989B LC/MS mass spectrometer.[24] Standard solutions of RDX and HMX were prepared in a methanol–water (50:50) solution. Figure 6.13 shows the APCI and ESI mass spectra of HMX. The adduct ions $[M + Cl]^-$ and $[M + TFA]^-$ are formed from background residues of TFA and a chlorinated compound.

A Finnigan LCQ ion trap was used for a study on ESI-MS and ESI-MS/MS of a series of explosives including RDX and HMX.[25] Samples were introduced as methanol–water (50:50) solutions. No additive was used. The major ions are at *m/z* 267 $[M + 45]^-$ (base peak), at *m/z* 268 $[M + 46]^-$, and at *m/z* 281 $[M + 59]^-$. A smaller ion appears at *m/z* 221 $[M - H]^-$. The ions at *m/z* 267 and *m/z* 281 were erroneously interpreted as $[M + NO_2 -H]^-$ and $[M + NNO_2 - H]^-$, respectively, because it was believed that the adduct ions were formed between RDX and its fragments, as in the gas phase ionization.[34–37]

A Finnigan LCQ ion trap mass spectrometer was used to study the ESI mass spectra of HMX, RDX and some of its degradation products, mononitroso-RDX (MNX) and trinitroso-RDX(TNX) (Scheme 6) in ground water.[38] Ring-labeled $^{15}N_3$-RDX was used as internal standard.

Samples were extracted and preconcentrated by Porapak RDX solid-phase extraction (SPE) cartridges (500 mg Sep-Pak, Waters Corp., Milford, MA).

Chromatographic separation was achieved with a Kromasil C8 reversed-phase column (250 × 2 mm, Eka Nobel, Bohus, Sweden), using an isocratic mobile phase of isopropanol–water–ammonium formate (20:78:2) at a flow rate of 0.2 ml/min and a column temperature of either 30° or 32°C. The concentration of the ammonium formate solution was 0.5 *M* and was adjusted to pH 8 with 10% ammonium hydroxide.

Several volatile buffers, including ammonium nitrate, ammonium acetate, ammonium formate, formic acid, acetic acid, and trifluoroacetic acid, were tested and evaluated for the ability to promote negative ion formation in the electrospray ion source. Ammonium formate was found to produce more stable negative ion adducts with the lowest background. All investigated nitramine compounds produced major formate adduct ions, $[M + CHOO]^-$.

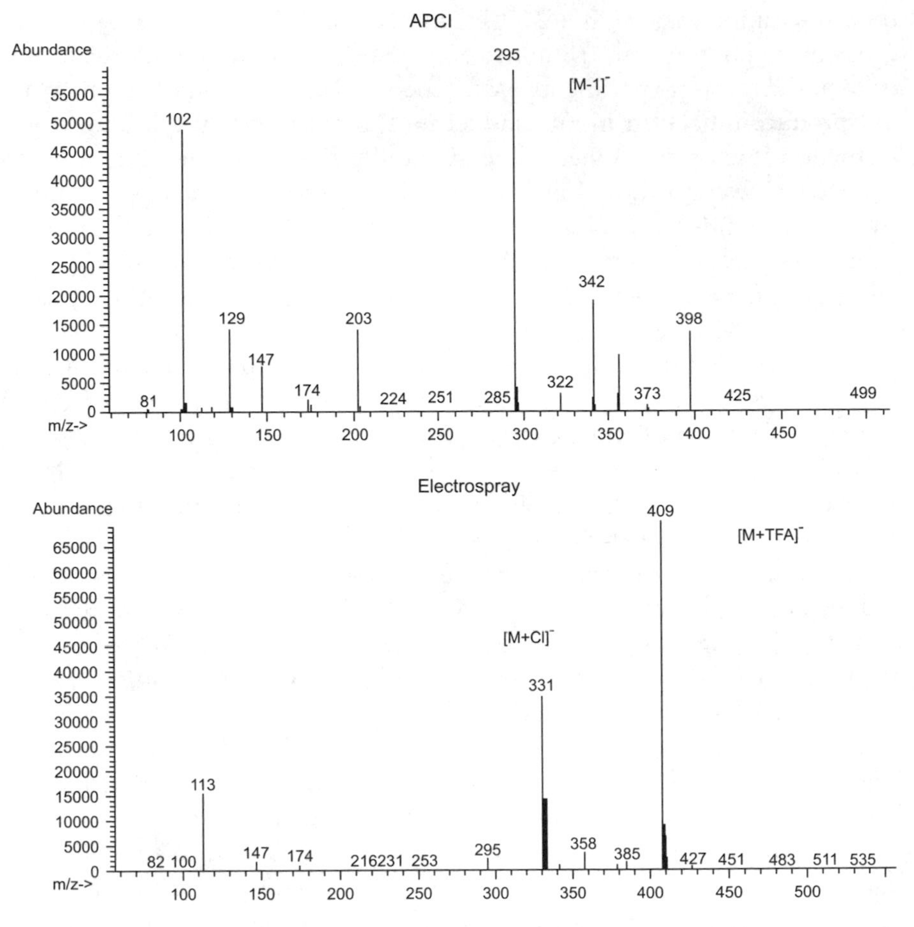

Figure 6.13 APCI and ESI mass spectra of HMX. (From Schilling, A.B., The analysis of explosives constituents and metabolites with electrospray and APCI LC/MS, presented at 44th ASMS Conference on Mass Spectrometry and Allied Topics, Portland, OR, 1996.)

NO, N, CH_2, CH_2, N, N, O_2N, CH_2, NO

DNX

NO_2, N, CH_2, CH_2, N, N, O_2N, CH_2, NO

MNX

Scheme 6

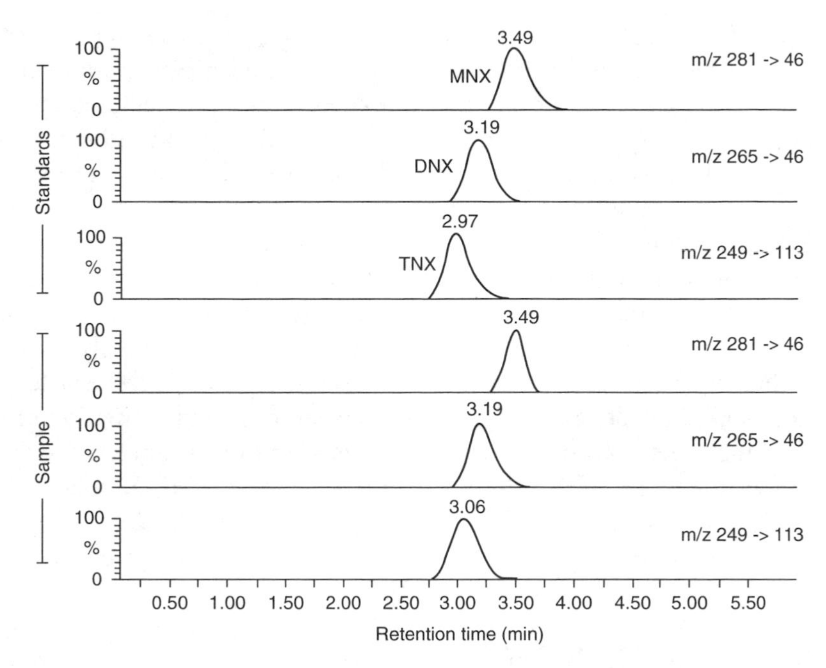

Figure 6.14 Mass chromatograms of LC/MS/MS in the selected reaction monitoring mode of the $[M + 75]^-$ adduct ions of MNX, DNX and TNX in a standard and in a groundwater sample. (Reprinted with permission from Beller, H.R. and Tiemeier, K., Use of liquid chromatography/tandem mass spectrometry to detect distinctive indicators of *in situ* RDX transformation in contaminated groundwater, *Environ. Sci. Technol.*, 36, 2060, 2002. © American Chemical Society.)

Detection limits were between 0.032 and 0.047 μg/l. Figure 6.14 shows the mass chromatograms of LC/MS/MS in the selected reaction monitoring mode of the $[M + 75]^-$ adduct ions of MNX, DNX, and TNX, in a standard and in a groundwater sample.

Another study of RDX and its degradation products in water was carried out using a Micromass Quattro LC (Z-spray) triple quadrupole mass spectrometer, operated in the electrospray, negative-ion mode.[39] The isocratic mobile phase was methanol–water (50:50) at a flow rate of 0.2 ml/min, the HPLC column was a Waters Nova-Pak C_{18} column (150 × 2.1 mm, 4 μm particle size). Argon was used as MS/MS collision gas. LC/MS mass spectra of standards of RDX, MNX, DNX, and TNX (Scheme 6), showed major $[M + 75]^-$ and $[M + 45]^-$ adduct ions and small $[M - H]^-$ ions in all 4 compounds. MS/MS of the $[M + 75]^-$ ions of all 4 compounds yielded daughter ions at *m/z* 46, NO_2^-. Detection limits of RDX and its degradation products in groundwater were 0.1 μg/l.

A series of explosives, including RDX and HMX, were analyzed by ESI-ion mobility spectrometry–mass spectrometry (ESI-IMS-MS).[40] The IMS was interfaced to an Extrel C50-Q quadrupole mass spectrometer. The electrospray needle was cooled by a flow of water-cooled nitrogen, in order to eliminate solvent evaporation inside the needle prior to electrospray. The mobile phase consisted of methanol–water (90:10) with an addition of 2 m*M* sodium chloride. Major ions in RDX and HMX were $[M + Cl]^-$, at *m/z* 257 and *m/z* 331, respectively.

Chloride adduct ion formation for a series of explosives, including RDX and HMX, was investigated in negative-ion APCI-MS.[41] It is important to maximize the number of gas-phase chloride ions formed under APCI conditions because their major reaction is to form the adduct ion $[M + Cl]^-$. The higher the electron attachment rate constant of a solvent, the faster the production of chloride ions and the higher the efficiency of formation of $[M + Cl]^-$ adduct ions. Carbon tetrachloride was found to produce the greatest increase in intensities of the $[M + Cl]^-$ ion of RDX and HMX, followed by chloroform and methylene chloride. These results were in agreement with the electron attachment rate constants of these chlorinated solvents.

A series of explosives, including RDX, were analyzed by negative-ion APCI mass spectrometry, using a Finnigan LCQ ion trap mass spectrometer.[42] Samples were introduced by infusion with a syringe pump into the LC stream, which consisted of methanol–water (50:50). Without any additive the mass spectrum contained weak ions at *m/z* 222, $[M]^-$, *m/z* 221, $[M - H]^-$, and prominent ions at *m/z* 267 and *m/z* 268, interpreted as $[M + NO_2 - H]^-$ and $[M + NO_2]^-$, respectively. Addition of ammonium chloride gave an improvement in response by a factor of 4, producing abundant $[M + Cl]^-$ ions at *m/z* 257 and 259.

A series of explosives, including RDX and HMX, were studied using negative-ion electrospray ionization Fourier transform ion cyclotron resonance (ESI-FTICR) mass spectrometry.[43] Samples were dissolved in acetonitrile and spiked with 0.5% ammonium acetate. The major ions in the mass spectra of RDX and HMX were $[M + 45]^-$ and $[M + 59]^-$. As a result of accurate mass measurements with the FTICR mass spectrometer, these ions were assigned as $[M + HCOO]^-$ and $[M + CH_3COO]^-$, respectively. ESI mass spectra of C4, a plasticized explosive based on RDX, contained adduct ions identified as $[RDX + NO_3]^-$, $[RDX + H_2CO + HCOO]^-$, and $[RDX + H_2CO + CH_3COO]^-$.[43] These ions were also observed in the mass spectrum of pure RDX but were more abundant in the mass spectrum of C4.

Table 6.2 shows a summary of the major adduct ions in ESI and APCI mass spectrometry of RDX. Both ESI and APCI seem to produce rather inconsistent mass spectra of RDX, resulting in the production of an array of adduct ions. Furthermore, relative abundances of the ions largely fluctuate, depending on analyte concentration, the presence of impurities in the mobile phase, and contamination of the LC/MS system. Consequently, a study was

Table 6.2 Adduct ions in LC/MS Mass Spectra of RDX

Mobile phase	Additive	Instrument	ESI or APCI	Ions	Ion Assignment	Reference
Acetonitrile–water (50:50)	2 m*M* ammonium acetate $(NH_4)COOCH_3$	Quadrupole	ESI	$[M + 59]^-$	$[RDX + CH_3COO]^-$	21
Methanol–water (75:25)	5 m*M* ammonium acetate $(NH_4)COOCH_3$	Quadrupole	ESI	$[M + 44]^-$		33
				$[M + 45]^-$		
				$[M + 59]^-$	$[RDX + CH_3COO]^-$	
				$[M + 75]^-$		
Methanol–water (50:50)	No additive	Triple quadrupole	ESI	$[M + 45]^-$	$[RDX+NO_2-H]^-$	39
				$[M + 75]^-$	$[RDX+NO_2+CH_3O+2H]^-$	
Methanol–water (50:50)	10 m*M* ammonium formate NH_4HCOO	Ion Trap	ESI	$[M + 45]^-$	$[RDX + CHOO]^-$	38
Methanol–water (50:50)	No additive	Ion Trap	ESI	$[M - 1]^-$	$[M - H]^-$	25
				$[M + 45]^-$	$[M + NO_2 - H]^-$	
				$[M + 46]^-$	$[M + NO_2]^-$	
				$[M + 59]^-$	$[M + NNO_2 - H]^-$	
Acetonitrile	0.5% ammonium acetate NH_4CH_3COO	FT ICR	ESI	$[M + 45]^-$	$[RDX + CHOO]^-$	43
				$[M + 59]^-$	$[RDX + CH_3CO_2]^-$	
Acetonitrile	0.5% ammonium hydroxide NH_4OH	FT ICR	ESI	$[M + 62]^-$	$[RDX + NO_3]^-$	43
				$[M + 75]^-$	$[RDX + H_2CO + HCO_2]^-$	
				$[M + 89]^-$	$[RDX+H_2CO+CH_3CO_2]^-$	
Methanol–water (50:50)	1 m*M* ammonium nitrate	Triple quadrupole	ESI	$[M + 62]^-$	$[RDX + NO_3]^-$	22, 23

Table 6.2 (*Continued*) Adduct ions in LC/MS Mass Spectra of RDX

Mobile phase	Additive	Instrument	ESI or APCI	Ions	Ion Assignment	Reference
Methanol–water (50:50)	0.5 m*M* ammonium chloride (NH_4Cl)	Ion Trap	APCI	$[M + 35]^-$	$[RDX + {}^{35}Cl]^-$	42
				$[M + 37]^-$	$[RDX + {}^{37}Cl]^-$	
Methanol–water (50:50)	No additive	Ion Trap	APCI	$[M - 1]^-$	$[RDX - H]^-$	42
				M^-	$[RDX]^-$	
				$[M + 45]^-$	$[RDX - NO_2 - H]^-$	
				$[M + 46]^-$	$[RDX - NO_2]^-$	
Methanol–water (65:35)	0.1% carbon tetrachloride	Ion trap	APCI	$[M + 35]^-$	$[RDX + {}^{35}Cl]^-$	41
				$[M + 37]^-$	$[RDX + {}^{37}Cl]^-$	

Source: Gapeev, A., Sigman, M. and Yinon, *J., Rap. Comm. Mass Spectrom.*, 17, 943, 2003. © John Wiley & Sons Ltd. Reproduced with permission.)

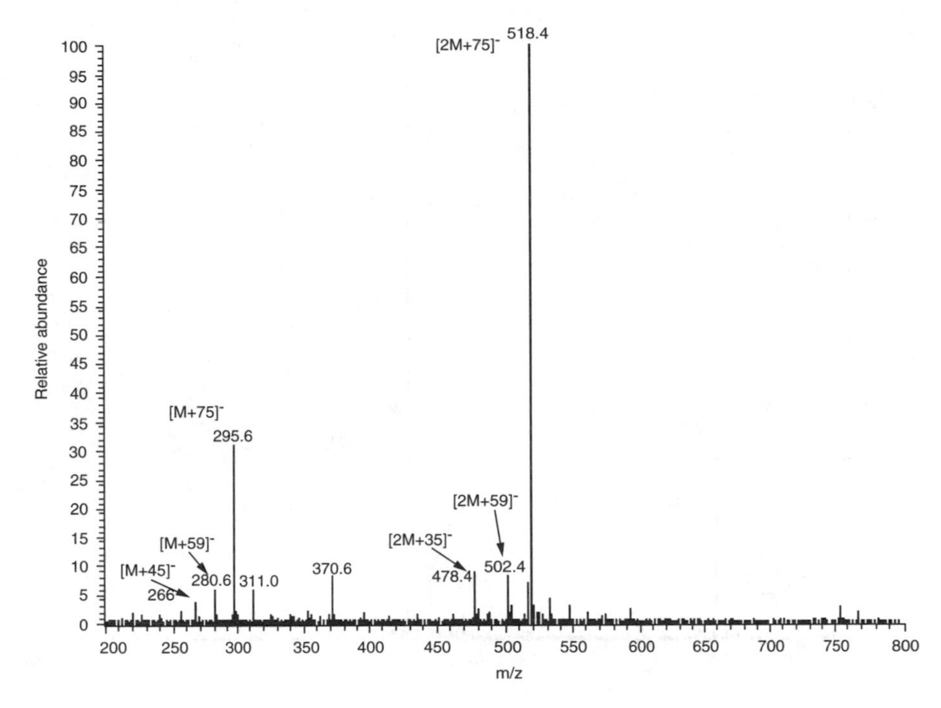

Figure 6.15 ESI mass spectrum of RDX. (From Gapeev, A., Sigman, M. and Yinon, J., LC/MS of explosives: RDX adduct ions, *Rap. Comm. Mass Spectrom.*, 17, 943, 2003. © John Wiley & Sons Ltd. Reproduced with permission).

undertaken to better understand the processes involved in the formation of these ions and their interpretation.[44]

In order to determine whether the clustering anions originate from self-decomposition of RDX in the source or from impurities in the mobile phase, isotopically labeled RDX ($^{13}C_3$-RDX and $^{15}N_6$-RDX) and isotopically labeled glycolic acid, acetic acid, ammonium formate, and formaldehyde were used in order to establish composition and formation route of RDX adduct ions produced in ESI and APCI sources. Figure 6.15 to Figure 6.17 show the ESI mass spectra of RDX, $^{13}C_3$-RDX, and $^{15}N_6$-RDX, respectively. MS analyses were carried out using a Thermo–Finnigan LCQ_{DUO} ion trap mass spectrometer. RDX was dissolved in methanol–water (1:1) and was injected into a 0.15 ml/min LC flow through a 5 μl sample loop.

When $^{13}C_3$-RDX was used (Figure 6.16), there was a shift of 3 mass units for $[M + 45]^-$, $[M + 59]^-$, $[M + 75]^-$ ions and a mass shift of 6 for $[2M + 35]^-$, $[2M + 37]^-$, $[2M + 59]^-$, and $[2M + 75]^-$ ions. When $^{15}N_6$-RDX was used (Figure 6.17), mass shifts of 6 were observed for $[M + 45]^-$, $[M + 59]^-$, $[M + 75]^-$ and mass shifts of 12 were observed for $[2M + 35]^-$, $[2M + 37]^-$, $[2M + 59]^-$, and $[2M + 75]^-$ ions. Results confirmed that $[2M + 35]^-$,

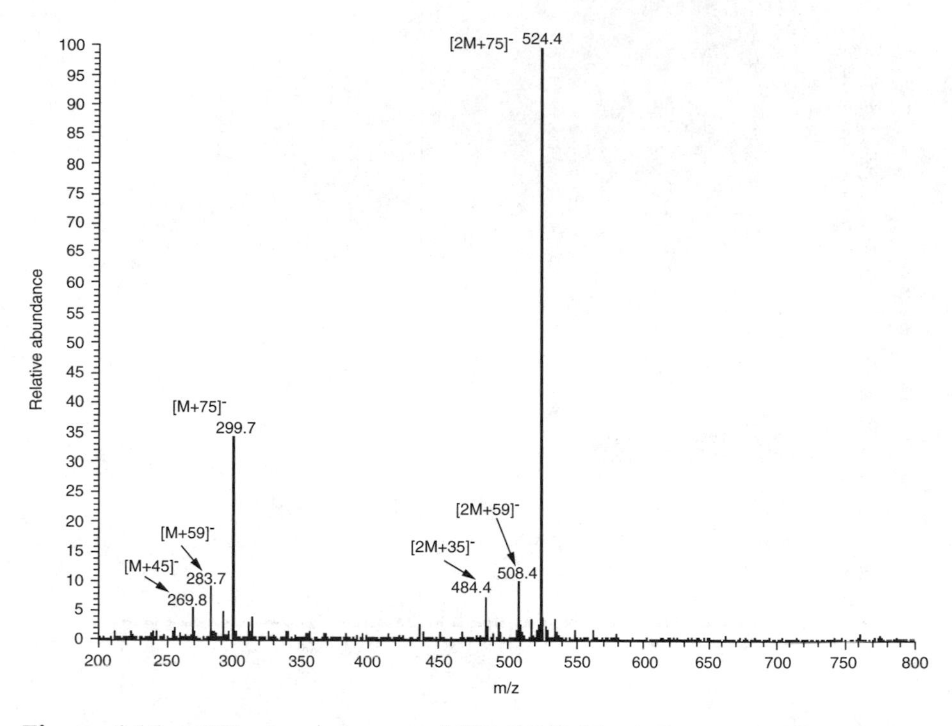

Figure 6.16 ESI mass spectrum of $^{13}C_3$-RDX. (From Gapeev, A., Sigman, M. and Yinon, J., LC/MS of explosives: RDX adduct ions, *Rap. Comm. Mass Spectrom.*, 17, 943, 2003. © John Wiley & Sons Ltd. Reproduced with permission).

$[2M + 37]^-$, $[2M + 59]^-$,and $[2M + 75]^-$ indeed contain two RDX molecules, and that RDX decomposition plays no role in ion formation in the ESI source.

The most abundant ions in the RDX concentration range of 25 ppb to 10 ppm were found to be $[2M + 75]^-$ and $[M + 75]^-$. Lower RDX concentrations lead to lower $[2M + 75]^-/[M + 75]^-$ ion ratios. At higher RDX concentrations, formation of the $[2M + 75]^-$ dimer becomes more favorable, so that the $[2M + 75]^-/[M + 75]^-$ ratio increases.

The elemental composition of the ion at *m/z* 297, $[M + 75]^-$, was determined by FT-ICR to be $[RDX + C_2H_3O_3]^-$.[43] Glycolic (hydroxyacetic) acid, $CH_2(OH)COOH$, having an anion at $[M - H]^-$, fits that elemental composition and was found to be an impurity at 8 ppm level in the methanol (HPLC grade) of the mobile phase.[44] The elemental composition of the ion at *m/z* 267, $[M + 45]^-$, was determined by FT-ICR to be $[RDX + HCOO]^-$.[43] Formate impurity in the methanol (HPLC grade) of the mobile phase, at an upper level of 100 ppm, was found to be the origin of this adduct ion.[44]

The elemental composition of the ion at *m/z* 281, $[M + 59]^-$, was determined by FT-ICR to be $[RDX + C_2H_3O_2]^-$.[43] Acetate impurity in the methanol (HPLC grade) of the mobile phase, at an upper level of 60 ppm was found

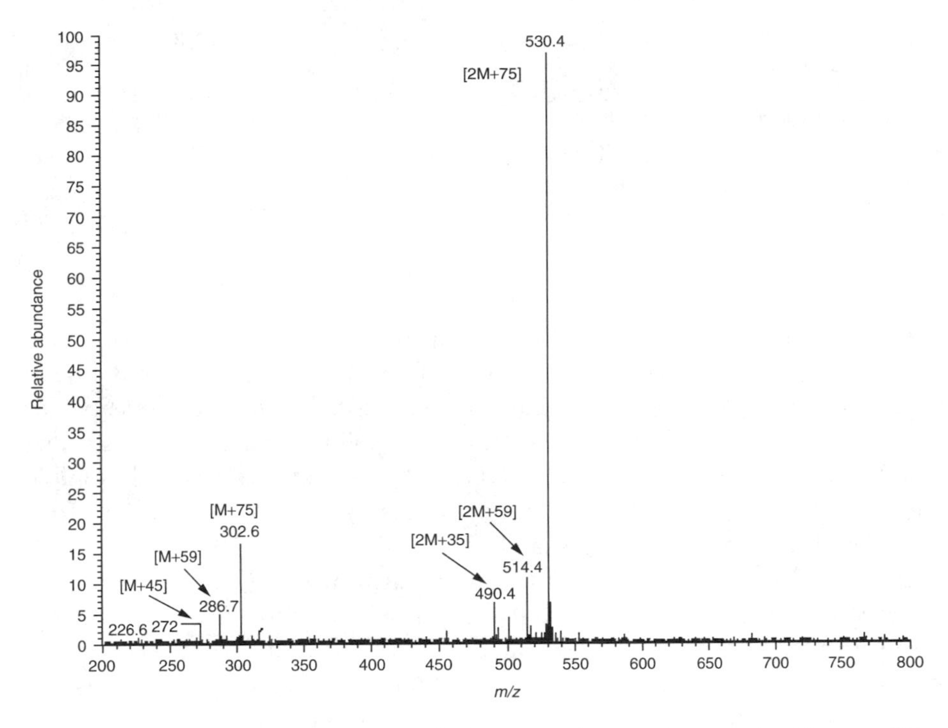

Figure 6.17 ESI mass spectrum of $^{15}N_6$-RDX. (From Gapeev, A., Sigman, M. and Yinon, J., LC/MS of explosives: RDX adduct ions, *Rap. Comm. Mass Spectrom.*, 17, 943, 2003. © John Wiley & Sons Ltd. Reproduced with permission).

to be the origin of this adduct ion.[44] When using purge-and-trap-grade distilled methanol, instead of HPLC grade, the abundances of the $[M + 45]^-$ and $[M + 59]^-$ ions were found to be lower.[44]

In APCI, which is a soft gas-phase ionization technique, part of the RDX molecules decompose, yielding NO_2^- species which in turn cluster with a second RDX molecule, producing abundant $[M + NO_2]^-$ cluster ions. This was confirmed by using $^{15}N_6$-RDX.[44] The ion $[M + 45]^-$ in the APCI mass spectrum of RDX was shifted by 7 mass units, when using $^{15}N_6$-RDX, which supports the $[M + NO_2]^-$ interpretation. Addition of 0.1 m*M* formate or chlorate in the mobile phase suppressed NO_2 formation, thus making possible the formation of a desired adduct ion.

6.3.3 Nitrate Ester Explosives

Three well-known explosives in this group are pentaerythritol tetranitrate (PETN), nitroglycerin (glycerol trinitrate, NG), and ethylene glycol dinitrate (EGDN) (Scheme 7).

PETN is a white, crystalline explosive used as a base charge in blasting caps and detonators, as the core explosive in commercial detonating cord, in

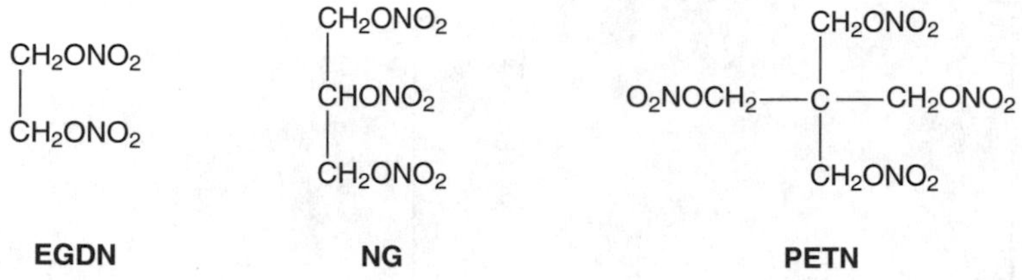

$$\begin{array}{c} CH_2ONO_2 \\ | \\ CH_2ONO_2 \end{array} \qquad \begin{array}{c} CH_2ONO_2 \\ | \\ CHONO_2 \\ | \\ CH_2ONO_2 \end{array} \qquad \begin{array}{c} CH_2ONO_2 \\ | \\ O_2NOCH_2—C—CH_2ONO_2 \\ | \\ CH_2ONO_2 \end{array}$$

EGDN **NG** **PETN**

Scheme 7

booster charges, and in plastic explosives (such as Semtex and Detasheet).[1,20,28] The chemical stability of PETN is very high. It is highly sensitive to impact and to initiation by explosion, but not very sensitive to friction. Its melting point is 141°C. It is most soluble in acetone (20 g/100 g at 20°C) and in ethyl acetate (6.3 g/100 g at 19°C), and very slightly soluble in water (0.2 mg/100 g at 20°C). Its vapor pressure at 25° C is 1.4×10^{-8} Torr.

NG is a colorless, odorless, and viscous liquid with a burning sweetish taste.[1,20,28] NG has been the main component in many dynamites and is an ingredient in multibase propellants.[28] It is very sensitive to shock, impact, and friction, and is used only when desensitized with other liquids or absorbent solids, or when mixed with nitrocellulose. NG is also used as a vasodilator in the treatment of angina pectoris. Its melting point is 13.2°C, and it is miscible with ether, acetone, chloroform, ethyl acetate, dichloroethylene, benzene, and nitrobenzene. Solubility in water is 0.15 g/100 g. Its vapor pressure is 2.5 m Torr at 20°C.

EGDN (also known as nitroglycol) is a transparent, colorless, liquid explosive. It is less sensitive to impact than NG and is considered to be more stable than NG. EGDN is very soluble in acetone, benzene, ethanol, diethyl ether, chloroform, and toluene. Its solubility in water at 20°C is 0.5 g/100 g. Its vapor pressure is 70 m Torr at 25° C. EGDN has been used in mixture with NG for low-temperature dynamites, as its melting point is –22.8° C.

Several additives were tested with a series of explosives, including EGDN, NG, and PETN.[22] Mobile phase of methanol–water (50:50) with 1 m*M* ammonium nitrate as additive produced best results, forming major $[M + NO_3]^-$ ions. EGDN formed nitrate adduct ions only for sample sizes of at least 500 ng. The base peak in the ESI mass spectrum of NG was at *m/z* 289, $[M + NO_3]^-$, but it had also an ion at *m/z* 226, $[M - H]^-$, having a 35% abundance.

Electrospray ionization coupled with tandem mass spectrometry was used for the analysis of a series of explosives, including NG and PETN.[21] Analyses were performed with a Perkin–Elmer Sciex API III-PLUS triple quadrupole mass spectrometer with electrospray ionization. Standard solutions of NG and PETN were made in acetonitrile–water (50:50), containing

Table 6.3 LC/MS-ESI Limits of Detection (pg/μL) of Nitrate Ester Explosives

Additive	NH_4NO_3	$NaNO_2$	Propionic acid	NH_4Cl
Adduct ion	$[M + NO_3]^-$	$[M + NO_2]^-$	$[M + CH_3CH_2CO_2]^-$	$[M + Cl]^-$
PETN	5	10	10	10
NG	5	20	25	10
EGDN	2000	10000	ND*	2500

* ND: not detected

Source: Reprinted from Zhao, X. and Yinon, J., Identification of nitrate ester explosives by liquid chromatography-electrospray ionization and atmospheric pressure chemical ionization mass spectrometry, *J. Chromatogr. A*, 977, 59, Copyright 2002, with permission from Elsevier.

2 m*M* of ammonium acetate. The solutions were introduced directly into the ion source by means of a syringe pump at a flow rate of 5 μl/min.

The base peak in the mass spectrum of PETN was at *m/z* 281, interpreted as $[M + CH_3COO]^-$, the additive originating from the ammonium acetate. Additional ions observed were $[M - H]^-$ at *m/z* 315 and a formate adduct ion, $[M + HCOO]^-$, which originated from formic acid background, used in previous experiments.

A comprehensive study on LC/MS of nitrate esters, including EGDN, NG, and PETN, was carried out using a Thermo–Finnigan LCQ_{DUO} ion trap mass spectrometer, with ESI and APCI in the negative-ion mode.[45] Postcolumn additives were investigated for the promotion of characteristic adduct ions. In ESI, additives were ammonium nitrate, sodium nitrite, propionic acid, and ammonium chloride. Additives were dissolved in methanol–water (70:30) and introduced postcolumn by a syringe pump through a T union into the LC flow before entering the ion source of the mass spectrometer. The characteristic adduct ions observed were $[M + NO_3]^-$, $[M + NO_2]^-$, $[M + CH_3CH_2CO_2]^-$, and $[M + Cl]^-$, respectively. Table 6.3 shows the limits of detection of the three explosives for each one of the additives.

Without additive, the major ions in the ESI mass spectrum of PETN are $[M + 45]^-$ at *m/z* 361, $[M + 59]^-$ at *m/z* 375, and $[M + 75]^-$ at *m/z* 391. In NG the major ions observed without additive are $[M + 45]^-$ at *m/z* 272 $[M + 75]^-$ at *m/z* 302, and $[2M + 75]^-$ at *m/z* 529. These adduct ions can probably be interpreted as originating from impurities in the solvent, as proven with RDX.[44]

Adduct ion formation in APCI of nitrate ester explosives was investigated using dichloromethane, chloroform, carbon tetrachloride, and ammonium chloride as additives.[45] In APCI, both vaporizer and heated capillary temperatures have great impact on the adduct ion formation. Except for dichloromethane, which required a vaporizer temperature of 330°C, the other additives produced intense $[M + Cl]^-$ ions at vaporizer temperature of 150°C. These adduct ions are easily recognizable by their isotopic composition: $[M + {}^{35}Cl]^-$ and $[M + {}^{37}Cl]^-$. Heated capillary temperatures were in the range

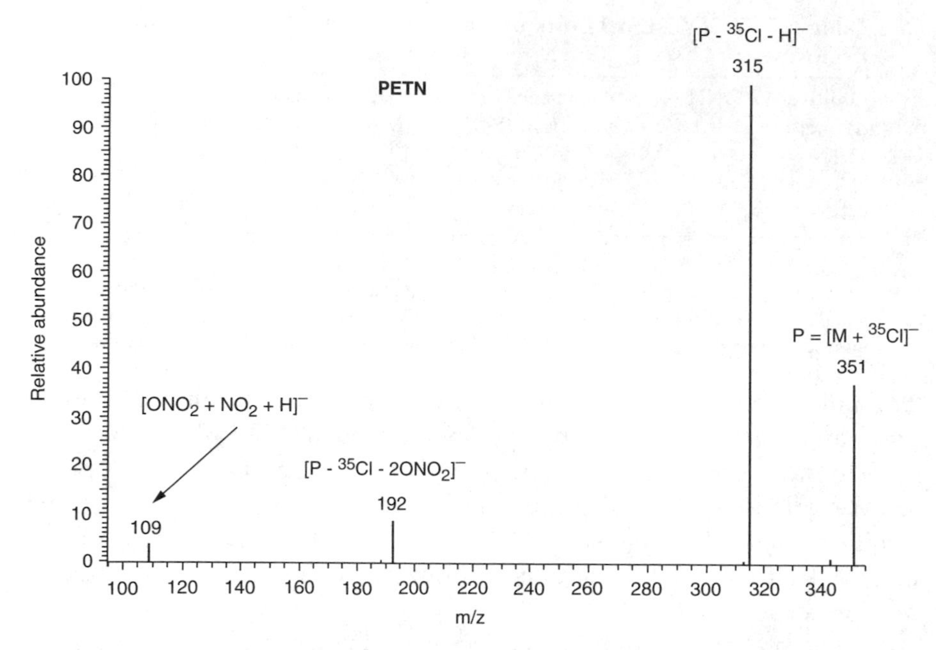

Figure 6.18 APCI-MS/MS mass spectrum of the $[M + {}^{35}Cl]^-$ ion of PETN (Reprinted from Zhao, X. and Yinon, J., Identification of nitrate ester explosives by liquid chromatography-electrospray ionization and atmospheric pressure chemical ionization mass spectrometry, *J. Chromatogr. A*, 977, 59, 2002. Copyright 2002, with permission from Elsevier Science.)

of 125 to 130°C. Chloroform and ammonium chloride produced the best sensitivity for adduct ion formation. Lowest detection limits for PETN, NG, and EGDN were 10, 20 pg/μl, and 2.5 ng/μl, respectively.

Figure 6.18 shows the APCI-MS/MS mass spectrum of the $[M + {}^{35}Cl]^-$ ion of PETN.

A series of explosives, including NG and PETN, were analyzed by negative-ion APCI mass spectrometry, using a Finnigan LCQ ion trap mass spectrometer.[42] Samples were introduced by infusion with a syringe pump into the LC stream, which consisted of methanol–water (50:50). Addition of ammonium chloride to the mobile phase at a final concentration of 0.5 m*M*, produced $[M + Cl]^-$ ions and improved the detection limits by one order of magnitude, compared to mass spectra without additive.

6.3.4 Various Explosives and Oxidizers

Electrospray ionization coupled with tandem mass spectrometry was used for the analysis of a series of explosives, including hexanitrostilbene (HNS) and hexanitrodiphenylamine (hexyl) (Scheme 8).[21,32] Analyses were performed with a Perkin–Elmer Sciex API III-PLUS triple quadrupole mass

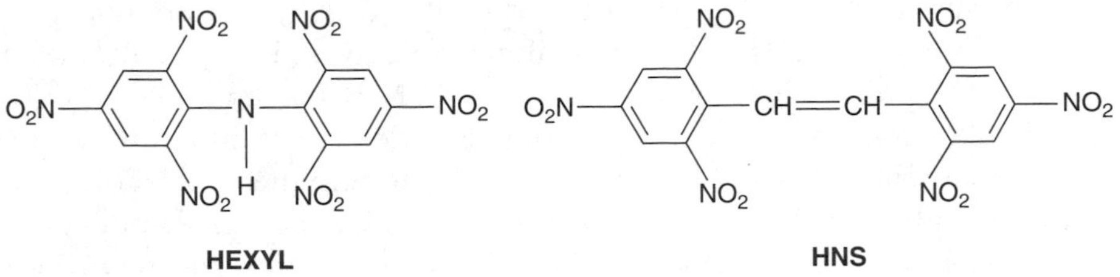

HEXYL HNS

Scheme 8

spectrometer with electrospray ionization in the negative-ion mode. Standard solutions of HNS and hexyl were made in dimethylformamide and diluted with acetonitrile. The solutions were introduced through the HPLC, using a gradient mobile phase consisting of 5 m*M* aqueous ammonium acetate and acetonitrile saturated with ammonium acetate, at a constant flow rate of 1.0 ml/min. The ESI mass spectra of HNS and hexyl contained mainly an $[M - H]^-$ ion at *m/z* 449 and *m/z* 438, respectively. The MS/MS mass spectra of the $[M - H]^-$ ions showed characteristic fragment ions at *m/z* 403 for HNS, assigned as $[M - H - NO_2]^-$, and at *m/z* 362 for hexyl, assigned as $[M - H - NO - NO_2]^-$.

Hexamethylenetriperoxidediamine (HMTD) (Scheme 9) is a white solid organic peroxide explosive. It is highly sensitive to initiation by impact, friction, and electrical discharge. APCI mass spectrometry in the positive-ion mode was used for the analysis of HMTD.[46] The instrument used was a Thermo–Finnigan Navigator quadrupole mass spectrometer. A solution of HMTD in acetone was injected through the HPLC, using a mobile phase of methanol–water (5:95) without an additive. The base peak at *m/z* 209 was assigned as the $[M + H]^+$ ion. Less intense fragment ions were observed at *m/z* 62, 90, 106, 179, and 207.

ESI-MS of ammonium nitrate and a group of additional inorganic oxidizers was studied using a Thermo–Finnigan LCQ_{DUO} ion trap mass spectrometer.[47,48]

Ammonium nitrate (AN) (NH_4NO_3) is an odorless, hygroscopic, white solid, used as a solid oxidizer in explosive mixtures such as ANFO, slurries, and emulsions.[1] Oxidizers included sodium nitrate ($NaNO_3$), potassium nitrate (KNO_3), ammonium sulfate [$(NH_4)_2SO_4$], potassium sulfate (K_2SO_4),

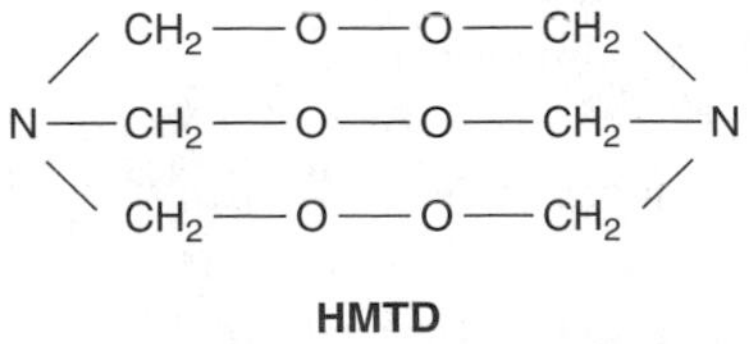

HMTD

Scheme 9

sodium chlorate ($NaClO_3$), potassium chlorate ($KClO_3$), ammonium perchlorate (NH_4ClO_4), and sodium perchlorate ($NaClO_4$). Isotopically labeled compounds [$^{15}NH_4NO_3$, $NH_4{}^{15}NO_3$, $NH_4N^{18}O_3$, $Na^{15}NO_3$, and $(^{15}NH_4)_2SO_4$] were used to confirm ion characterization. Solutions of ammonium nitrate were prepared in methanol–water (50:50) and in pure methanol. Solutions of other oxidizers were prepared in water and diluted in methanol–water (50:50). Solutions were introduced into the mass spectrometer by syringe pump infusion at a flow rate of 5 μl/min. Heated capillary temperatures were in the range of 55 to 250°C.

It was found that in the ESI mass spectrum of AN, at heated capillary temperatures of 55 to 150°C, in the positive-ion mode, cluster ions of the type $[(NH_4NO_3)_nNH_4]^+$ (n = 1–3) were dominant, which enabled characterization and identification of the integral ammonium nitrate molecule. Figure 6.19 [47] shows the full-scan positive-ion ESI mass spectra at heated capillary of 100°C, of (a) 1 m*M* NH_4NO_3, (b) 1 m*M* $NH_4{}^{15}NO_3$, (c) 1 m*M* $^{15}NH_4NO_3$, and (d) 1 m*M* $NH_4N^{18}O_3$ in methanol–water (50:50) introduced by syringe pump infusion.

Characteristic ESI mass spectra of other oxidizers could be obtained in both positive- and negative-ion mode, and over a wider range of temperatures of the heated capillary.[48] An exception is ammonium sulfate for which only the positive-ion mass spectrum produced characteristic cluster ions. Figure 6.20 [48] shows the full-scan positive-(upper trace) and negative-ion (lower trace) ESI mass spectra, at heated capillary of 220°C, of 1 m*M* ammonium perchlorate in methanol–water (50:50). Table 6.4 shows the characteristic ESI ions, in both positive- and negative-ion modes of explosive oxidizers.

6.3.5 Military, Commercial, and Other Explosive Mixtures

ESI and APCI in the negative-ion mode were compared for the analysis of a mixture of explosives, using a Hewlett–Packard 5989B LC/MS mass spectrometer with a HP 1090L HPLC.[24] The mixture contained 14 standard compounds (Table 6.5). HPLC column was HP Hypersil BDS, 100 mm × 4 mm ID, 3 μm particle size. Mobile phase consisted of methanol–water with a gradient of 26 to 40% methanol at 10 min to 53% methanol at 20 min, at a flow rate of 0.72 ml/min. Figure 6.21 shows the total ion chromatograms obtained in APCI and ESI. As can be seen, not all the components could be detected in either mode of ionization.

Table 6.5 shows the base peak ions in ESI and APCI. In ESI of RDX, HMX and tetryl, the major ion is the adduct ion $[M + TFA]^-$, the TFA probably originating from residue background.

In APCI of tetryl, the major ions are at *m/z* 242, $[M^* - H]^-$ and *m/z* 212, $[M^* - NO]^-$, where M* represents *N*-methylpicramide, a hydrolysis product of tetryl.

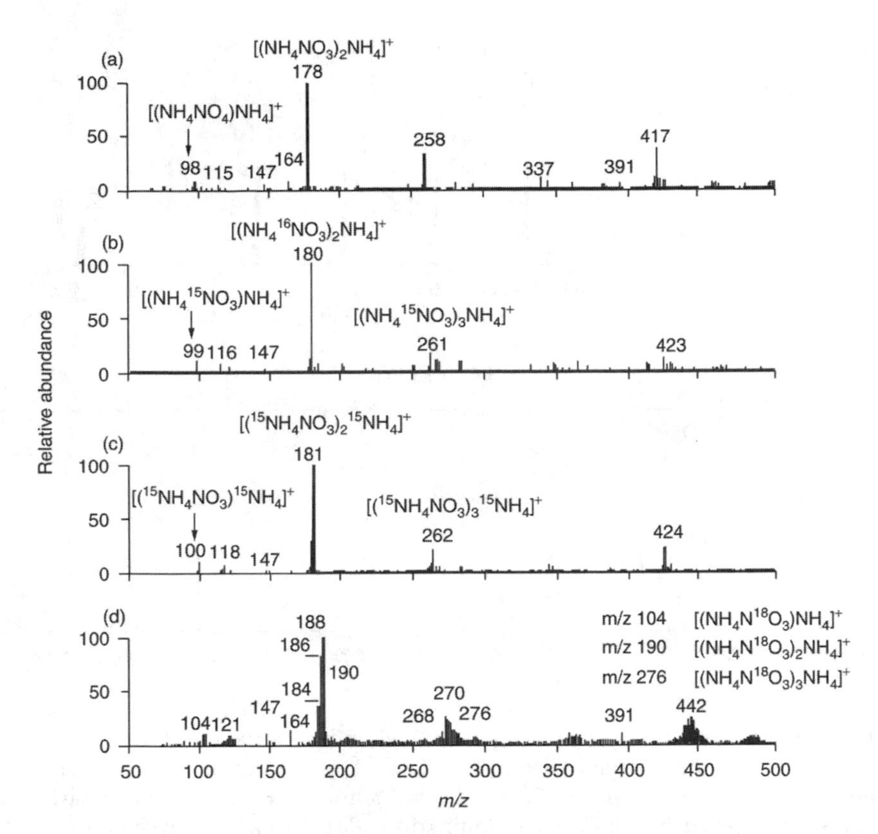

Figure 6.19 Full scan positive-ion ESI mass spectra at heated capillary of 100° C, of (a) 1 mM NH_4NO_3, (b) 1 mM $NH_4{}^{15}NO_3$, (c) 1 mM $^{15}NH_4NO_3$ and (d) 1 mM $NH_4N^{18}O_3$ in methanol–water (50:50) introduced by syringe pump infusion. (From Zhao, X. and Yinon, J., Characterization of ammonium nitrate by electrospray ionization tandem mass spectrometry, *Rap. Comm. Mass Spectrom.*, 15, 1514, 2001. © John Wiley & Sons Ltd. Reproduced with permission).

Chloride adduct ion formation for a series of explosives was investigated in negative-ion APCI-MS, using a Finnigan LCQ ion trap mass spectrometer.[41] Chloride adduct ion formation was studied in 5-nitro-2,4-dihydro-3H-1,2,4-triazol-3-one (NTO), nitroguanidine (NQ), RDX, HMX, 2,4,6,8,10,12-hexanitro-2,4,6,8,10,12-hexazaisowurtzitane (CL-20), TNT, tetryl, and PETN. Limits of detection obtained were 40, 60, 12, 52, 10, 19, 25, and 12 ppb, respectively.

A Finnigan LCQ ion trap mass spectrometer was used to study the ESI mass spectra of RDX and its degradation products MNX and TNX, as well as HMX, TNB, TNT, DNT, and 2-A-4,6-DNT in ground water.[38] Solid phase extraction (SPE) cartridges were used to concentrate samples. HPLC separation was achieved with a Kromasil C8 reversed-phase column (250 × 2 mm ID), using an isocratic mobile phase of isopropanol–water–ammonium

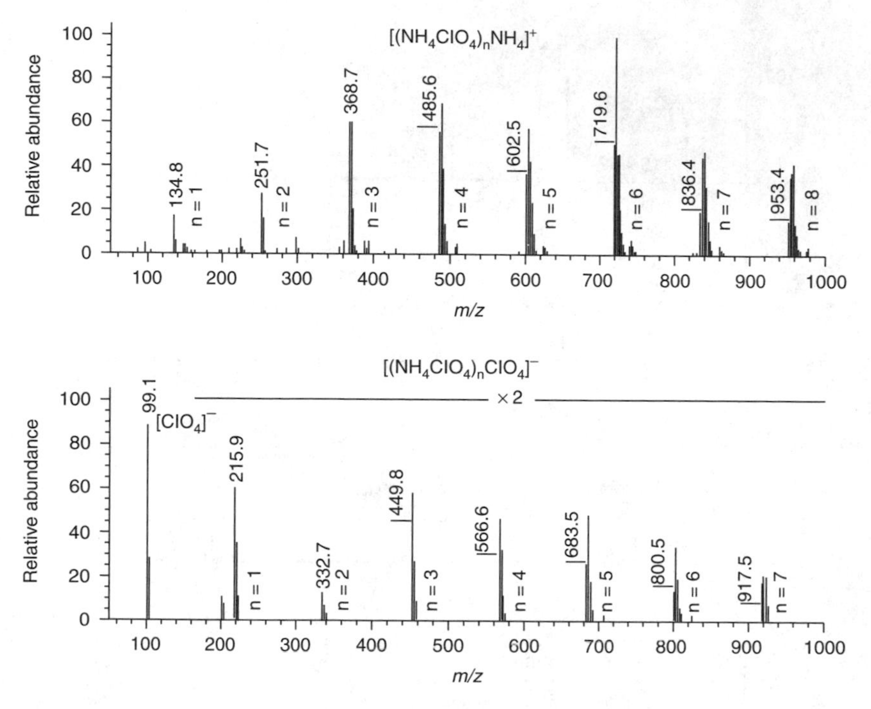

Figure 6.20 Full-scan positive-(upper trace) and negative-ion (lower trace) ESI mass spectra, at heated capillary of 220° C, of 1 mM ammonium perchlorate in methanol–water (50:50) (From Zhao, X. and Yinon, J., Forensic identification of explosive oxidizers by electrospray ionization, *Rap. Comm. Mass Spectrom.*, 16, 1137, 2002. © John Wiley & Sons Ltd. Reproduced with permission).

Table 6.4 Characteristic Ions in ESI Mass Spectra of Explosive Oxidizers

Oxidizer	Positive-ion	Negative-ion
Sodium nitrate	$[(NaNO_3)_nNa]^+$	$[(NaNO_3)_nNO_3]^-$
Potassium nitrate	$[(KNO_3)_nK]^+$	$[(KNO_3)_nNO_3]^-$
Ammonium sulfate	$[(NH_4HSO_4)_nNH_4]^+$	
Potassium sulfate	$[(K_2SO_4)_nK]^+$	$[(K_2SO_4)_nKSO_4]^-$
Sodium chlorate	$[(NaClO_3)_nNa]^+$	$[(NaClO_3)_nClO_3]^-$
Potassium chlorate	$[(KClO_3)_nK]^+$	$[(KClO_3)_nClO_3]^-$
Ammonium perchlorate	$[(NH_4ClO_4)_nNH_4]^+$	$[(NH_4ClO_4)_nClO_4]^-$
Sodium perchlorate	$[(NaClO_4)_nNa]^+$	$[(NaClO_4)_nClO_4]^-$

Table 6.5 List of Compounds in Mixture and Base Peak Ions

Compound #	Compound name	M. W.	Base Ion in ESI	Base Ion in APCI
1	HMX	296	409 $[M + TFA]^-$	295 $[M - H]^-$
2	RDX	222	335 $[M + TFA]^-$	102
3	1,3,5-TNB	213	183 $[M - NO]^-$	213 $[M]^-$
4	1,3-DNB	168	ND	168 $[M]^-$
5	Nitrobenzene	123	ND	ND
6	Tetryl	287	400 $[M + TFA]^-$	212, 241
7	TNT	227	226 $[M - H]^-$	227 $[M]^-$
8	2-A-4,6-DNT	197	196 $[M - H]^-$	196 $[M - H]^-$
9	4-A-2,6-DNT	197	196 $[M - H]^-$	196 $[M - H]^-$
10	2,4-DNT	182	ND	181 $[M - H]^-$
11	2,6-DNT	182	ND	182 $[M]^-$
12	2-NT	137	ND	ND
13	4-NT	137	ND	136 $[M - H]^-$
14	3-NT	137	ND	ND

Source: Schilling, A.B., The analysis of explosives constituents and metabolites with electrospray and APCI LC/MS, Presented at 44th ASMS Conference on Mass Spectrometry and Allied Topics, Portland, OR, 1996.

formate (20:78:2), at a flow rate of 0.2 ml/min and a column temperature of 30 to 32°C. The concentration of the ammonium formate solution was 0.5 *M* and was adjusted to pH 8 with 10% ammonium hydroxide. Characteristic ions observed in the LC/MS-ESI ion chromatograms of a standard mixture of the investigated explosives were formate adduct ions, $[M + 45]^-$, for the nitramine compounds, and alkoxy adduct ions, $[M + 59]^-$, for the nitroaromatic compounds.

A study on LC/MS of nitrate ester explosives was carried out using a Thermo–Finnigan LCQ_{DUO} ion trap mass spectrometer, with ESI and APCI, in the negative-ion mode.[45] In LC/MS-ESI, HPLC separation was achieved with a Restek Allure C_{18} column (100 × 2.1 mm I.D., 5 μm particle size), using an isocratic mobile phase of methanol–water (70:30), at a flow rate of 150 μl/min. In LC/MS-APCI, a Restek Allure C_{18} column (150 × 3.2 mm ID, 5 μm particle size) was used with an isocratic mobile phase of methanol–water (70:30) at a flow rate of 400 μl/min.

Sample injection volume in both ionization modes was 10 μl. Postcolumn additives were introduced with a syringe pump (flow rate 5 μl/min) through a T union into the LC flow, before entering the mass spectrometer.

Figure 6.22 [45] shows the LC/MS-APCI mass chromatograms of a 50 pg/μl Booster DYNO sample, with postcolumn introduction of chloroform. Figure 6.23 [45] shows the LC/MS-ESI mass chromatograms of a 25 pg/μl Semtex sample, with postcolumn introduction of ammonium nitrate. Mixtures containing various inorganic oxidizers and commercial explosives containing

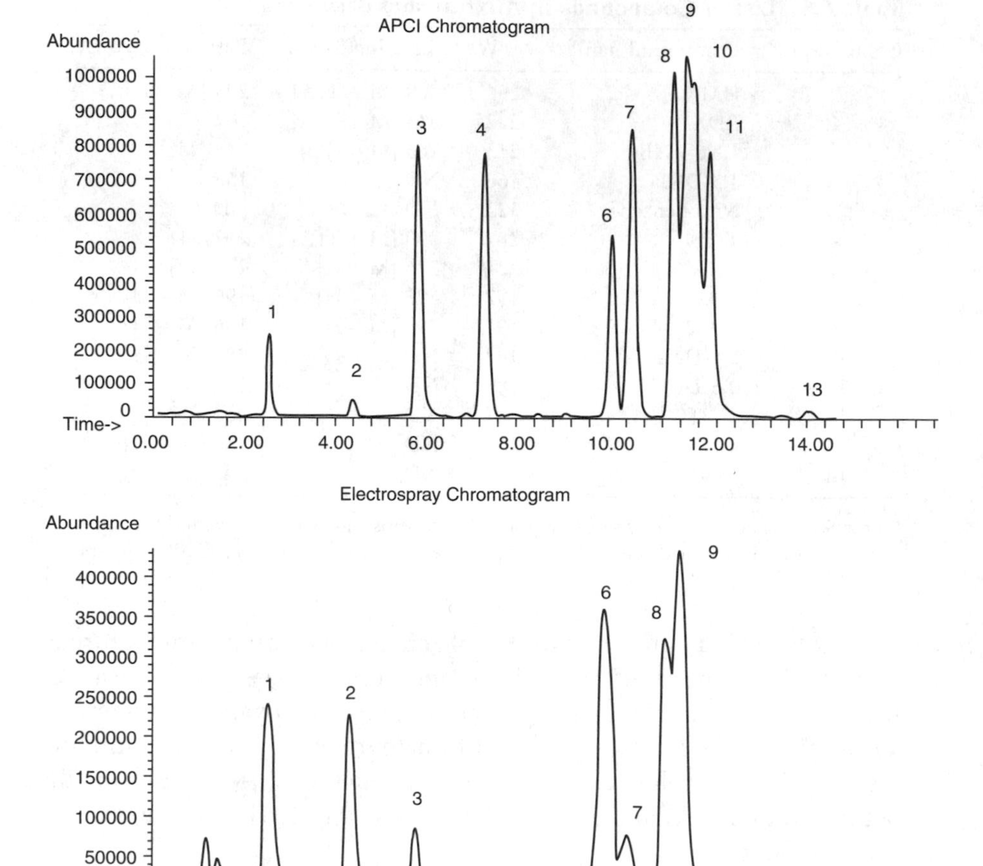

Figure 6.21 LC/MS total ion chromatograms of a mixture of explosives in APCI and ESI. Peak identification in Table 6.5. (From Schilling, A. B., The analysis of explosives constituents and metabolites with electrospray and APCI LC/MS, 44th ASMS Conference on Mass Spectrometry and Allied Topics, Portland, OR, 1996. With permission)

oxidizers were analyzed by ESI-MS, using a Thermo–Finnigan LCQ_{DUO} ion trap mass spectrometer. [48]

Solutions were prepared in water and diluted in methanol–water (50:50) and introduced into the mass spectrometer by syringe pump infusion at a flow rate of 5 μl/min. Heated capillary temperatures were in the range of 55 to 250°C. Figure 6.24 shows the positive-ion ESI mass spectrum of Power-mite, a commercially available slurry explosive. The mass spectrum obtained at 100° C shows the presence of ammonium nitrate and sodium nitrate in the sample. In addition to ions specific to each one of the oxidizers, ions

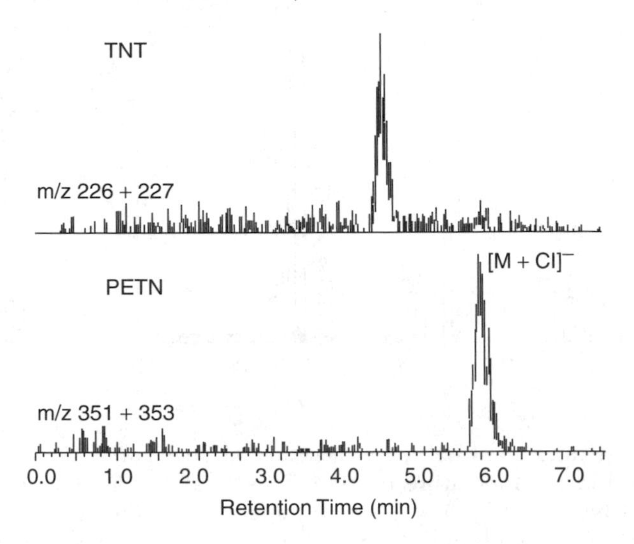

Figure 6.22 LC/MS-APCI mass chromatograms of a 50 pg/μL Booster DYNO sample, with post-column introduction of chloroform. (Reprinted from Zhao, X. and Yinon, J., Identification of nitrate ester explosives by liquid chromatography-electrospray ionization and atmospheric pressure chemical ionization mass spectrometry, *J. Chromatogr. A*, 977, 59, 2002. Copyright 2002, with permission from Elsevier Science.)

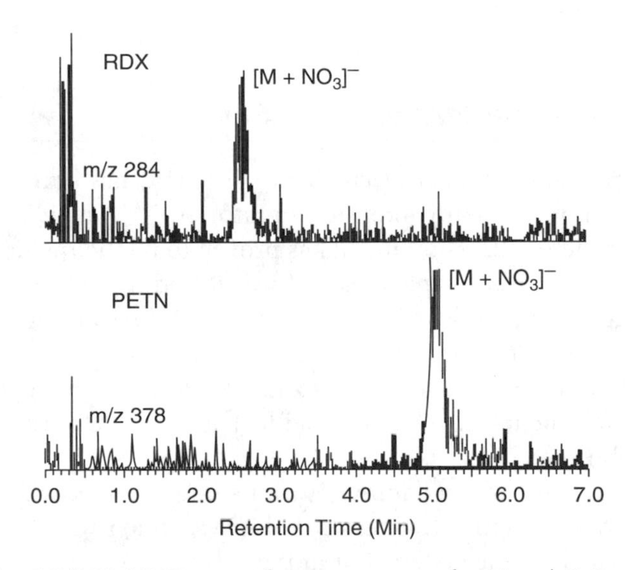

Figure 6.23 LC/MS-APCI mass chromatograms of a 25 pg/μL Semtex sample, with post-column introduction of ammonium nitrate. (Reprinted from Zhao, X. and Yinon, J., Identification of nitrate ester explosives by liquid chromatography-electrospray ionization and atmospheric pressure chemical ionization mass spectrometry, *J. Chromatogr. A*, 977, 59, 2002. Copyright 2002, with permission from Elsevier Science.)

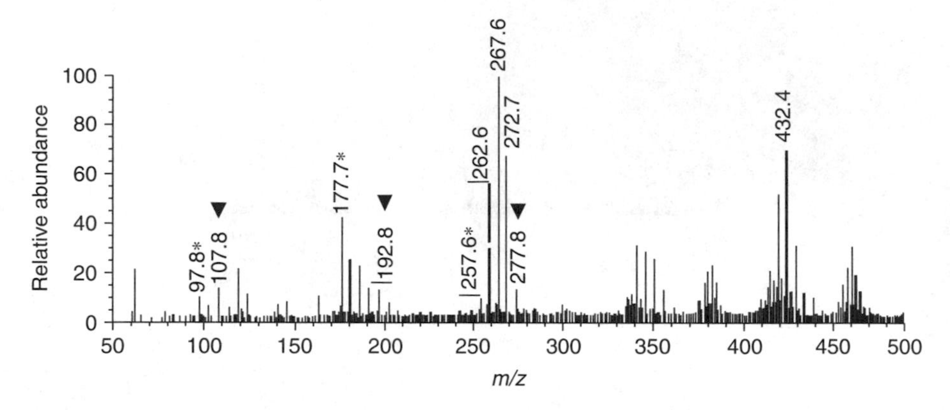

Figure 6.24 Positive-ion ESI mass spectrum of Powermite, a slurry explosive, at heated capillary temperature of 100° C. * cluster ions due to ammonium nitrate, $[(NH_4NO_3)_nNH_4]^+$, n = 1,2,3; ▼ cluster ions due to sodium nitrate, $[(NaNO_3)_nNa]^+$, n = 1,2,3 (From Zhao, X. and Yinon, J., Forensic identification of explosive oxidizers by electrospray ionization, *Rap. Comm. Mass Spectrom.*, 16, 1137, 2002. © John Wiley & Sons Ltd. Reproduced with permission).

result from ion-molecule interactions between the two compounds. For example, the ion at *m/z* 268 is probably due to $[(NH_4NO_3)_2(NaNO_3)Na]^+$ or $[(NH_4NO_3)(NaNO_3)_2NH_4]^+$.

6.4 Summary and Future Directions

Liquid chromatography/mass spectrometry (LC/MS) has become a routine analytical technique in many forensic laboratories.

LC/MS in the negative-ion mode has proven to be the method of choice for trace analysis of most explosives. ESI was found to be most suitable for nitramine explosives, while APCI is better suited for nitroaromatic compounds. For nitrate esters, both ionization techniques are adequate. Best results are obtained when using a postcolumn additive, which will result in formation of abundant characteristic adduct ions. Lowest detection limits are in the ppb range.

Without the use of an additive, various adduct ions will be formed, mainly in nitramine and nitrate ester explosives, which will depend on the impurities present in the system and in the solvent.

Future directions indicate more use of MS/MS techniques, in combination with LC/MS, for characterization of explosive residues.

The forensic community is now ready for a database of LC/MS chromatograms and mass spectra, which will be of great value to identify and characterize explosive residues in bombing cases.

References

1. Yinon, J. and Zitrin, S., *Modern Methods and Applications in Analysis of Explosives*, John Wiley, Chichester, U.K., 1993.
2. Abian, J., The coupling of gas and liquid chromatography with mass spectrometry, *J. Mass Spectrom.*, 34, 157, 1999.
3. Niessen, W.M.A., State-of-the-art in liquid chromatography-mass spectrometry, *J. Chromatogr. A*, 856, 179, 1999.
4. Lim, C.K. and Lord, G., Current developments in LC-MS for pharmaceutical analysis, *Biol. Pharm. Bull.*, 25, 547, 2002.
5. Rustichelli, C., Ferioli, V., Vezzalini, F., Rossi, M.C., and Gamberini, G., Simultaneous separation and identification of hashish constituents by coupled liquid chromatography-mass spectrometry (HPLC-MS), *Chromatographia*, 43, 129, 1996.
6. Van Bocxlaer, J.F., Clauwaert, K.M., Lambert, W. E., Deforce, D.L., Van den Eeckhout, E.G., and De Leenheer, A.P., Liquid chromatography-mass spectrometry in forensic toxicology, *Mass Spectrom. Rev.*, 19, 165, 2000.
7. Rivier, L., New trends in doping analysis, *Chimia*, 56, 84, 2002.
8. Parker, C.E., Voyksner, R.D., Tondeur, Y., Henion, J.D., Hass, J.R., and Yinon, J., Analysis of explosives by liquid chromatography-negative ion chemical ionization mass spectrometry, *J. Forensic Sci.*, 27, 495, 1982.
9. Yinon, J. and Hwang, D.-G., High-performance liquid chromatography-mass spectrometry of explosives, *J. Chromatogr.*, 268, 45, 1983.
10. Yinon, J., Forensic analysis of explosives by LC/MS, *Forensic Sci. Rev.*, 13, 19, 2001.
11. Voyksner, R.D. and Yinon, J., Trace analysis of explosives by thermospray high-performance liquid chromatography-mass spectrometry, *J. Chromatogr.*, 354, 393, 1986.
12. Berberich, D.W., Yost, R.A., and Fetterolf, D.D., Analysis of explosives by liquid chromatography/thermospray/mass spectrometry, *J. Forensic Sci.*, 33, 946, 1988.
13. Cappiello, A., Famiglini, G., Lombardozzi, A., Massari, A., and Vadala, G.G., Electron capture ionization of explosives with a microflow rate particle beam interface, *J. Am. Soc. Mass Spectrom.*, 7, 753, 1996.
14. Yamashita, M. and Fenn, J.B., Electrospray ion source. Another variation on the free-jet theme., *J. Phys. Chem.*, 88(20), 4451, 1984.
15. Fenn, J.B., Mann, M., Meng, C.K., and Wong, S.F., Electrospray ionization for mass spectrometry of large biomolecules, *Science*, 246, 64, 1989.
16. Gaskell, S.J., Electrospray — principles and practice, *J. Mass Spectrom.*, 32, 677, 1997.
17. Whitehouse, C.R., Dreyer, R.N., Yamashita, M., and Fenn, J.B., Electrospray interface for liquid chromatographs and mass spectrometers, *Anal. Chem.*, 57, 675, 1985.

18. Horning, E.C., New picogram detection system based on a mass spectrometer with an external ionization source at atmospheric pressure, *Anal. Chem.*, 45, 936, 1973.

19. Urbanski, T., *Chemistry and Technology of Explosives*, Vol. 4, Pergamon Press, Oxford, 1984.

20. Yinon, J., *Forensic and Environmental Detection of Explosives*, John Wiley & Sons, Chichester, U.K., 1999.

21. Casetta, B. and Garafolo, F., Characterization of explosives by liquid chromatography/mass spectrometry and liquid chromatography/tandem mass spectrometry using electrospray ionization and parent-ion scanning techniques, *Org. Mass Spectrom.*, 29, 517, 1994.

22. Miller, M.L., Leibowitz, J., and Martz, R., Additive enhancement for ESI of nitrated explosives, Presented at 44th ASMS Conference on Mass Spectrometry and Allied Topics, Portland, OR, 1996, 1389.

23. Miller, M.L., The analysis of nitrated organic explosives by LC/MS: additive enhancement, Presented at 45th ASMS Conference on Mass Spectrometry and Allied Topics, Palm Springs, CA, 1997, 52.

24. Schilling, A.B., The analysis of explosives constituents and metabolites with electrospray and APCI LC/MS, presented at 44th ASMS Conference on Mass Spectrometry and Allied Topics, Portland, OR, 1996.

25. Yinon, J., McClellan, J.E., and Yost, R.A., Electrospray ionization tandem mass spectrometry collision-induced dissociation study of explosives in an ion trap mass spectrometer, *Rap. Comm. Mass Spectrom.*, 11, 1961, 1997.

26. Zhao, X. and Yinon, J., Characterization and origin identification of 2,4,6-trinitrotoluene through its by-product isomers by liquid chromatography-atmospheric pressure chemical ionization mass spectrometry, *J. Chromatogr. A*, 946, 125, 2002.

27. Yinon, J., Mass spectrometry of explosives: nitro compounds, nitrate esters, and nitramines, *Mass Spectrom. Rev.*, 1, 257, 1982.

28. Urbanski, T., *Chemistry and Technology of Explosives*, Vols. 1–3, McMillan, New York, 1964.

29. Tervo, J., Personal communication, 2001.

30. McClellan, J.E., Fundamentals and applications of electrospray ionization-quadrupole ion trap mass spectrometry for the analysis of explosives, Ph.D. thesis, University of Florida, Gainesville, 2000.

31. Yinon, J., Zitrin, S., and Tamiri, T., Reactions in the mass spectrometry of 2,4,6-*N*-tetranitro-*N*-methylaniline (tetryl), *Rap. Comm. Mass Spectrom.*, 7, 1051, 1993.

32. Garafolo, F., Longo, A., Migliozzi, V., and Tallarico, C., Quantitative analysis of thermostable explosive compounds by combined liquid chromatography/tandem mass spectrometry, *Rap. Comm. Mass Spectrom.*, 10, 1273, 1996.

33. Schreiber, A., Efer, J., and Engewald, W., Application of spectral libraries for high-performance liquid chromatography-atmospheric pressure ionization mass spectrometry to the analysis of pesticide and explosive residues in environmental samples., *J. Chromatogr. A*, 869, 2000.
34. Yinon, J., Identification of explosives by chemical ionization mass spectrometry using water as reagent, *Biom. Mass Spectrom.*, 1, 393, 1974.
35. Zitrin, S. and Yinon, J., Chemical ionization mass spectrometry of explosives, in *Advances in Mass Spectrometry in Biochemistry and Medicine,* Vol. 1, Frigerio, A. and Castagnoli, N., Eds., Spectrum Publications, New York, 1976, 369.
36. Yinon, J., Analysis of explosives by negative ion chemical ionization mass spectrometry, *J. Forensic Sci.*, 25, 401, 1980.
37. Yinon, J., Direct exposure chemical ionization mass spectra of explosives, *Org. Mass Spectrom.*, 15, 637, 1980.
38. Cassada, D.A., Monson, S.J., Snow, D.D., and Spalding, R.F., Sensitive determination of RDX, nitroso-RDX metabolites, and other munitions in ground water by solid-phase extraction and isotope dilution liquid chromatography-atmospheric pressure electrospray ionization mass spectrometry, *J. Chromatogr. A*, 844, 87, 1999.
39. Beller, H.R. and Tiemeier, K., Use of liquid chromatography/tandem mass spectrometry to detect distinctive indicators of *in situ* RDX transformation in contaminated groundwater, *Environ. Sci. Technol.*, 36, 2060, 2002.
40. Asbury, G.R., Klasmeier, J., and Hill, H.H., Jr., Analysis of explosives using electrospray ionization/ion mobility spectrometry (ESI/IMS), *Talanta*, 50, 1291, 2000.
41. Reich, R.F. and Yost, R.A., Trace detection and quantitation of explosives using LC/APCI-MS, 49th Annual ASMS Conference on Mass Spectrometry and Allied Topics., Chicago, IL, 2001.
42. Evans, C.S., Sleeman, R., Luke, J., and Keely, B.J., A rapid and efficient mass spectrometric method for the analysis of explosives, *Rap. Comm. Mass Spectrom.*, 16, 1883, 2002.
43. Wu, Z., Hendrickson, C.L., Rodgers, R.P., and Marshall, A.G., Composition of explosives by electrospray ionization Fourier transform ion cyclotron resonance mass spectrometry, *Anal. Chem.*, 74, 1879, 2002.
44. Gapeev, A., Sigman, M. and Yinon, J., LC/MS of explosives: RDX adduct ions, *Rap. Comm.. Mass Spectrom.*, 17, 943, 2003.
45. Zhao, X. and Yinon, J., Identification of nitrate ester explosives by liquid chromatography-electrospray ionization and atmospheric pressure chemical ionization mass spectrometry, *J. Chromatogr. A*, 977, 59, 2002.
46. Crowson, A. and Beardah, M.S., Development of an LC/MS method for the trace analysis of hexamethylenetriperoxidediamine (HMTD), *Analyst*, 126, 1689, 2001.

47. Zhao, X. and Yinon, J., Characterization of ammonium nitrate by electrospray ionization tandem mass spectrometry, *Rap. Comm. Mass Spectrom.*, 15, 1514, 2001.

48. 48. Zhao, X. and Yinon, J., Forensic identification of explosive oxidizers by electrospray ionization, *Rap. Comm. Mass Spectrom.*, 16, 1137, 2002.

Index

A

B

C

D

E

F

G

O

P

R

S